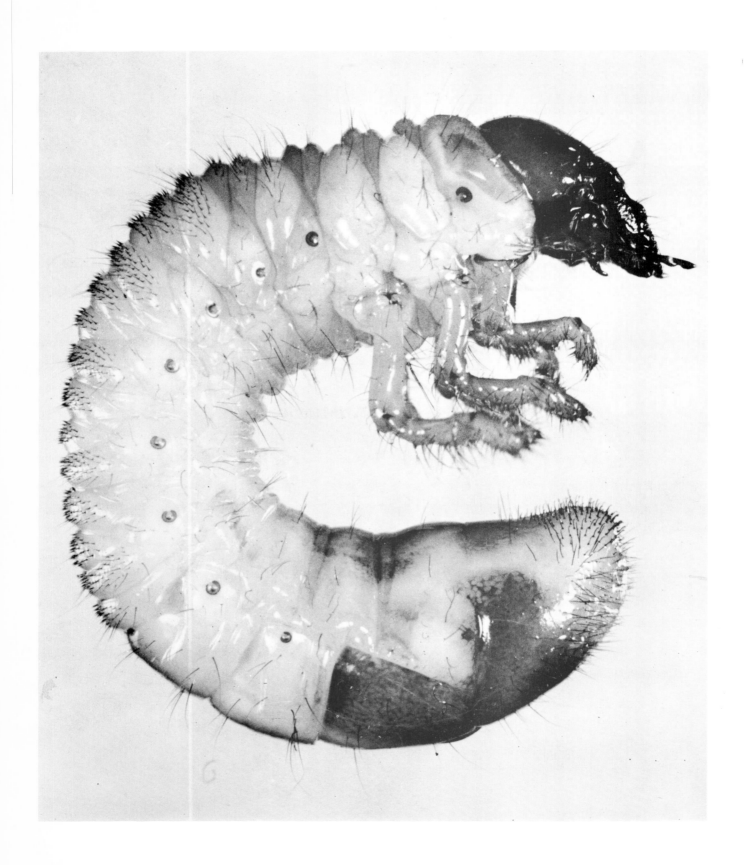

WHITE GRUBS

AND

THEIR ALLIES

A Study of North American Scarabaeoid Larvae

STUDIES IN ENTOMOLOGY NUMBER FOUR

By Paul O. Ritcher

Corvallis, Oregon
OREGON STATE UNIVERSITY PRESS

OREGON STATE MONOGRAPHS
STUDIES IN ENTOMOLOGY

JOHN D. LATTIN, *Consulting Editor*

NUMBER ONE
A Review of the Genus *Eucerceris* (Hymenoptera: Sphecidae)
By HERMAN A. SCULLEN

NUMBER TWO
The Scolytoidea of the Northwest:
Oregon, Washington, Idaho, and British Columbia
By W. J. CHAMBERLAIN

NUMBER THREE
Stoneflies of the Pacific Northwest
By STANLEY G. JEWITT, JR.

NUMBER FOUR
White Grubs and Their Allies
By PAUL O. RITCHER

© 1966
Oregon State University Press
Library of Congress Catalog Card number: 66-63008
Printed in the United States of America
By the Department of Printing, Oregon State University

Author's Acknowledgments

THE INFORMATION published in this book represents work done over the past thirty years while the writer was on the staffs of the Kentucky Agricultural Experiment Station (1936-1949), North Carolina State College (1949-1952), and Oregon State University (1952-1966). I am especially indebted to the Kentucky Agricultural Experiment Station for permission to reproduce much of the material contained in my Kentucky Bulletins 401, 442, 467, 471, 476, 477, 506, and 537, which have long been out of print.

For loan or gift of specimens, for identification of adults, for advice or for other favors in connection with these studies, I am greatly indebted to the following people in this country: W. H. Anderson, the late A. G. Böving, E. A. Chapin, O. L. Cartwright, and D. M. Anderson of the United States National Museum; the late T. R. Chamberlin of the Madison, Wisconsin, White Grub Laboratory; E. M. Searles and the late C. L. Fluke of the University of Wisconsin; H. H. Ross and M. W. Sanderson of the Illinois State Natural History Survey; J. J. Davis of Purdue University; the late Philip Luginbill of the Federal Bureau of Entomology and Plant Quarantine; W. P. Hayes of the University of Illinois; L. P. Wehrle, Floyd Werner, and George Butler of the University of Arizona; C. H. Hadley of the Federal Japanese Beetle Laboratory; C. F. Smith, T. B. Mitchell, and the late B. B. Fulton of North Carolina State College; H. G. Reinhard of the Texas Agricultural Experiment Station; C. H. Hoffman, United States Department of Agriculture; Melville Hatch, University of Washington; Hugh Leech and E. S. Ross of the California Academy of Science; Larry Saylor, Los Gatos, California; E. Gorton Linsley, Ray Smith, John MacSwain, and Paul Hurd of the University of California; Mont Cazier of Arizona State College; Robert Potts and Peter Ting, California Department of Agriculture;

Mrs. Patricia Vaurie, American Museum of Natural History; Bernard Benesh, Sunbright, Tennessee; E. C. Cole, University of Tennessee; W. A. Price, the late H. H. Jewett, L. H. Townsend, and other members of the Kentucky Department of Entomology and Botany; J. D. Lattin, Louis Gentner, and other entomologists at Oregon State University; D. Elmo Hardy, University of Hawaii; W. F. Barr of the University of Idaho; Joe Schuh of Klamath Falls, Oregon; Kenneth Fender and Dorothy McKay-Fender of McMinnville, Oregon; William Gibson of the Rockefeller Foundation; Ferd Butt, Lopez Island, Washington; Roger Friend and J. Peter Johnson, Connecticut Agricultural Experiment Station; C. B. Dominick, Chatham, Virginia; H. J. Franklin, Massachusetts Agricultural Experiment Station; George Gyrisco and Henry Dietrich, Cornell University; C. E. Heit, New York; Vincent Roth, American Museum of Natural History, Southwest Research Station; Robert Lauderdale, University of Nevada; Ray Hutson, Michigan State University; H. R. Bryson, Kansas State University; Vernon Olney, formerly at The Dalles, Oregon, Mid-Columbia Experiment Station; Frank Beer of the General Science Department at Oregon State University; Forrest Peifer of Hood River, Oregon, and a host of others.

Workers abroad to whom I am indebted for material or other help pertaining to this study include the late Fritz van Emden, the late Gilbert Arrow and E. B. Britton of the British Museum of Natural History; E. S. Narayanan and T. V. Venkatraman of the Indian Agricultural Research Institute, New Delhi, India; Philip Carne of the C.S.I.R.O., Canberra, Australia; Niilo Virkki of Finland; R. A. Crowson, University of Glasgow; and Kenneth Rachie, of the Rockefeller Foundation, New Delhi, India.

A number of former graduate students have given me a great deal of help with my studies and their assistance is gratefully acknowledged. These include the following, all of whom completed a thesis on some phase dealing with Scarabaeidae: Henry Howden, Science Service, Ottawa, Canada (biology and taxonomy of Geotrupinae, and so forth); Manohar Jerath, Nigeria (Aphodiinae); Sutharm Areekul, Thailand (comparative morphology of the alimentary canal and nervous system of Scarabaeidae); Aphirat Arunin, Thailand (comparative musculature of scarabaeoid larvae); and the late Floyd Ellertson, Mid-Columbia Experiment Station, Hood River, Oregon (biology of *Pleocoma*).

The assistance of Gerald Kraft, Western Washington State College, Bellingham; Everett Burts, United States Department of Agriculture, Wenatchee, Washington; and David Fellin, United States Forest Service, Missoula, Montana, is also acknowledged.

The writer is especially indebted to his two capable assistants, Charles Baker and Mrs. Nandini Rajadhyaksha, who have assisted in many ways. Most of the Figures on Plates IV to VII, XIV, XV, XX, XXXI, and XXXVI to XLI were drawn by Mrs. Rajadhyaksha.

The entire manuscript has been reviewed critically by John D. Lattin of Oregon State University and Henry Howden of the Canadian Science Service, resulting in many improvements and additions.

Financial support from the Kentucky Agricultural Experiment Station, the Oregon Agricultural Experiment Station, the General Research Fund of Oregon State University, and the National Science Foundation (Grant NSF-G17935) is gratefully acknowledged.

Paul O. Ritcher
September 1965

Contents

FRONTISPIECE: Third-stage larva of *Phyllophaga hirticula.*

List of Plates

Introduction

"WHITE GRUBS" is the common name applied in many countries to larvae of beetles belonging to the family Scarabaeidae, and in particular to those of economic importance to agricultural crops. While most "white grubs" are white with yellowish or reddish brown heads, some have orange, yellow, or bluish bodies.

General biology. In general, grubs of Scarabaeidae, and of Lucanidae as well, are curled (C-shaped), stout bodied, and have three pairs of well-developed legs. Grubs of Passalidae, however, are elongate and have only two pairs of well-developed legs. The head has stout, downward projecting (hypognathous) mandibles and a pair of three- or four-segmented antennae. There are three thoracic segments and ten abdominal segments. The thorax and the first eight abdominal segments bear spiracles on each side which are usually C-shaped or kidney shaped.

Adults of white grubs and related forms have lamellate antennae and are sluggish beetles ranging from one-eighth inch to six inches in length. Many are an inconspicuous brown or black color; but some are bright green, blue, or yellow, with metallic lustre. Many species are nocturnal and rarely seen by the average person except when the beetles are attracted to light. Others are diurnal and may be found on flowers or overripe fruits.

The superfamily Scarabaeoidea includes three families, the Scarabaeidae, Lucanidae, and Passalidae. According to Peterson (1951) about 1,250 species occur in the United States and Canada. The bulk of our species belong to the family Scarabaeidae.

The Lucanidae and Passalidae inhabit decaying wood. The family Scarabaeidae, which includes fourteen commonly recognized subfamilies, has often been divided into the lamellicorn scavengers and the lamellicorn leaf chafers (Comstock, 1950). This is a generalization which has numerous exceptions, especially in the so-called scavenger group. Food habits of both adults and larvae are summarized in Table 1 taken from a review of their biology (Ritcher, 1958).

TABLE 1. FOOD HABITS OF SCARABAEIDS
(A = Adults   L = Larvae)

SUBFAMILY	SAPROPHAGOUS							PHYTOPHAGOUS	
	Carrion	Dung	Humus	Decaying vegetable matter	Duff	Litter	Wood	Fungi	Seed plants
Scarabaeinae	A L	A L		A				A	
Aphodiinae	A L	A L	L	A L				A	L
Geotrupinae	A	A L	L		L	L		A	
Acanthocerinae							L		
Pleocominae									L
Glaphyrinae				L					L
Troginae	A L								
Melolonthinae			L					A	L
Rutelinae			L				L	A	L
Dynastinae		L	L	L	L	L	L	A	L
Cetoniinae		L	L				L	A	

7

More details of their feeding habits are given in the introductory remarks for each chapter and following many of the larval descriptions. In general, most of the species of economic importance belong to the more specialized subfamilies Melolonthinae, Rutelinae, Dynastinae, and Cetoniinae. Adults of several species of *Phyllophaga* belonging to the Melolonthinae frequently defoliate deciduous forest and shade trees in the midwestern states. Larvae of a number of genera of Melolonthinae are especially destructive underground to grasses of lawns and pastures, strawberries, vegetable crops, and tree seedlings.

Scarabaeoidea usually have life cycles of from one to three years but that of Pleocominae is nine years or longer (Ellertson and Ritcher, 1959). Except for Pleocominae, white grubs have three larval stages (instars). Descriptions in the following pages are of the third-stage larva (except for *Pleocoma*) but will usually apply as well to second-stage larvae.

Previous work. European workers were the first to publish classifications of scarabaeoid larvae (DeHaan, 1836; Mulsant, 1842; Burmeister, 1842; Erickson, 1848). These were followed by the classical works of Schiödte (1874) and Perris (1877) on coleopterous larvae which both include large sections dealing with scarabaeoid larvae.

During the past thirty years, many workers in Europe and other parts of the world have published comprehensive studies of their scarabaeoid fauna which include keys for separating larvae to family, subfamily, and genus (VanEmden, 1941, Great Britain; Gardner, 1935, India; Golovianko, 1936, Russia; Janssens, 1947, Belgium; Korschefsky, 1940, Germany; Panin, 1957, Roumania; Schaerffenberg, 1941, Germany; Paulian, 1941 and 1945, France; Viado, 1939, Philippine Islands; Medvedev, 1952, Russia; Garcia-Tejero, 1947, Spain; and so forth). Many of these keys can be used to separate the families and subfamilies of North American Scarabaeoidea since the larger groups are nearly worldwide in distribution. Van Emden's keys to the British fauna (1941) are the most applicable for our fauna and include many of our genera.

For the United States fauna, only two general works are available. Böving and Craighead (1931), in their illustrated synopsis of the principal larval forms of the order Coleoptera, give a key to the sub-families and tribes of scarabaeoid larvae but with only an occasional mention of the genera or groups of genera. Hayes (1929) published a monograph of the morphology, taxonomy, and biology of larval Scarabaeoidea of the United States which contains keys to families, subfamilies, genera, and, in some groups, to species.

Scope of this monograph. The present work contains most of the descriptions, drawings, and notes on biology of Scarabaeidae published previously by the writer in Kentucky Agricultural Experiment Station Bulletins 442, 467, 471, 476, 477, 506, and 527, and in other papers. In addition, new descriptions are included of species in the following genera of Scarabaeidae: *Onthophagus, Bolboceras, Lichnanthe, Cloeotus, Serica, Rutela, Ancognatha, Dynastes, Phileurus,* and *Valgus.* Entirely new sections on the subfamilies Hybosorinae, Glaphyrinae, Acanthocerinae, and Troginae, and on the families Lucanidae and Passalidae, are also included.

This monograph does not mention all of the North American species of scarabaeoid larvae which have been described. Descriptions of most of these, which are mainly in the subfamilies Geotrupinae, Aphodiinae, and Melolonthinae, can be found by referring to papers of Howden, Jerath, and Böving, cited in the text and bibliography. A few exotic genera and species are included to point out relationships and because they were included in the author's previous papers.

Methods and materials. Positively identified larvae are a necessity for larval taxonomy, but they are much more difficult to secure than adults since rearing is often involved. Part of the larvae used in this study were named specimens borrowed from other workers, from museums, and from university collections. Many well-associated larvae or their cast skins were obtained by the writer and his assistants in the following ways:

1. Third-stage grubs, collected from the soil, from decaying wood, or elsewhere, were segregated as to kind, part were preserved, and the rest kept (usually individually) in two or three ounce metal salve boxes, with suitable food and host media, until they transformed to the adult stage. Identification of the adults was then made or obtained from a specialist. For taxonomic studies, the cast larval skins so obtained can be used almost as well as preserved larvae.

2. Adult scarabs were collected on food plants, from the soil or manure, from carrion, or at lights. They were separated as to species and maintained in egg-laying cages. In the case of leaf-feeding forms, adults were usually confined in a gallon can half filled with sifted soil and containing a vial holding a few fresh leaves. The soil was sifted through screens at intervals of several days to remove the eggs. Eggs were then incubated in a cool place in individual cavities made in packed soil in a two-ounce salve box (Ritcher, 1940). First-stage larvae were placed individually in two-ounce salve boxes containing soil and a few kernels of wheat. To rear grubs quickly, the boxes were kept at room temperature and the soil and wheat changed every two weeks. By this method, third-stage larvae, even of species of *Phyllophaga* with two- or three-year life cycles, could be obtained within two months after hatching.

Aphodius larvae were obtained by confining the adults in pint jars of fresh manure. Larvae of other dung beetles (Scarabaeinae) were obtained by isolating adults in wooden screen-top boxes, nearly full of sifted soil and sunk in the ground. Some Scarabaeinae which roll balls of dung, or carrion, such as *Canthon* and *Deltochilum,* can best be reared within screened, four by six foot wooden barriers sunk in undisturbed soil.

Trox larvae are easily reared in two quart glass jars half filled with sifted soil. Feathers and bits of fur and animal skin, mixed in the top of the soil and laid on the soil surface, provide the necessary food for both adults and larvae.

3. Pupae or adults in pupal cells were collected from the soil or decaying wood, together with the associated cast larval skins. In some cases, larvae found at the same time could be identified by comparing their characters with those of the cast skins. Pupae were reared to the adult stage or in some cases could be identified later in their development by characters of the male genitalia.

4. Adults of Geotrupinae, some Scarabaeinae, and a few Dynastinae provision their burrows with food for their larvae. Frequently, the adults and eggs or larvae may be taken from the same burrow, giving a positive association. Adults of several Scarabaeinae construct a brood chamber in the soil where they remain with the balls of dung containing their larvae. Thus both may be collected at the same time. Passalid adults and larvae are found together since the adults prepare food for the larvae.

Larvae were preserved by dropping them into near boiling water and, in about three minutes removing them to 70 percent ethyl alcohol. Cast skins were dropped directly into 70 percent alcohol.

Larvae were studied with a binocular dissecting microscope using 10X and 15X oculars and 1X and 3X objectives, giving a range of magnification of from 10X to 45X. Mouth parts and spiracles were dissected and mounted on microscope slides in Hoyer's medium. A compound microscope was used to study structural details.

Drawings were prepared using a squared grid in the ocular of the microscope. Drawings were made large enough to reduce by a factor of 3:1.

Explanation of characters used in keys and descriptions. The keys are admittedly artificial and not intended to show phylogenetic relationships, although such are frequently apparent. Distinguishing characters that are readily seen with a binocular microscope and subject to the least variation were selected for use in the keys. Where minute characters are used, they were selected from necessity. The writer attempted to keep the terminology to a minimum and still retain the exactness of description made possible by Böving's naming of the various structures (Böving, 1936 and 1942). The most valuable distinguishing characters are found on the head and mouthparts and on the ventral surface of the last abdominal segment.

The mandibles of larvae of several of the more generalized scarabaeoid genera have a scissorial region consisting of three to five teeth. Of these, the distal two or three teeth, on the right and left mandibles, respectively, are usually larger and similar in shape. In certain genera of the subfamily Dynastinae (and in many other groups as well) the scissorial region of the mandible consists of a distal blade-like portion separated from a small proximal tooth by a distinct notch, called the scissorial notch (SN, Fig. 317), caudo-laterad of which is found a prominent seta. Judging by the position of the scissorial notch and the caudo-laterad seta, it appears that the anterior, blade-like portion of the left mandible of dynastid and melolonthid larvae represents a fusion of the two distal teeth characteristic of lucanid larvae and larvae of several primitive scarabaeid genera (S_{1+2}, Figs. 341, 342, and 344) while the narrower blade of the right mandible has developed from a single tooth (S_1, Fig. 343). The tooth proximad of the scissorial notch on the left mandible of dynastid larvae would then represent a third tooth (S_3, Figs. 317, 341, 342, and 344), while the similar tooth of the right mandible would represent a second tooth (S_2, Fig. 343).

In other dynastid genera and in certain cetoniid genera there is, in addition, a second tooth posterior to the scissorial notch which may be more or less remote from the first tooth. This tooth is better developed on the left mandible (S_4, Figs. 342, 344, and 345) and is frequently less well developed or absent on the right mandible (S_3, Fig. 343). This additional tooth on the inner margin of the mandible is probably homologous to the small, proximal, scissorial teeth of *Dorcus* (Fig. 459) and to the prominent, bifurcate tooth on the inner margin of the mandible of *Geotrupes* (Fig. 108). In the subfamily Dynastinae it is a good character for separation of groups of genera.

In using the keys, care should be taken to locate the scissorial notch of the left mandible before deciding whether or not the fourth scissorial tooth is absent or present. This is necessary since mandibles are subject to considerable wear. Mandibles of cast skins often lack much of the terminal blade-like scissorial portion.

The number of sensory spots on the last segment of the antenna also offers an excellent character for separating genera (Figs. 335-337). Use of this character simplifies the separation of many tribes and subfamilies of the Scarabaeidae. In this paper use is made only of the dorsal sensory spots which are the spots on the surface of the side opposite from the distal projection of the subapical antennal segment (Fig. 325).

The relative size of the spiracles on the thoracic and abdominal segments is quite constant for the various species and genera. Frequently, the relative size is much the same for all the species of a given genus. There is considerable variation among genera in the shapes of the "holes" of the respiratory plates of their spiracles and considerable variation among genera and species in the number of "holes." Since the "holes" usually occur in rather irregular, transverse rows, some indication of their comparative number is given in the species descriptions by mention of the maximum number found along any diameter of the respiratory plate of each thoracic spiracle. In the description of *Cyclocephala immaculata* Oliv., for example, it is stated that the spiracular respiratory plate has a maximum of fourteen to twenty-two "holes" along any diameter. The smaller number of "holes," at least fourteen in this case, usually occurs along diameters across the middle part of the plate, while a larger number of "holes," but not over twenty-two, usually occurs across the arms of the plate. There is also considerable variation among species in the shapes of the respiratory plate. Where the distance between the two lobes of the respiratory plate

of the thoracic spiracle ranges from 100 to 80 percent of the dorso-ventral diameter of the bulla, the writer has used "equal to or slightly less than" in describing the degree of constriction. "Somewhat less than" (Fig. 352) is used where the distance is from less than 80 percent to 60 percent; "much less than" (Fig. 348) where it is from less than 60 percent to 20 percent; and "almost contiguous" (Fig. 354) where it is less than 20 percent.

Except for the setation of the head and venter of the last abdominal segment, little detailed mention of body setation has been made by most writers. Study of larvae of the tribe Anomalini and of the subfamily Dynastinae has shown many quite constant differences in the body setation of the various genera and species which are worthy of description and use in keys. The writer has made considerable use of differences in the setation of the dorsal portions of the abdominal segments.

Superfamily Scarabaeoidea

Key to the Families of the Superfamily Scarabaeoidea, Based on Characters of the Larvae

1. Mandible without a ventral process (Fig. 492). Antenna 2-segmented (Fig. 493). Maxillary palpus 3-segmented (Fig 489). Thoracic spiracles with emarginations of respiratory plates facing anteriorly, those of abdominal segments facing posteriorly (Fig. 488). Metathoracic legs greatly reduced in size, unsegmented (Fig. 488). .... family **Passalidae**
page 199

   Mandible with a ventral process (Fig. 189). Antenna with 3, 4, or apparently 5 segments (Figs. 138, 176, and 51). Maxillary palpus 4-segmented (Fig. 179). Thoracic spiracles with emarginations of respiratory plates facing anteriorly (Fig. 453), cephaloventrally, posteriorly (Fig. 322), caudoventrally or ventrally (Fig. 3); those of abdominal segments facing anteriorly (Fig. 453), cephaloventrally or ventrally (Fig. 48) but never posteriorly. Metathoracic legs normal (Fig. 322) or reduced in size; if reduced in size, with 2 or more segments (Fig. 89) 2

2. Maxillary stridulatory teeth usually absent (Fig. 466); if present, then the distal segment of the antenna greatly reduced in size. Anal opening longitudinal (Fig. 484) or Y-shaped (Fig. 485), usually lying between 2 oval lobes (Fig. 484). Stridulatory organs present on mesothoracic and metathoracic legs (Figs. 469-478). Legs always normal in size .... family **Lucanidae**
page 185

   Maxillary stridulatory teeth usually present (Fig. 179); if absent, then the distal segment of the antenna not greatly reduced in size. Anal opening transverse (Figs. 366-370), angulate (Fig. 145), or Y-shaped (Fig. 210). Stridulatory organs present or absent on legs. Legs may be reduced in size .... family **Scarabaeidae**
page 11

Family Scarabaeidae

Key to Subfamilies of Scarabaeidae Based on Characters of the Larvae

1. Maxilla with galea and lacinia distinctly separated (Fig. 21) 2

   Maxilla with galea and lacinia entirely fused (Fig. 413), or fused proximally but distally free, or (rarely) not fused but fitting tightly together (Fig. 179) 10

2. Distal segment of antenna little if any reduced in diameter compared to the other antennal segments (Figs. 49 and 156).................................. 3

   Distal segment of antenna much reduced in size, often appearing as an appendage of the penultimate segment (Figs. 27, 138, and 166) 5

3. Anterior margin of labrum trilobed or broadly rounded; without serrations or truncate lobes .... a few **Aphodiinae**
page 26

   Anterior margin of labrum serrate or with truncate lobes (Figs. 77 and 156) 4

4. Anterior margin of labrum with 3 truncate lobes (Figs. 77 and 84). Prothoracic and mesothoracic legs with stridulatory organs (Figs. 80 and 82) .... **Hybosorinae**
page 37

   Anterior margin of labrum strongly serrate (Figs. 156 and 165). Stridulatory organs may be present on mesothoracic and metathoracic legs ...... **Acanthocerinae**
page 67

5. Epipharynx with tormae united mesally, anterior phoba present (Figs. 32 to 38). Anal opening usually surrounded by fleshy lobes (Figs. 42 and 131) 6

   Tormae of epipharynx not united mesally, anterior phoba absent (Figs. 137 and 149). Last abdominal segment without fleshy lobes (Figs. 145 and 153) 9

Subfamily Scarabaeinae (Coprinae)

The scarabaeid subfamily Scarabaeinae contains numerous genera and species of small to very large, round or oval dung-feeding beetles. Many species are black or purplish-black, but a few have brilliant metallic colors. The head, especially of the male, often bears a prominent median horn.

Contrary to common belief, dung beetles are of considerable economic importance. They are agriculturally beneficial in destroying the breeding places of such important cattle pests as the horn fly and in incorporating organic matter into the soil (Lindquist, 1935). They are harmful in that many species serve as intermediate hosts for parasites of domestic animals (Hall, 1929).

Adults of the subfamily Scarabaeinae are commonly called "tumble bugs." Strictly speaking, in the United States and Canada, this name should apply only to the genera *Canthon* and *Deltochilum*, inasmuch as other genera occurring here do not roll balls of dung. Species of *Copris* and *Phanaeus* form balls of dung only in underground chambers, while species of *Pinotus, Onthophagus,* and *Ateuchus* pack dung into the ends of their underground burrows.

Part of the dung buried by Scarabaeine beetles is used as food; in other balls or pellets of dung a single, large egg is deposited. The larva develops inside the ball or pellet, and when full grown forms a round pupal cell within which it transforms to the adult. In some genera the parent beetles remain for some time in the burrows with their progeny.

There are several excellent accounts of the biology of our native Scarabaeinae by Warren (1917), Lindquist (1933 and 1935), and Cooper (1938), and Mathews (1961 and 1963). Howden and Cartwright (1963) have summarized what is known about the biology of the genus *Onthophagus* and have given new information about the habits of a number of species. Little has been known, however, of the taxonomy and morphology of the larvae. Hayes (1928 and 1929) partially described the larvae of *Pinotus carolinus* (Linn.), *Canthon pilularius* (Linn.), *Copris tullius* Oliv. (= *C. fricator* [Fab.]), and *Onthophagus* sp., separating the latter three in a key.[1] Howden and Ritcher (1952), were the first to describe the larva of *Deltochilum gibbosum* (Fab.). Böving and Craighead (1931), in their key to families and subfamilies of the Scarabaeoidea, separated the genera *Canthon, Copris,* and *Onthophagus,* using differences in leg structure.[2]

In this section the larvae of eight genera[3] and eight species of the subfamily Scarabaeinae are described in detail and keys are given for their separation. A brief description of the adult and the biology of each species is appended to the description of the larva. This material is a reprint of my 1945d paper, with emendations, plus a description of the larva of *Deltochilum gibbosum* (Fab.) (reprinted from Howden and Ritcher, 1952) and an expanded section on the genus *Onthophagus* which includes new descriptions of the larvae of several species, based on reared material loaned by Howden.

Larva hump-backed. Labrum symmetrical, distal margin trilobed. Epipharynx with tormae united mesally; plegmata, proplegmata, clithra, epizygum, zygum, and nesia absent. Pedium of epipharynx surrounded laterally and anteriorly by phobae.[4] Chaetopariae each consisting of a very few stout setae or a patch of stout setae. Anterior part of haptolachus with a transverse, sinuate mesophoba. Scissorial area of left mandible bearing 3 teeth (S_1, S_2, and S_3), that of the right mandible bearing 2 teeth (S_{1+2} and S_3). Galea and lacinia of maxilla widely separated, not fused. Maxillary stridulatory area consisting of a row of short, conical teeth; anterior process absent. Hypopharynx with 2 asymmetrical, sclerotized processes termed "oncyli."[5] Antenna 4-segmented with the distal segment so reduced in diameter that it appears to be an appendage of the third segment. Third segment of antenna with a distal, oval sensorial spot or a conical sensorial appendage. Concavities of spiracular respiratory plates, when present, directed ventrally; arms of respiratory plates not constricted.

---

[1] Through the kindness of Dr. Hayes, the writer has examined the specimen of *Canthon pilularius* (Linn.) which was used in making the drawings for his 1928 and 1929 papers under the name *Canthon laevis* (Drury). The epipharynx of this specimen, which was drawn from an oblique angle, is torn; therefore the central portion bearing the phobae, haptolachus, and much of the tormae is missing (see Hayes, 1929, Fig. 55, Plate 6).

[2] According to Dr. Böving, Dr. Craighead and he were handicapped in their study of coprinid larvae by having only a scanty amount of poorly preserved material available for study.

[3] *Megathopa* is the only genus known to occur in the United States which is not included.

[4] The new term "protophoba" is suggested for the anterior phoba; the term "dexiophoba" for the right, lateral phoba; and the term "laeophoba" for the left, lateral phoba.

[5] At Dr. Böving's suggestion, the new term "oncylus" (plural oncyli) is proposed for the hypopharyngeal sclerite (or sclerites) regularly found in beetle larvae with masticating mandibles. The term is derived from the Greek word "oncylos" meaning something bulky and swollen, a welt or protuberance.

Last abdominal segment with several fleshy, lateral and caudal lobes. Raster with a teges or paired polystichous or monostichous palidia. Anal slit transverse. Legs without stridulatory structures, 2-segmented and bearing 1 or 2 terminal setae. Claws minute or absent; when present, each bearing a terminal seta. Metathoracic legs not conspicuously smaller than prothoracic and mesothoracic legs.

Key to Known Genera of the Subfamily Scarabaeinae, Based on Characters of the Third-Stage Larvae

1. Prothoracic shield without anteriorly projecting processes (Figs. 1 and 2) 2

   Prothoracic shield with an anteriorly projecting, angular process on each side (Figs. 3 and 23) 4

2. Third abdominal segment bearing a prominent conical, dorsal gibbosity covered with numerous short, stout setae (Fig. 1) ....*Onthophagus*

   Third abdominal segment without a prominent, conical, dorsal gibbosity (Figs. 2 and 3) 3

3. Venter of last abdominal segment with 2 monostichous, longitudinal palidia (Fig. 43) ....*Ateuchus*

   Venter of last abdominal segment with 2 patches of short, spine-like setae ....*Oniticellus*[6]

4. Legs with a pair of terminal setae, claws absent (Figs. 13, 17, and 18) 5

   Legs with a single terminal seta which may be set on a small blunt claw (Figs. 14, 16, 19, and 20) 6

5. Venter of last abdominal segment with a pair of polystichous palidia ....*Pinotus*

   Venter of last abdominal segment covered with numerous (200 or more) short, stout, caudally directed setae ....*Deltochilum*

6. Venter of last abdominal segment with paired, median, caudal lobes (Fig. 42) or a cleft median lobe (Fig. 41). Legs with small, blunt claws each bearing a terminal seta (Fig. 14) ....*Copris*

   Venter of last abdominal segment with a single, broad, caudal, median lobe (Figs. 40 and 46). Claws absent (Figs. 16 and 20) 7

7. Median portion of venter of last abdominal segment covered with a large quadrate patch of stout, caudally directed, spine-like setae (Fig. 40)..............*Phanaeus*

   Median portion of venter of last abdominal segment with 2 inconspicuous patches of very short setae (Fig. 46) ....*Canthon*

---

[6] Based on larvae of *Oniticellus cinctus* F., from India. Loaned by the United States National Museum. See Gardner (1929).

TRIBE Scarabaeini

Genus *Canthon* Hoffmansegg
Canthon pilularius (Linn.),[7] Third-Stage Larva
(Figs. 15, 16, 23, 26, 29, 35, and 46)

This description is based on the following material and reprinted from my 1945d bulletin:

Twenty-eight third-stage larvae obtained by confining adults with fresh cow dung, during the summer of 1944, at Lexington, Kentucky.

Maximum width of head capsule 3.48 to 3.6 mm. Surface of cranium light yellow-brown, faintly reticulate. Frons on each side with 2 to 5 posterior frontal setae, one seta or rarely 2 setae in each anterior angle, a group of 4 to 9 exterior frontal setae and a group of 8 to 17 anterior frontal setae (Fig. 29). Pedium of epipharynx (Fig. 35) bare except for a very few setae in the anterior, median portion. Each chaetoparia consisting of about 7 to 9 setae. Tormae symmetrical. Mesophoba monostichous. A group of 4 isolated macrosensilla in the center of the haptolachus. Maxillary stridulatory area with a more or less irregular row of 12 to 17 sharp-pointed, conical, anteriorly pointing teeth (Fig. 15). Uncus of lacinia without a proximal tooth (Fig. 26). Glossa, anterior to the oncyli, with a transverse row of spine-like setae. Third antennal segment with a distal subconical sensory organ.

Prothoracic shield with an anterior, angular projection on each side (Fig. 23). Dorsum of third abdominal segment without a projecting wart. Dorsa of abdominal segments 6 to 8 inclusive, each with 2 widely separated, narrow transverse bands of about 3 irregular rows of setae; setae longer and stouter in posterior band. Dorsum of abdominal segment 9 with anterior band absent or represented by a few very short setae. Venter of last abdominal segment (Fig. 46) with a pair of inconspicuous, polystichous palidia easily overlooked. Each palidium consists of a patch of very small, short, spine-like setae directed caudolaterally. Posterior to the palidia is a broad median lobe flanked on each side by 2 other fleshy lobes. Legs each with a long, slender terminal seta surrounded by a circlet of 8 to 9 short, stout setae (Fig. 16). Claws absent.

Canthon pilularius (Linn.) is the very common, ball-rolling "tumble bug" familiar to everyone who has lived in the country in the eastern United States. According to Leng (1920), it occurs throughout the Atlantic states and as far west as Arizona and southern California. The beetle is flattened dorsally, ranging from 12 to 19 mm in length and from 7.5 to 11.5 mm in width. The head, pronotum, and elytra are purplish-

---

[7] *Canthon laevis* (Drury) is a synonym of this name.

black, finely shagreened, and free of protuberances, pits, and prominent striae. The venter and long legs are black. Adults of *Canthon pilularius* form balls of cow, horse, or sheep dung which they roll about the pasture or along lanes until they find a suitable spot for burial. They seem to avoid loose soil for this purpose, burying balls 3 to 5 inches deep, each in a separate chamber. Here a single, large egg is laid in a cavity near the surface of each ball and covered over, giving the ball a pyriform shape. Completed balls average 21 to 27 mm in length and 19 to 24 mm in diameter. Ball rolling in Kentucky is most common in May and June, the activity being renewed in the summer, after rains. The presence of a larva within a ball can be detected by a warty covering of dried excrement on the outside of the ball. Lindquist (1935) gives an account of the biology of this species as observed in Texas.

Genus *Deltochilum* Escholtz
Deltochilum gibbosum (Fab.), Third-Stage Larva

The following description is taken from the paper by Howden and Ritcher (1952):

"Maximum width of head capsule 5.3 to 5.6 mm. Surface of cranium light yellow-brown, faintly reticulate. Body hump-backed, grayish-white in color.

"Frons with from 1 to 3 small, posterior frontal setae on each side, a single seta in each anterior angle, and a combined transverse patch of 7 to 10 anterior and exterior frontal setae on each side. Labrum symmetrical, strongly trilobed. Pedium of epipharynx bare. Each chaetoparia covered with from 20 to 25 setae. Tormae symmetrical. Mesophoba monostichous. Anterior epitorma absent. Haptolachus with 4 macrosensilla just caudad of the mesophoba. Maxillary stridulatory area with a row of 5 to 11 conical teeth. Lacinia with a large, distal, sclerotized, blade-like uncus; uncus with a small, proximal, tooth-like projection. Galea with a small conical uncus. Glossa, anterior to the oncyli, with a transverse row of spine-like setae. Third antennal segment with a small, distal, subconical sensory organ.

"Prothoracic shield with an anterior, angular process on each side. Dorsum of third abdominal segment without a projecting wart. Abdominal segments 6 to 8 inclusive, each with 2 dorsal annulets, each annulet with a sparsely set transverse row of short, slender setae. Venter of last abdominal segment covered with numerous (200 or more), short, rather stout, caudally directed setae. Ventral anal lobe entire but with a deep median cleft which, on some larvae, joins a transverse cleft forming an inverted Y or T.

"Legs each with 1 or 2 apical setae (usually 2) surrounded by a circlet of 8 setae. Claws absent."

Deltochilum gibbosum is a large, round, rather flat, black scarab, with long legs, which occurs in the southeastern United States. The adults are attracted to ani-

mal fur, feathers, and human excrement (Cartwright, 1949). Adults were observed rolling balls of feathers. The completed egg-ball is coated with mud and leaves and is shaped like a narcissus bulb (Howden and Ritcher, 1952). Egg-balls are set in cup-shaped depressions in the ground.

TRIBE Coprini

Genus *Ateuchus* Weber
Ateuchus histeroides (Web.), Third-Stage Larva
(Figs. 2, 4, 12, 19, 30, 37, and 43)

This description is based on the following material and reprinted from my 1945d bulletin:

Four third-stage larvae obtained by confining adults with fresh cow dung, during the summer of 1944, Lexington, Kentucky.

Maximum width of head capsule 2.0 to 2.1 mm. Surface of cranium light yellow-brown; smooth. Frons on each side with 1 or 2 posterior frontal setae, one seta in each anterior angle, one exterior frontal seta and 2 or 3 anterior frontal setae (Fig. 30). Pedium of epipharynx (Fig. 37) bare. Each chaetoparia with about 4 setae. Tormae asymmetrical (left pternotorma absent). Mesophoba polystichous on left side. A pair of macrosensilla is borne on each side of the median line, just behind the mesophoba. Maxillary stridulatory area with a row of 8 to 12 short, conical teeth (Fig. 12). Uncus of lacinia with a proximal tooth. Glossa, anterior to the oncyli, with a transverse fringe of spine-like setae. Third antennal segment with a distal, conical sensory organ (Fig. 4).

Prothoracic shield without anterior projections (Fig. 2). Dorsum of third abdominal segment without a projecting wart. Dorsa of abdominal segments 6 to 8 inclusive, each with an anterior, sparsely set, transverse patch of fairly long, slender setae and a posterior, transverse band of similar setae. Dorsum of abdominal segment 9, with setation limited to a posterior patch of sparsely set, slender setae. Venter of last abdominal segment with 2 longitudinal palidia (Fig. 43). Each palidium consisting of a single curved, closely set row of 27 to 33 slender pali. Septula oval. Venter, caudal of the palidia, with a broad, median, fleshy lobe flanked on each side by a single fleshy lobe. Legs each with a long terminal seta surrounded by a circlet of 4 or 5 shorter setae (Fig. 19). Claws absent.

Ateuchus histeroides (Web.) is a small woodland species occurring in a number of states east of the Mississippi River (Leng, 1920 and Brimley, 1938). The adult is a dark, shiny, iridescent purple, ranging from 6 to 7 mm in length and 4 to 5 mm in width. The head and pronotum are smooth with many small, shallow punctures. The smooth elytra have a series of shallow, longitudinal striae which are sparsely punctate. Adults are found in cow manure in woodlands. They excavate vertical tunnels beneath dung to a depth of at least 10 or 12 inches. A single egg is laid in a cell near the upper end of the vertical pellet of dung packed in the lower end of each tunnel. Eggs are laid from late April into September.

Genus *Pinotus* Erichson
Pinotus carolinus (Linn.), Third-Stage Larva
(Figs. 3, 6, 9, 10, 13, 17, 18, 21, 27, 32, and 39)

This description is based on the following material and reprinted from my 1945d bulletin:

Cast skins of 2 third-stage larvae collected September 23, 1941, at Lexington, Kentucky, from the soil of a bluegrass pasture, and reared to the pupal stage.

Six third-stage larvae collected February 7, 1944, at Lexington, Kentucky, behind the plow in a bluegrass pasture.

Maximum width of head capsule 5.36 to 6.1 mm. Surface of cranium light yellow-brown; smooth, not pitted. Frons on each side with 1 posterior frontal seta, 1 seta in each anterior angle and a combined row or transverse patch of 3 to 8 anterior and exterior frontal setae (Fig. 27). Pedium of epipharynx (Fig. 32) with a cluster of small spine-like setae in the angle between the protophoba and the dexiophoba. Chaetopariae each consisting of 15 to 20 setae. Tormae symmetrical. Two pairs of macrosensilla are located just behind the mesophoba. Maxillary stridulatory area with a row of 6 to 11 short, conical teeth. Uncus of lacinia with a small proximal tooth. Glossa with a transverse, irregular row of granules anterior to the oncyli (Fig. 21). Third antennal segment with a distal, oval sensory spot (Fig. 6).

Prothoracic shield with an anterior, angular projection on each side (Fig. 3). Dorsum of third abdominal segment without a projecting wart. Dorsa of abdominal segments 6 to 9 inclusive, each with a single, posterior, transverse row of setae. Venter of last abdominal segment with a pair of polystichous palidia (Fig. 39). Each palidium consisting of a dense patch of caudomesally directed, spine-like setae, not arranged in definite rows. Legs each with a pair of terminal setae surrounded by a circlet of 9 or 10 setae (Figs. 17 and 18). Claws absent.

Pinotus carolinus (Linn.) is a very large, common dung beetle widely distributed over the eastern, southern, and southwestern states. The adult is strongly convex dorsally, ranging from 20 to 29 mm in length and 14 to 18 mm in width. The head, pronotum, and elytra are black and the venter and legs are reddish-black. The surface of the head is rugulose and there is often a short median horn between the eyes. The elytra have shallow, longitudinal striae. Adults of *Pinotus caro-*

linus make branched burrows which enter the soil at an oblique angle, near or under cattle droppings. The egg is laid in a mass of manure packed in the burrow. At Lexington, Kentucky, the winter is passed both in the full-grown larval and adult stages. The adult is strongly attracted to lights. Lindquist (1933) describes the biology of this species as observed in Kansas.

Genus *Copris* Geoffroy

Key to Known Larvae of the Genus *Copris*

Tormae asymmetrical (left pternotorma much longer than right pternotorma) (Fig. 36). Dorsa of abdominal segments 7 to 9 each with two widely separated transverse rows of setae ....*C. minutus* (Drury)
Tormae symmetrical (Fig. 34). Dorsa of abdominal segments 7 to 9 each with a single transverse row of setae ....*C. fricator* (Fab.)

Copris minutus (Drury), Third-Stage Larva
(Figs. 36 and 41)

This description is based on the following material and reprinted from my 1945d bulletin:

Two third-stage larvae collected June 23, 1943, at Lexington, Kentucky. Larvae removed from balls of dung found in a brood chamber in garden soil; associated with adult female.

Three third-stage larvae collected in April, 1939, in Woodford County, Kentucky; associated with 2 adults.

Cast skins of 2 third-stage larvae collected July 17, 1933, at Lexington, Kentucky. Skins removed from 2 balls, one containing a pupa and the other an adult; balls found in a brood chamber beneath cow manure.

Maximum width of head capsule 2.6 to 2.9 mm. Surface of cranium light straw-colored, faintly reticulate. Frons on each side with 1 or 2 posterior frontal setae, one seta in each anterior angle, 2 or 3 exterior frontal setae, and 1 or 2 anterior frontal setae. Pedium of epipharynx (Fig. 36) bare. Each chaetoparia with 2 to 4 setae. Tormae asymmetrical. Mesophoba polystichous on the left side. Four macrosensilla on the haptolachus with 2 borne against the posterior phoba and a transverse pair borne considerably behind these. Maxillary stridulatory area with a row of 5 to 7 conical teeth. Uncus of lacinia with a prominent proximal tooth. Glossa, anterior to the oncyli, with a transverse row of close-set, spine-like setae. Third antennal segment with a conical sensory organ.

Prothoracic shield with an anterior, angular projection on each side. Dorsum of third abdominal segment without a projecting wart. Dorsa of abdominal segments 7 to 9, each with an anterior, inconspicuous, transverse row of short setae and a conspicuous posterior row of fairly long setae. Venter of last abdominal segment with only a very few small, scattered setae; practically bare (Fig. 41). Venter of last abdominal segment posteriorly with a cleft median lobe, flanked on each side by 2 other fleshy lobes. Legs each with a short, blunt, lightly sclerotized claw bearing a terminal seta. Claw surrounded by a circlet of 7 or 8 fairly short setae.

Copris minutus (Drury) is a small, common dung beetle occurring from Canada to Florida and westward to Texas (Leng, 1920). The black adult ranges from 8 to 10.5 mm in length and 4.5 to 5.5 mm in width. The head of male has a short median horn. The elytra of both sexes have prominent, longitudinal, punctate striae. Adults of *C. minutus* construct a brood chamber several inches deep in the soil beneath cattle droppings. Here several balls are formed from an unshaped mass of dung. Balls with eggs have a slightly pyriform shape, measuring 13 to 15 mm in length and 13 to 14 mm in width. Adults are usually found in the brood chambers with the balls, even after the larvae within have pupated.

Copris fricator (Fab.), Third-Stage Larva[8]
(Figs. 8, 14, 28, 34, and 42)

This description is based on the following material:
Three third-stage larvae obtained by confining adults with fresh cow dung, during the summer of 1944, at Lexington, Kentucky.
Three cast skins of 3 third-stage larvae associated with one pupa and 2 adults of *C. fricator,* dug from pasture soil beneath cow dung, July 18, 1944, at Lexington, Kentucky.

Maximum width of head capsule 3.6 to 3.9 mm. Surface of cranium light yellow-brown, faintly reticulate. Frons on each side with 1 or 2 posterior frontal setae, one seta in each anterior angle, 1 to 3 exterior frontal setae, and 2 to 4 anterior frontal setae (Fig. 28). Pedium of epipharynx bare (Fig. 34). Each chaetoparia with 1 to 3 setae. Tormae symmetrical. Mesophoba polystichous on the left side. Four macrosensilla on the haptolachus with 2 just behind the posterior phoba and the 2 others considerably behind these. Maxillary stridulatory area with a row of 5 to 9 conical teeth. Uncus of lacinia with a proximal tooth. Glossa, anterior to the oncyli, with a transverse row of close-set, fairly short, spine-like setae. Third antennal segment with a subconical sensory organ (Fig. 8).

Prothoracic shield with an anterior angular projection on each side. Dorsum of third abdominal segment without a projecting wart. Dorsa of abdominal segments 7 to 9 with a single posterior, transverse row of

[8] Described previously (Ritcher, 1945d) under the name of *Copris tullius* Oliver, now considered a synonym (Mathews, 1961).

short setae; other setae absent. Venter of last abdominal segment with a few, very small, scattered setae (Fig. 42). Venter posteriorly with a fleshy lobe on each side of the median line, and each of these flanked laterally by 2 other fleshy lobes. Legs each with a short, blunt claw bearing a short terminal seta (Fig. 14). Claw surrounded by a circlet of 7 to 8 setae.

Copris fricator (Fab.) is a fairly large species occupying much the same range as *C. minutus* (Mathews, 1961). The black adult ranges from 14 to 17 mm in length and 8.5 to 9.5 mm in width. The head of the male bears a short to fairly long median horn. The elytra of both sexes have prominent, longitudinal, punctate striae. Adults of *C. fricator* have habits similar to those of *C. minutus*. Burrows are 3 or 4 inches deep, each terminating in a brood chamber containing 2 to 4 balls of dung. Balls are slightly pyriform in shape, ranging from 23 to 35 mm in length and 19 to 27 mm in width. Adults remain in the brood chamber with the balls even after the larvae within have become adults. Mathews (1961) gives a good account of the biology of *C. fricator* (Fab.). Details of the biology of *Copris remotus* Lec. are given by Lindquist (1935).

Genus *Phanaeus* MacLeay
Phanaeus vindex McL., Third-Stage Larva
(Figs. 7, 20, 24, 31, 33, and 40)

This description is based on the following material and reprinted from my 1945d bulletin:
Eight third-stage larvae obtained by confining 5 adults with fresh cow dung during the summer of 1944, at Lexington, Kentucky. Adults identified by E. A. Chapin of the United States National Museum.

Maximum width of head capsule 4.2 to 4.6 mm. Surface of cranium light yellow-brown, smooth. Frons on each side with 1 to 4 posterior frontal setae, one seta in each anterior angle, 1 or 2 exterior frontal setae, and an irregular oblique row of 6 to 9 anterior frontal setae (Fig. 31). Pedium of epipharynx bare (Fig. 33). Each chaetoparia consisting of about 11 setae. Tormae symmetrical. Mesophoba monostichous. A pair of macrosensilla on each side of the median line just behind the posterior phoba. Maxillary stridulatory area with a row of 6 to 9 sparse-set conical teeth. Uncus of lacinia with a small proximal tooth. Glossa, anterior to the oncyli, with a close-set fringe of spine-like setae (Fig. 24). Third antennal segment with oval sensory spot (Fig. 7).

Prothoracic shield with an anterior angular projection on each side. Thoracic spiracle considerably larger than the abdominal spiracles. Dorsum of third abdominal segment without a projecting wart. Dorsa of abdominal segments 6 to 9 each with a posterior, transverse single or double row of setae. Dorsa of abdominal segments 6 to 8 each also with an anterior, transverse, single row of much smaller sparse-set setae. Venter of last abdominal segment with a broad, quadrate patch of caudally directed, spine-like setae separated on each side from a longitudinal, subtriangular patch of similar setae by a sunken, longitudinal, bare area (Fig. 40). Caudad of these patches of setae with a broad, fleshy lobe with a smaller fleshy lobe on each side. Legs each with a long, terminal seta within a circlet of 9 to 11 much shorter setae (Fig. 20). Claws absent.

Phanaeus vindex McL. is a large, rather uncommon, brilliantly colored dung beetle occurring from New York to Florida and as far west as Colorado (Leng, 1920). The adult is somewhat flattened dorsally, ranging from 15 to 19 mm in length and 9 to 12.5 mm in width. The beetle is highly iridescent with predominantly bright green elytra and a pink and yellow-green pronotum. Fully developed males have a long, slender, posteriorly curved horn on the head and an angular, posterior process on each side of the pronotum. Adults of *P. vindex* burrow vertically in the soil at the edge of or under cattle droppings. Dung is carried to the bottom of the burrows at a depth of 8 or 10 inches and a single large ball is formed in the terminal cavity of each burrow. The completed ball, containing an egg near one end, is slightly pyriform, ranging from 34 to 44 mm in length and 31 to 43 mm in width. Adults do not stay with the young as do the adults of *Copris*. In rearing cages at Lexington, Kentucky, larvae were obtained during June and July. Lindquist (1935) gives an excellent account of the biology of *P. triangularis* (Say).

Genus *Onthophagus* Latreille
(Figs. 1, 5, 11, 22, 25, 38, 44, and 45)

Larvae of this genus may be distinguished by the following characters: Cranium straw-colored, smooth, shiny. Frons on each side with 1 or 2 posterior frontal setae, 1 seta in each anterior angle and 2 or 3 anterior frontal setae (Fig. 22). Exterior frontal setae absent. Pedium of epipharynx bare, surrounded by anterior and lateral phobae (Fig. 38). Each chaetoparia with 2 to 4 setae. Tormae symmetrical. One macrosensillum on each side behind mesophoba. Maxillary stridulatory area with a row of 5 to 11 conical teeth. Uncus of lacinia with a prominent proximal tooth (Fig. 25). Hypopharyngeal glossa, anterior to the oncyli, with a dense transverse row of spine-like setae. Third antennal segment with a conical sensory organ (Fig. 5).

Prothoracic shield without anterior projection. Dorsum of third abdominal segment with a prominent, setose gibbosity (Fig. 1). Dorsa of abdominal segments 4 to 9, each with 2 widely separated, transverse rows of setae. Venter of last abdominal segment with 1 or 2 patches of short, stout setae, not arranged in definite rows. Posterior to the raster is a broad, median anal lobe

flanked on each side by 1 or 2 other lobes. Legs each having a long terminal seta surrounded by a circlet of 8 or 9 shorter setae. Claws absent.

Key to Known Species of *Onthophagus* Larvae

1. Raster set with more than 60 short, stout setae which may be in 1 or 2 patches; maxilla with a row of 7 to 11 stridulatory teeth 2

   Raster set with less than 60 short, stout setae which are usually in one patch; maxilla with a row of 4 to 7 stridulatory teeth 3

2. Gibbosity on dorsum of third abdominal segment with 2 patches of 34 to 40 setae. Raster with 2 distinct patches of 25 to 40 short, stout setae ....*O. hecate hecate*

   Gibbosity on dorsum of third abdominal segment with 2 patches of 17 to 22 setae. Raster with 2 patches of 34 to 38 short, stout setae which may be confluent posteriorly ....*O. striatulus*

3. Gibbosity on dorsum of third abdominal segment with 2 patches of 25 to 28 setae ....*O. alluvius*

   Gibbosity on dorsum of third abdominal segment with 2 patches of fewer than 21 setae .... *O. oklahomensis, O. pennsylvanicus,* and *O. landolti texanus.*

Onthophagus alluvius Howden and Cartwright, Third-Stage Larva

This description is based on the following material:

Twenty-four third-stage larvae reared by H. H. Howden at Knoxville, Tennessee, in 1954, from adults confined with fresh cow dung in soil-filled flower pots.

Maximum width of head capsule 1.48 to 1.58 mm. Maxillary stridulatory area with a row of 4 to 7 conical teeth. Gibbosity on dorsum of third abdominal segment with 2 patches of 25 to 28 curved setae. Venter of last abdominal segment with a patch of 36 to 42 caudally or caudomesally directed, short hamate setae.

Onthophagus alluvius is a common species in lowland areas of eastern Texas and eastern Mexico (Howden and Cartwright, 1963). Information on its biology in Texas was published by Lindquist in 1935 under the name *anthracinus*.

Onthophagus hecate hecate (Panz.), Third-Stage Larva
(Figs. 1, 5, 11, 22, 25, 38, and 44)

This description is based on the following material and reprinted from my 1945d bulletin:

Sixteen third-stage larvae obtained by confining adults of *O. hecate* with fresh cow dung, during the summer of 1944, at Lexington, Kentucky.

Maximum width of head capsule 1.88 to 2.1 mm. Maxillary stridulatory area with a row of 6 to 11 short, conical teeth. Gibbosity on dorsum of third abdominal segment with 2 patches of 34 to 40 curved setae. Venter of last abdominal segment with 2 patches of short setae. Each patch consisting of about 25 to 40 caudomesally directed, spine-like setae not arranged in definite rows.

Onthophagus hecate hecate (Panz.) is a small, common dung beetle occurring from Canada to northern Florida and westward into Wyoming, Utah, and Arizona (Howden and Cartwright, 1963). Adult somewhat flattened above, ranging from 5.5 to 8.5 mm in length and from 3.5 to 4.5 mm in width. Head, pronotum, and elytra a dull purplish-black, covered with short whitish hairs. Male with the anterior median part of the head bent upward to form a flat triangular process; pronotum of male with a prominent, bifurcate, anterior process. At Lexington, Kentucky, adults of *O. hecate* are abundant in fresh cow dung during May, June, and July. They dig tunnels beneath the droppings to a depth of 2 to 9 inches and pack the lower end of each tunnel with a wad of dung. Tunnels are nearly vertical for most of their length, but each turns near its lower end so that the manure pellet lies nearly horizontal. The farther end of the pellet, at the end of the burrow, is bulbous; the egg is laid in a small cell in the other, neck-like end. When full grown, the larva constructs an elliptical, pupal cell within the remains of the old pellet.

Onthophagus landolti texanus Schaeffer, Third-Stage Larva

This description is based on the following material:
Eleven third-stage larvae reared by H. H. Howden at Knoxville, Tennessee, in 1954, from adults confined with fresh cow dung in soil-filled pots.

Maximum width of head capsule 1.22 to 1.52 mm. Maxillary stridulatory area with a sparsely set row of 5 to 7 conical teeth. Gibbosity on dorsum of third abdominal segment with 2 patches of 15 to 21 curved setae. Venter of last abdominal segment with a sparsely set patch of 28 to 32 short, caudally directed setae.

Onthophagus landolti texanus is fairly common in southern Texas, especially in the lower Rio Grande Valley (Howden and Cartwright, 1963). The same writers give details of the biology as observed in the laboratory.

Onthophagus oklahomensis Brown, Third-Stage Larva

This description is based on the following material:
Three third-stage larvae reared by H. H. Howden at Knoxville, Tennessee, in 1954, from adults collected at Bastrop State Park in Texas.

Maximum width of head capsule 1.11 to 1.29 mm. Maxillary stridulatory area with a row of 5 to 7 conical teeth. Gibbosity on dorsum of third abdominal segment with 2 sparsely set patches of 15 to 18 short, slender, curved setae. Venter of last abdominal segment with a sparsely set patch of 33 to 36 short setae.

Onthophagus oklahomensis is a very small species, closely related to *O. pennsylvanicus,* occurring in sandy areas of the southeastern coastal plain and in Oklahoma and Texas (Howden and Cartwright, 1963). According to the same writers, this species makes shallow 1- to 3-inch deep burrows under or beside cow dung.

Onthophagus pennsylvanicus Harris, Third-Stage Larva
(Fig. 45)

This description is based on the following material and reprinted from my 1945d bulletin:
Thirteen third-stage larvae obtained by confining adults of *O. pennsylvanicus* with fresh cow manure, during the summer of 1944, at Lexington, Kentucky.

Maximum width of head capsule 1.23 to 1.4 mm. Maxillary stridulatory area with a row of 5 to 7 conical teeth. Gibbosity on dorsum of third abdominal segment with two patches of 10 to 18 curved setae. Venter of last abdominal segment with a subquadrate patch of 25 to 45 short, caudally directed, spine-like setae (Fig. 45) and little or no suggestion of a septula.

Onthophagus pennsylvanicus Harris is a very small dung beetle occurring from Ontario to central Florida and westward to Colorado and central Texas (Howden and Cartwright, 1963). Adult somewhat flattened above, ranging from 2 to 4 mm in length and 2 to 2.5 mm in width. Head, pronotum, and elytra of adult a dull purplish-black and covered with short, whitish hairs. Pronotum with numerous shallow pits. Sexes not dimorphic. The biology of *O. pennsylvanicus* is very similar to that of *O. hecate.* Larvae have been reared successfully in pint jars of soil topped with fresh cow dung.

Onthophagus striatulus striatulus Palisot de Beauvois, Third-Stage Larva

This description is based on the following material:
Two third-stage larvae reared by H. H. Howden in 1954, from adults from Bastrop State Park, Texas.

Maximum width of head capsule 2.1 mm. Maxillary stridulatory area with a row of 8 to 11 conical teeth. Gibbosity on dorsum of third abdominal segment with 2 patches of 17 to 22 long, slender, curved setae. Venter of last abdominal segment with 2 patches of 34 to 38 short, stout, caudally directed setae, which may be confluent posteriorly.

Onthophagus striatulus striatulus is a variable species so far as adult characters are concerned. The male major may be recognized by the two long horns on the head (Howden and Cartwright, 1963). According to the same writers, adults prefer rotting fungi as food, but the larval burrow is provisioned with dung.

Plate I

FIGURE 1. *Onthophagus hecate hecate* (Panz.). Third-stage larva, left lateral view.

FIGURE 2. *Ateuchus histeroides* Web. Third-stage larva, left lateral view.

FIGURE 3. *Pinotus carolinus* (Linn.). Third-stage larva, left lateral view.

FIGURE 4. *Ateuchus histeroides* (Web.). Distal portion of right antenna.

FIGURE 5. *Onthophagus hecate hecate* (Panz.). Distal portion of right antenna.

FIGURE 6. *Pinotus carolinus* (Linn.). Distal portion of right antenna.

FIGURE 7. *Phanaeus vindex* McL. Distal portion of right antenna.

FIGURE 8. *Copris fricator* (Fab.). Distal portion of right antenna.

FIGURE 9. *Pinotus carolinus* (Linn.). Left mandible, dorsal view.

FIGURE 10. *Pinotus carolinus* (Linn.). Scissorial area of right mandible, dorsal view.

FIGURE 11. *Onthophagus hecate hecate* (Panz.). Left mandible, dorsal view.

FIGURE 12. *Ateuchus histeroides* (Web.). Teeth of stridulatory area on left maxilla.

FIGURE 13. *Pinotus carolinus* (Linn.). Left mesothoracic leg, side view.

FIGURE 14. *Copris fricator* (Fab.). Tip of left mesothoracic leg.

FIGURE 15. *Canthon pilularius* (Linn.). Teeth of stridulatory area on left maxilla.

FIGURE 16. *Canthon pilularius* (Linn.). Tip of left mesothoracic leg, side view.

FIGURE 17. *Pinotus carolinus* (Linn.). Tip of left mesothoracic leg, end view.

FIGURE 18. *Pinotus carolinus* (Linn.). Tip of left mesothoracic leg, side view.

FIGURE 19. *Ateuchus histeroides* (Web.). Tip of left mesothoracic leg, side view.

FIGURE 20. *Phanaeus vindex* McL. Tip of left mesothoracic leg, side view.

Symbols Used

Cl—Claw S_{1-3}—Scissorial teeth   SE—Sensory appendage
R—Raster

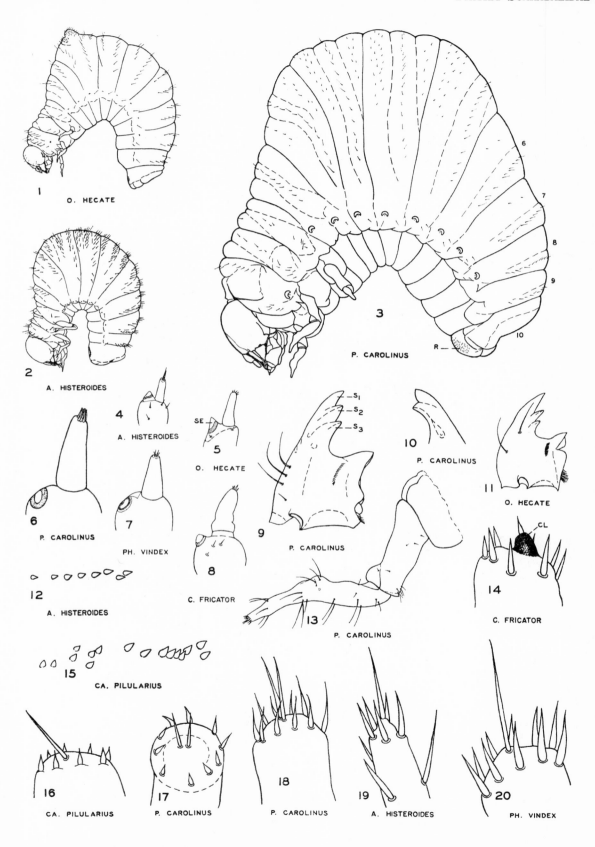

1   O. HECATE

2

A. HISTEROIDES

3   P. CAROLINUS

6

7

8

9

10

R

4   A. HISTEROIDES

5   O. HECATE

SE

S_1
S_2
S_3

10   P. CAROLINUS

6   P. CAROLINUS

7   PH. VINDEX

8   C. FRICATOR

9   P. CAROLINUS

11   O. HECATE

12   A. HISTEROIDES

13   P. CAROLINUS

14   C. FRICATOR
CL

15   CA. PILULARIUS

16   CA. PILULARIUS

17   P. CAROLINUS

18   P. CAROLINUS

19   A. HISTEROIDES

20   PH. VINDEX

Plate II

FIGURE 21.   *Pinotus carolinus* (Linn.). Left maxilla, labium, and hypopharynx, ental view.

FIGURE 22.   *Onthophagus hecate hecate* (Panz.). Head, dorsal view.

FIGURE 23.   *Canthon pilularius* (Linn.). Head and prothorax, dorsal view.

FIGURE 24.   *Phanaeus vindex* McL. Labium and hypopharynx, ental view.

FIGURE 25.   *Onthophagus hecate hecate* (Panz.). Uncus of lacinia of left maxilla, ental view.

FIGURE 26.   *Canthon pilularius* (Linn.). Uncus of lacinia of left maxilla, ental view.

FIGURE 27.   *Pinotus carolinus* (Linn.). Head, dorsal view.

FIGURE 28.   *Copris fricator* (Fab.). Head, dorsal view.

FIGURE 29.   *Canthon pilularius* (Linn.). Head, dorsal view.

FIGURE 30.   *Ateuchus histeroides* (Web.). Head, dorsal view.

FIGURE 31.   *Phanaeus vindex* McL. Head, dorsal view.

FIGURE 32.   *Pinotus carolinus* (Linn.). Epipharynx.

FIGURE 33.   *Phanaeus vindex* McL. Epipharynx.

Symbols Used

A—Antenna
AA—Seta of anterior frontal angle
ACP—Acanthoparia
ACR—Acroparia
ACS—Anterior clypeal setae
AFS—Anterior frontal setae
AP—Anterior process of thoracic shield
CAR—Cardo
CPA—Chaetoparia
DES—Dorsoepicranial setae
DPH—Dexiophoba
DX—Dexiotorma
E—Epicranium
ECS—Exterior clypeal setae

EFS—Exterior frontal setae
ES—Epicranial stem
ETA—Anterior epitorma
ETP—Posterior epitorma
F—Frons
FS—Frontal suture
G—Galea
GL—Glossa
GP—Gymnoparia
H—Haptomerum
L—Labrum
LA—Lacinia
LP—Labial palpus
LPH—Laeophoba

LT—Laeotorma
MP—Maxillary palpus
MPH—Mesophoba
MS—Macrosensilla
O—Oncylus (=hypopharyngeal sclerite)
PC—Preclypeus
PE—Pedium
PF—Palpifer
PES—Posterior frontal seta
PPH—Protophoba
PSC—Postclypeus
PTT—Pternotorma
ST—Maxillary stridulatory area

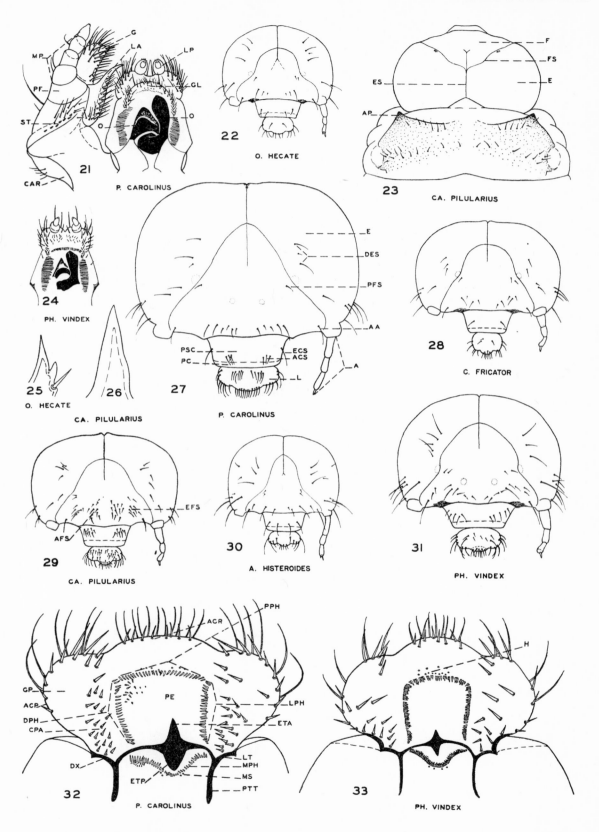

21 P. CAROLINUS

22 O. HECATE

23 CA. PILULARIUS

24 PH. VINDEX

25 O. HECATE

26 CA. PILULARIUS

27 P. CAROLINUS

28 C. FRICATOR

29 CA. PILULARIUS

30 A. HISTEROIDES

31 PH. VINDEX

32 P. CAROLINUS

33 PH. VINDEX

Plate III

FIGURE 34. *Copris fricator* (Fab.). Epipharynx.

FIGURE 35. *Canthon pilularis* (Linn.). Epipharynx.

FIGURE 36. *Copris minutus* (Drury). Epipharynx.

FIGURE 37. *Ateuchus histeroides* (Web.). Epipharynx.

FIGURE 38. *Onthophagus hecate hecate* (Panz.). Epipharynx.

FIGURE 39. *Pinotus carolinus* (Linn.). Venter of tenth abdominal segment.

FIGURE 40. *Phanaeus vindex* McL. Venter of tenth abdominal segment.

FIGURE 41. *Copris minutus* (Drury.). Venter of tenth abdominal segment.

FIGURE 42. *Copris fricator* (Fab.). Venter of tenth abdominal segment.

FIGURE 43. *Ateuchus histeroides* (Web.). Venter of tenth abdominal segment.

FIGURE 44. *Onthophagus hecate hecate* (Panz.). Venter of tenth abdominal segment.

FIGURE 45. *Onthophagus pennsylvanicus* Harris. Venter of tenth abdominal segment.

FIGURE 46. *Canthon pilularius* (Linn.). Venter of tenth abdominal segment.

Symbols Used

CR—Crepis   S—Septula   T—Teges
PLA—Palidium

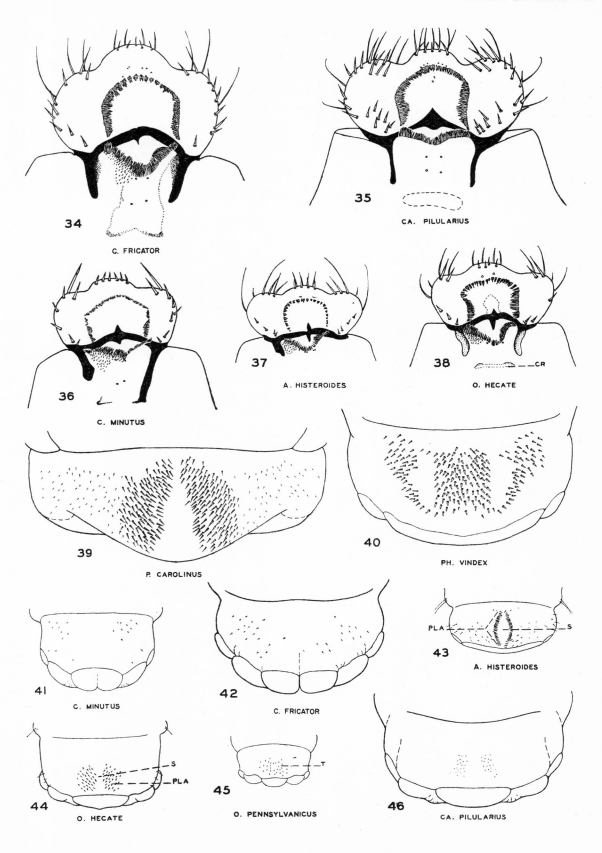

34   C. FRICATOR

35   CA. PILULARIUS

36   C. MINUTUS

37   A. HISTEROIDES

38   O. HECATE   CR

39   P. CAROLINUS

40   PH. VINDEX

41   C. MINUTUS

42   C. FRICATOR

43   PLA   S   A. HISTEROIDES

44   S   PLA   O. HECATE

45   T   O. PENNSYLVANICUS

46   CA. PILULARIUS

Subfamily Aphodiinae

This subfamily contains a large number of species of small, inconspicuous beetles commonly found in dung or in the soil. Many of the species are coprophagous both as adults and as larvae, but the larvae of some species, mostly of the genus *Aphodius,* are known to be injurious to roots of grasses and other plants (Jerath and Ritcher, 1959). A few species of *Aphodius* and *Euparia* are associated with ants and several *Aphodius* are parasitic on other scarabaeids (Howden, 1955a).

Larvae of 36 species belonging to nine genera of Aphodiinae have been recently described by Jerath (1960) and keys given separating many of the species occurring in the United States. The keys presented here are largely taken from his paper.

Larvae of Aphodiinae may be characterized as follows: Antenna 4- (or apparently 5) segmented with last segment reduced in size. Third antennal segment usually with apical process. Epipharynx trilobed, pedium surrounded by phobae. Tormae united, with prominent epitorma. Galea and lacinia of maxilla distinctly separate but often close together. Maxillary stridulatory teeth sometimes absent. Dorsa of thoracic and abdominal segments plicate. Concavities of respiratory plates of thoracic spiracles facing posteriorly, those of abdominal segments facing ventrally or cephaloventrally. Legs 4-segmented, with well-developed claws. Anal lobes whitish or yellowish, bare of setae.

Key to Genera and Tribes of Aphodiinae Based on Characters of the Larvae
(From Jerath, 1960, with some changes)

1. Lower anal lobes emarginate (Fig. 72) or entire (Fig. 75) 2
   With two anal lobes (Fig. 74) 4

2. Lower anal lobe entire (Fig. 75) ....*Aegialia* (Aegialiini)
   page 26
   Lower anal lobe emarginate (Fig. 72).......**Aphodiini**........ 3
   page 27

3. Scissorial area of left mandible with blade-like anterior portion and a single tooth posterior to scissorial notch (Fig. 63); galia dorsally with 5 or more stout setae ....*Aphodius*
   Scissorial area of left mandible with bladelike anterior portion and 2 teeth posterior to scissorial notch; galea dorsally with 4 stout setae ....*Oxyomus*

4. Galea ventrally with a long seta and a row of 2 or 3 short setae 5
   Galea ventrally with a long seta and a row of 4 to 7 short setae ....most **Eupariini**........ 6
   page 28

5. Maxillary stridulatory area without teeth; lacinia without apical uncus ....*Saprosites* (**Eupariini**)
   Maxillary stridulatory area with a patch of minute teeth; lacinia with apical uncus (Fig. 57) ....
   .... **Psammodiini**........ 8
   page 29

6. Clypeus with 1 seta on each side; frons without posterior frontal and anterior frontal setae; maxillary stridulatory area with 5 or fewer teeth; raster with fewer than 24 tegillar setae ....*Aphotaenius*
   Clypeus with 3 setae on each side; frons with 2 posterior frontal setae and 1 anterior frontal seta; maxillary stridulatory area with 7 or more teeth; raster with more than 25 tegillar setae (Fig. 70) 7

7. Maxilla with a row of 6 or 7 setae on ventral surface of galea ....*Euparia*
   Maxilla with a row of 4 or 5 setae on ventral surface of galea ....*Ataenius*

8. Maxilla with 1 long seta and a row of 3 short setae on ventral surface of galea; each abdominal spiracle-bearing area with 6 to 8 setae ventrally and 2 dorsal setae ....*Psammodius*
   Maxilla with 1 long seta and 2 short setae on ventral surface of galea; each abdominal spiracle-bearing area with 2 setae ventrally and 1 dorsal seta..*Pleurophorus*

TRIBE Aegialiini

Study of *Aegialia* larvae showed that this group, previously considered in a separate subfamily Aegialiinae, has many morphological characters in common with larvae of other Aphodiinae (Jerath, 1960a).

Genus *Aegialia* Latreille
(Figs. 47, 52, 56, 60, 68, and 75)

Larvae of this genus may be distinguished by the following characters: Frons of head, on each side, with 1 anterior angle seta, 1 anterior frontal seta, 1 exterior frontal seta, and a transverse pair of posterior frontal setae. Epipharynx with subquadrate pedium surrounded by phobae. Tormae united, with a long median epitorma, and produced cephalad and caudad on the right side. Maxilla with an irregular row of conical stridulatory teeth and 1 to 3 teeth on the palpifer. First 3 antennal segments subequal, third segment with an apical, conical process. Dorsa of abdominal segments 2 to 5 each with 3 annulets, each sparsely set with a transverse row of short, stiff, spine-like setae. Dorsa of abdominal segments 6 to 8 each with 3 transverse, sparsely set rows of long, slender setae. Raster with a patch of hamate setae, truncate and slightly flared at their tips. Lower anal lobe entire.

Key to Known *Aegialia* Larvae

(From Jerath, 1960)

1. Galea dorsally with 5 stout setae (Fig. 56); raster
   with 48 to 58 hamate setae ....*A. lacustris*
   Galea dorsally with 4 stout setae; raster with 25 to 37
   hamate setae (Fig. 68) ....*A. blanchardi*

Aegialia lacustris LeConte is a rather uncommon species which, according to Brown (1931), occurs across southern Canada and in Maine, Colorado, and Washington. In Oregon, larvae and adults were found at shallow depths under willows near streams (Jerath and Ritcher, 1959, and Jerath, 1960b). The larvae were feeding on decaying organic matter. The adult is a dark brown or reddish-brown beetle 3.7 to 5.3 mm in length (Brown, 1931).

Aegialia blanchardi Horn is a sand dune species common along the west coast from California to British Columbia. It also occurs in the Willamette Valley of Oregon (Jerath and Ritcher, 1959) and is known from Massachusetts and North Carolina (Brown, 1931). Larvae and pupae were found only in the spring, at Waldport, Oregon, at depths of 6 to 8 inches in dune sand beneath vegetation (Jerath and Ritcher, 1959). The adult is a blackish, oblong oval, robust beetle about 4.5 mm long (Brown, 1931).

TRIBE Aphodiini

Three genera of this tribe occur in America north of Mexico. These are *Oxyomus*, represented only by one introduced European species, the genus *Xeropsamobeus* from California (larva unknown) and the very large genus *Aphodius* containing many common species.

Genus *Oxyomus* Cast.

Oxyomus silvestris (Scop.) has become established near New York City and Philadelphia, Pennsylvania (Horn, 1887). The larva was described by Jerath (1960a) from specimens collected in Luxembourg.

Genus *Aphodius* Illiger

(Figs. 48, 50, 51, 54, 55, 61-66, 69, 72, and 73)

Jerath (1960a) characterized *Aphodius* larvae as follows: "Frons, on each side, with 2 short posterior frontal setae with a microsensilla (in *A. erraticus* 2 or 3 setae with 2 microsensillae), a short anterior frontal seta and a microsensilla, a long exterior frontal seta, and a microsensilla and a long seta at each anterior angle. First antennal segment apparently subdivided. Scissorial area of left mandible with S1+2 and S3 and of right mandible with S1+2 and S3+4. Each mandible dorsally with 2 or 3 setae and ventrally with 3 or 4 short setae. Galea dorsally with more than 4 setae. Abdominal segments 1-8 each with 3 dorsal annulets; segments 9 and 10 each with 2 rows of dorsal setae, dorsa not divided. Lower anal lobe emarginate, partially subdivided into 2 lobes (or into 4 as in *A. erraticus*)."

My studies of *Aphodius* larvae show that some species have a different setal pattern than indicated above, on the dorsa of the posterior abdominal segments. For example, *A. lividus* has 3 transverse rows of setae on the dorsa of abdominal segments 6 to 8.

More than 100 species belonging to this genus are known to occur in America north of Mexico. Of these, the larvae of 15 species were described by Jerath in 1960 and a modified key to the same group is presented below. Adults of most species of *Aphodius* are dung feeders, but some also feed in decaying fungi or in decaying organic matter in or on the soil. The larvae include feeders on dung, organic matter, and live roots. Some of the latter are of economic importance. A few species are found in ant nests.

Key to *Aphodius* Larvae

1. Venter of last abdominal segment with definite palidia
   (Fig. 66) 2
   Venter of last abdominal segment with a teges only;
   palidia absent (Fig. 65) 3

2. Maxilla with a longitudinal row of 6 or 7 setae on ventral surface of galea ....*A. granarius*
   Maxilla with a longitudinal row of 11 or 12 setae on
   ventral surface of galea ....*A. pardalis*

3. Maxilla with a longitudinal row of 11 to 20 setae on
   ventral surface of galea (Fig. 62) 4
   Maxilla with a longitudinal row of 4 to 10 setae on
   ventral surface of galea 7

4. Teges with from 32 to 38 hamate setae arranged in 5
   or 6 rows (Fig. 65); dorsal folds of abdominal segments 2 to 5 each with two irregular transverse rows
   of setae ....*A. hamatus*
   Teges with 50 or more hamate setae; dorsal folds of
   abdominal segments 2 to 5 each with a single, sparse,
   transverse row of setae 5

5. Maxillary stridulatory area with a row of 6 to 10 teeth
   on the stipes ....*A. aleutus*
   Maxillary stridulatory area with a row of 18 to 24 teeth
   on the stipes 6

6. Teges consisting of 55 to 90 hamate setae (Fig. 69)......
   .... *A. fimetarius*
   Teges consisting of 125 to 165 hamate setae ....*A. fossor*

7. Teges with fewer than 50 hamate setae 8
   Teges with more than 50 hamate setae 13

8. With 1 to 3 stridulatory teeth on palpifer 9
   Without stridulatory teeth on palpifer 10

9. Frons with 2 depressions on each side; maximum width
   of head capsule 1.862 mm ....*A. sparsus*
   Frons with 4 depressions on each side; maximum width
   of head capsule 0.86 to 0.92 mm ....*A. neotomae*

10. Ventral surface of galea with a sparsely set row of 4
    to 6 setae ....*A. prodromis*
    Ventral surface of galea with a row of 6 to 12 setae 11

11. Teges set with 18 to 21 hamate setae arranged in 4 ir-
    regular, longitudinal rows ....*A. stercososus*
    Teges set with 25 to 35 hamate setae not arranged in
    rows 12

12. Dorsa of abdominal segments 6 to 8 each with 2 trans-
    verse rows of setae ....*A. troglodytes*
    Dorsa of abdominal segments 6 to 8 each with 3 trans-
    verse rows of setae ....*A. lividus*

13. Labrum with a broad truncate, median process and a
    small conical process on each side ....*A. erraticus*
    Labrum without conspicuous processes 14

14. With 1 or 2 stridulatory teeth on palpifer...............*A. vittatus*
    Without stridulatory teeth on palpifer 15

15. Dorsa of abdominal segments 6 to 8 each with 2 trans-
    verse rows of setae ....*A. pectoralis*
    Dorsa of abdominal segments 6 to 8 each with 3 trans-
    verse rows of setae ....*A. haemorrhoidalis*

The following notes on the biology and distribution of *Aphodius* species have been compiled from Jerath and Ritcher (1959), Jerath (1960a and 1960b), Leng's Catalogue (1920 and supplements), and personal observations, unless otherwise noted.

1. *Aphodius aleutus* Esch. is abundant in deer droppings in Oregon at high altitudes. Known from Alaska, California, Colorado, New Mexico, Oregon, and Washington.

2. *Aphodius erraticus* (L.) is an introduced species occurring in mid-United States and Maryland. Larvae are found in soil beneath cow dung.

3. *Aphodius fimetarius* (L.) is an introduced, very abundant, widespread species that breeds in horse manure and older cow dung. The distinctive adult has reddish elytra.

4. *Aphodius fossor* (L.) is a large introduced species occurring in eastern Canada and northeastern United States, westward to Iowa. Larvae are found in cow dung.

5. *Aphodius granarius* (L.) is an introduced species which is now widespread in the United States and Canada. The black adults are sometimes abundant in cow dung, but the larvae are usually found in soil where they may feed on grass roots. Lugger (1899) observed larvae eating sprouting seeds of corn in Minnesota.

6. *Aphodius hamatus* Say is found from Maine to Oregon and in New Mexico. In the west it often occurs at higher altitudes. The distinctively marked adults, which have black and whitish elytra, are found in cow dung and are sometimes attracted in large numbers to oily spots on highways. The larvae were found injuring grass roots in Nevada.

7. *Aphodius haemorrhoidalis* (L.) is a common introduced species found in New York, New Jersey, Massachusetts, and Kentucky. This species breeds in cow dung.

8. *Aphodius lividus* (Oliv.) is an introduced species which occurs in our southern states and westward to California. Both adults and larvae are found in cow dung.

9. *Aphodius neotomae* Fall is a very small, uncommon species found in dung in wood-rat nests in California and Oregon.

10. *Aphodius pardalis* Lec. is a west coast species found in turf of lawns, golf courses, and bowling greens where the larvae may cause severe damage by feeding on roots and rhizomes (see Downes, 1928, and Ritcher and Morrison, 1955).

11. *Aphodius pectoralis* Lec. is found in the Pacific Northwest. In western Oregon it is common in deer droppings in wooded areas.

12. *Aphodius prodromus* (Brahm) is an introduced species occurring in the northeastern United States. Adults are found in horse manure. The author obtained mature larvae in Kentucky by confining adults in pint jars of soil containing horse manure.

13. *Aphodius sparsus* Lec. is a rather large species common in dung in wood-rat nests in western California and Oregon.

14. *Aphodius troglodytes* Hub. is a Florida species found in burrows of the Florida land tortoise (Hubbard, 1894).

15. *Aphodius vittatus* Say is a common, widely distributed species. Both adults and larvae are found in cow manure. The small black adult is distinctly marked with red on the elytra.

TRIBE Eupariini
Genus *Aphotaenius* Cartwright

Aphotaenius carolinus (Van Dyke) is the only species of this genus known to occur in the United States (Cartwright, 1963). The larva was described by Jerath (1960a) from specimens reared by O. L. Cartwright. According to Cartwright (1952), the species is common in deer droppings in western North Carolina, was taken in sheep droppings and old cow dung in South Carolina, and was found in old cow dung in Georgia. The black adult is 2.75 mm long (Cartwright, 1952).

Genus *Euparia* Serville

Euparia castanae Serville, which occurs in Florida, Alabama, and Louisiana, is the only species of this genus known to occur in the United States. The larva was described by Jerath (1960a). Both adults and larvae are found in ant nests (Schmidt, 1911).

Genus *Ataenius* Harold
(Figs. 58, 59, 70, and 74)

Larvae of this genus may be distinguished by the following characters: Maxilla with a row of 4 or 5 stout setae; lacinia with a row of 5 stout setae, decreasing in size posteriorly, near the mesal edge. Maxillary stridulatory area on stipes consisting of a row or elongate patch of blunt teeth. Epipharynx with tormae united. Dexiotorma with projections cephalad and caudad; laeotorma with a blunt pternotorma. Venter of last abdominal segment with a patch of hamate setae with 2 separate lower anal lobes.

This large genus contains more than 50 species. Jerath (1960a) characterized larvae of the genus and presented a key to larvae of 8 known species and 4 others not definitely known as to species, part of which were reared by O. L. Cartwright.

Information on the biology of this genus is fragmentary. Hoffman (1935) described the life history of *Ataenius spretulus* (Hald), (under the name of *A. cognatus* (Lec.)). Larvae of this species damaged turf of golf courses. Adults hibernated in and under dry cow dung. They sometimes frequent fungi and are attracted to lights (Hoffman, 1935).

TRIBE Psammodiini

Four genera of this tribe are represented in the United States, *Psammodius, Rhyssemus, Trichiorhyssemus,* and *Pleurophorus.*

Genus *Psammodius* Fallen
(Figs. 49, 53, 57, 67, and 76)

Jerath (1960a) characterized larvae of this genus as follows: "Clypeus divided into smaller preclypeus and large postclypeus. First and third antennal segments subequal, second shorter than first and third. Epipharynx with dexiophoba and laeophoba monostechous; mesophoba incomplete in the middle and polystechous on the sides. Dexiotorma produced cephalad and caudad into an armlike structure; laeotorma small and produced caudad with end blunt.

"Galea ventrally with a long seta and a row of 3 short setae. Lacinia dorsally with a row of 5 long setae

near the mesal edge and a short seta posteriorly. Each abdominal spiracle-bearing area with 6 to 8 setae ventrally and 2 setae dorsally. Lower anal lobe divided into 2 sublobes remote from each other."

Jerath (1960a) described the larvae of *Psammodius oregonensis* Carter and *Psammodius hydropicus* Horn., but his description of the latter was based only on a single specimen. Adults and larvae of *P. oregonensis* can be found throughout the year in dune areas along the Oregon coast. The larvae feed on live roots of vegetation.

Genus *Pleurophorus* Mulsant
(Fig. 71)

Jerath (1960a) characterized larvae of this genus as follows: "Each antennal base with 2 long setae and a short seta exterolaterally and one long seta dorsally. Clypeus not distinctly divided into preclypeus and postclypeus. Lacinia dorsally with a row of 4 or 5 long setae near the mesal edge and one short seta posteriorly. Galea ventrally with a long seta and 2 short setae. Spiracular concavity facing ventrally. Each abdominal spiracle-bearing area with 2 setae ventrally and one seta dorsally. Lower anal lobe divided into 2 sublobes placed adjacent to each other."

To this description I would add the following characters: First antennal segment longer than second; second and third segments subequal. Maxilla with a patch of stridulatory teeth on palpifer and a patch of similar teeth on stipes. Raster with a patch of hamate setae.

Nine species belonging to this genus are known to occur in America, north of Mexico (Cartwright, 1948). Jerath (1960a) described the larvae of two species, *Pleurophorus caesus* (Creutz.) and *Pleurophorus longulus* (Cartw.).

Pleurophorus caesus (Creutz.) is an introduced species (Hatch, 1946) which occurs from coast to coast (Cartwright, 1948). The adult is a very small, very slender scarab 2.7 to 3.3 mm long and 0.9 to 1.1 mm wide (Cartwright, 1948). In the Yakima Valley of Washington, it often infests potato seed pieces in April, soon after planting (Landis, 1959). The adults are sometimes abundant in western Oregon in the fall, under cantaloupes. Larvae have been collected from the soil of potato fields (Henny, 1942) and in the soil around roots of *Zinnia* (Jerath and Ritcher, 1959).

Plate IV

FIGURE 47. *Aegialia lacustris* Lec. Head.

FIGURE 48. *Aphodius sparsus* Lec. Third-stage larva, left lateral view.

FIGURE 49. *Psammodius oregonensis* Cartw. Head.

FIGURE 50. *Aphodius stercorosus* Melsh. Last two segments of antenna.

FIGURE 51. *Aphodius hamatus* Say. Head.

FIGURE 52. *Aegialia lacustris* Lec. Epipharynx.

FIGURE 53. *Psammodius oregonensis* Cartw. Epipharynx.

FIGURE 54. *Aphodius fossor* (L.). Epipharynx.

FIGURE 55. *Aphodius stercorosus* Melsh. Epipharynx.

Symbols Used

A—Antenna
AA—Seta of anterior frontal angles
AFS—Anterior frontal seta
C—Clypeus
CL—Clithrum
DES—Dorsoepicranial setae

DPH—Dexiophoba
EFS—Exterior frontal seta
ETA—Epitorma
F—Frons
FS—Frontal suture
GP—Gymoparia

L—Labrum
LPH—Laeophoba
PFS—Posterior frontal seta
PPH—Protophoba
PTT—Ptermotorma

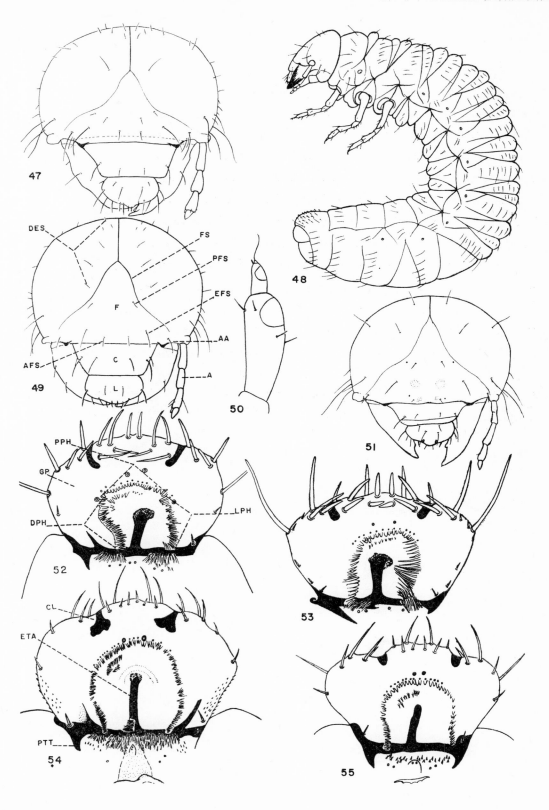

Plate V

FIGURE 56. *Aegialia lacustris* Lec. Left maxilla, dorsal view.

FIGURE 57. *Psammodius oregonensis* Cartw. Left maxilla and hypopharynx, dorsal view.

FIGURE 58. *Ataenius* sp. Right maxilla, dorsal view.

FIGURE 59. *Ataenius falli* Hinton. Left mandible, dorsal view.

FIGURE 60. *Aegialia lacustris* Lec. Left mandible, dorsal view.

FIGURE 61. *Aphodius hamatus* Say. Left mandible, dorsal view.

FIGURE 62. *Aphodius hamatus* Say. Left mandible, ventral view.

FIGURE 63. *Aphodius fimetarius* (L.). Left mandible, dorsal view.

FIGURE 64. *Aphodius fimetarius* (L.). Right mandible, dorsal view.

FIGURE 65. *Aphodius hamatus* Say. Venter of last abdominal segment.

FIGURE 66. *Aphodius pardalis* Lec. Venter of last abdominal segment.

FIGURE 67. *Psammodius oregonensis* Cartw. Venter of last abdominal segment.

Symbols Used

AC—Acia
G—Galea
L—Lacinia
LAL—Lower anal lobe

MO—Molar area
PF—Palpifer
PLA—Palidium

SA—Scissorial area
ST—Stipes
UN—Uncus

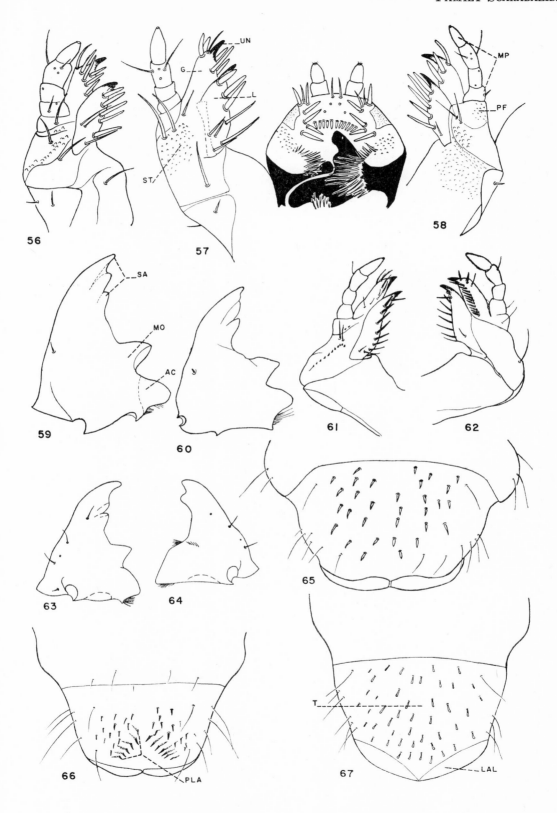

Plate VI

FIGURE 68.   *Aegialia blanchardi* Horn. Venter of last abdominal segment.

FIGURE 69.   *Aphodius fimetarius* (L.). Venter of last abdominal segment.

FIGURE 70.   *Ataenius falli* Hinton. Venter of last abdominal segment.

FIGURE 71.   *Pleurophorus caesus* (Creutz). Left maxilla and hypopharynx, dorsal view.

FIGURE 72.   *Aphodius fimetarius* (L.). Caudal view of last abdominal segment.

FIGURE 73.   *Aphodius haemorrhoidalis* (L.). Caudal view of last abdominal segment.

FIGURE 74.   *Ataenius falli* Hinton. Caudal view of last abdominal segment.

FIGURE 75.   *Aegialia lacustris* Lec. Caudal view of last abdominal segment.

FIGURE 76.   *Psammodius oregonensis* Cartw. Caudal view of last abdominal segment.

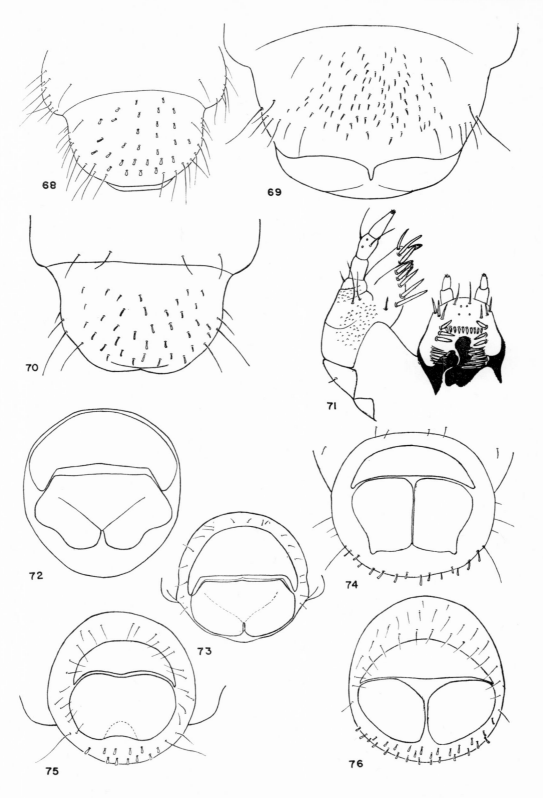

Subfamily Hybosorinae

Only two species belonging to this subfamily are known to occur in the United States. These are the cosmopolitan *Hybosorus illigeri* Reiche, which occurs in some of our southern states, in southern Europe, Asia, and Africa and *Pachyplectrus laevis* Lec. which occurs in southern California and Arizona (Leng, 1920). I have collected a few adults of *H. illigeri* at light in Raleigh, N. C., but nothing is known of the biology.

Gardner (1935) described the larva of *Phaeochrous emarginatus* Cast., an Indian species, and figured the epipharynx, portions of the prothoracic and mesothoracic legs, and the venter of the last abdominal segment. He stated that Fletcher (1919) has given habitus figures of the larva and pupa of *Hybosorus orientalis* West and that his figure of the raster shows a resemblance to that of *Phaeochrous*. Gardner characterized larvae of Hybosoridae[9] as differing from known scarabaeoid larvae in possessing stridulatory organs on the prothoracic legs. He also stated that the structure of the labrum-epipharynx is distinctive, the lacinia and galea are distinctly separate, and the setae in the two lines on the raster are flattened and widened towards their outwardly directed apices.

My studies of both *Phaeochrous* and *Hybosorus* larvae from India indicate that larvae of this subfamily possess characters showing relationships with the Acanthocerinae. Larvae of Hybosorinae may be characterized as follows: Anterior margin of labrum with 3 truncate lobes. Epipharynx with a row of setae on each chaetoparia, a blunt tooth in the haptomeral region, and united tormae. Maxilla with galea and lacinia distinctly separate. Maxillary stridulatory area consisting of a sparse row of conical teeth. Antenna 3-segmented (resulting from fusion of third and fourth segments). Spiracles with concavities of respiratory plates facing ventrally. Anterior abdominal terga plicate. Venter of last abdominal segment with 2 palidia each consisting of a curved row of outwardly directed setae. Legs 4-segmented, with well-developed claws. Prothoracic and mesothoracic legs with stridulatory organs.

Genus *Hybosorus* MacLeay
Hybosorus orientalis West, Third-Stage Larva
(Figs. 77 to 83)

This description is based on 3 third-stage larvae, associated with one pupa and reared adults. Collected at

[9] He considered them as belonging to a separate family.

Dehra Dun (U. P.), India. Loaned by the United States National Museum.

Frons with a transverse patch of about 15 anterior frontal setae and, on each side, with a single exterior frontal seta, a single seta in each anterior frontal angle, and a single posterior frontal seta (Fig. 77). Dorsoepicranial setae 2 or 3 in number on each side. Labrum with 3 apical, truncate lobes. Scissorial area of each mandible with an anterior, blade-like portion and a posterior tooth. Maxilla with separate galea and lacinia (Fig. 81). Maxillary stridulatory area consisting of a sparsely set row of about 9 rather blunt, conical teeth and 2 similar teeth located on the palpifer. Epipharynx (Fig. 79) with 3 apical, truncate lobes flanked on either side with a bulbous lobe bearing 7 setae. Haptomeral region with a blunt, beak-like process located dextrad of the median line. Each chaetoparia with an irregular row of 8 or 10 setae. Tormae united. Antenna 3-segmented, resulting from the fusion of the third and fourth segments. Last antennal segment with a single, dorsal sensory area which continues around to the ventral surface, also with a sensory appendage.

Abdominal segments 1 to 6 each with 3 dorsal folds. Spiracles with concavities of respiratory plates facing ventrally. Raster consisting of 2 bent rows of about 29 short, caudolaterally pointing setae which converge caudally (Fig. 83).

Legs 4-segmented, each with a well-developed claw. Prothoracic and mesothoracic legs with stridulatory structures; femur and tibiotarsus of each mesothoracic leg with a row of 8 conical stridulatory teeth, coxa of prothoracic leg with a large patch of granules on its outer surface.

Key to Genera of Hybosorinae
Based on Characters of the Larvae

1. Frons, on each side, with a single posterior frontal seta (Fig. 77). Palidia of raster set with short spinelike setae (Fig. 83) ....*Hybosorus*

   Frons, on each side, with a small patch of 5 or 6 posterior frontal setae (Fig. 84). Palidia of raster set with flat, blunt setae (Fig. 88) ....*Phaeochrus*[10]

[10] Extra limital; occurs in India.

Plate VII

Symbols Used

COX—Coxa	H—Haptomerum	SD—Stridulatory teeth
CPA—Chaetoparia	PLA—Palidium	TTS—Tibiotarsus
FEM—Femur	PTT—Pternotorma	

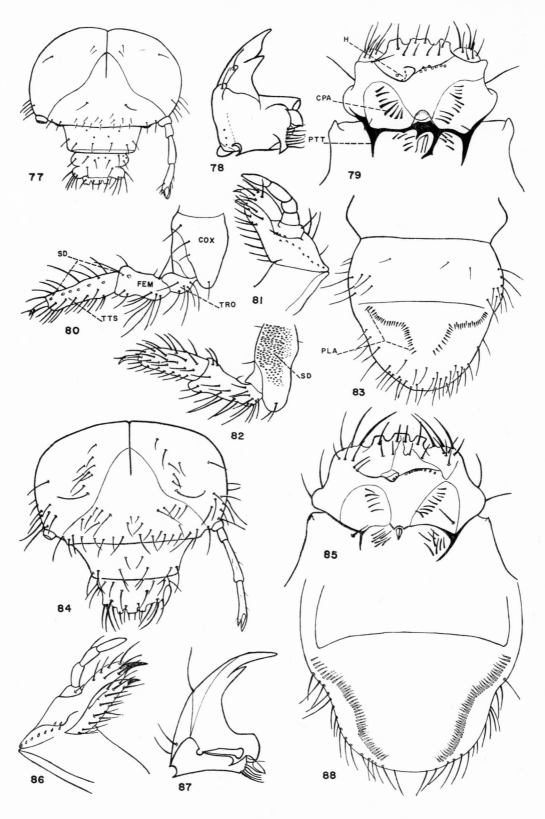

Subfamily Geotrupinae

The scarabaeid subfamily Geotrupinae includes four tribes and a number of very distinct genera of laparostict beetles. The group consists of the tribes Geotrupini and Lethrini, which show affinities with the Scarabaeinae, and the tribes Bolboceratini and Athyreini.

Several geotrupid genera, such as *Geotrupes,* are nearly world-wide in distribution, while the tribe Lethrini is confined almost entirely to southeastern Europe and southwestern Asia (Boucomont, 1912). The tribe Athyreini, recently erected by Howden and Martinez (1963), occurs in South America and Africa. In the United States, the subfamily is represented by some 65 species belonging to 10 genera.

Adult geotrupids range from 5 to about 45 mm in length, are oval to round, and are often strongly convex dorsally. The head often bears a median horn on the clypeus. Color of the adults usually ranges from yellowish, brown, orange-brown, reddish-brown, and brown to black but some species are blue, green, purple, or metallic bronze.

The biologies of the various genera and species are quite diverse and until the work of Howden (1952, 1954, and 1955) were not too well known. Some are saprophagous, some are coprophagous, some are mycetophagous, and a few feed on the shoots of living plants. Howden (1955) reports that the larval food of Bolboceratini is surface humus provisioned in a cell by the adult. Adults of many genera are nocturnal and frequently attracted to lights. Much of the adult life of all geotrupids is spent in burrows in the soil. Burrows of a number of species extend to a depth of 6 feet or more (Howden, 1955).

The diversity of structure both in adults and larvae has led to differences of opinion as to the systematic position of the group and the genera assigned to it. Most writers agree that the tribes Bolboceratini, Lethrini, and Geotrupini belong together. Some writers (Paulian, 1941, and Ritcher, 1947b) have also included the genus *Pleocoma.* My present thinking is that *Pleocoma* belongs in a separate subfamily.

The bizarre larva of *Geotrupes,* with its peculiar legs and mandibles, has been known for many years. Study of this odd genus has influenced all subsequent treatment of the entire group, and several writers have even made a separate family Geotrupidae (Böving and Craighead, 1930-31, and Paulian, 1939). The fact that larvae of several other geotrupid genera, and even of one genus in the same tribe as *Geotrupes,* have neither greatly reduced metathoracic legs nor stridulatory organs on the legs (van Emden, 1941) makes such separation from the Scarabaeidae untenable.

The writer, in 1947, described the larvae of several genera and species of Geotrupinae and presented keys for their separation. Later, Howden, one of my graduate students, became interested in the group and published several papers (1952 and 1954) and a monograph (1955) which contained much new information about both larvae and adults. The material on Geotrupinae presented in this section is largely reprinted from my 1947 bulletin, with emendations. In addition, a description of the previously unknown larva of *Bolboceras* (*Odontaeus*) *obesus* (Lec.) is included, plus some information from Howden's papers concerning several genera and species of Geotrupini and larvae of species of *Bolboceras* (*Odontaeus*) other than *obesus.*

Larvae of the subfamily Geotrupinae may be characterized as follows: Antenna 3-segmented with penultimate segment bearing 1 or more distal sense organs and last segment reduced in diameter. Epipharynx usually trilobed with symmetrical tormae. Maxilla with galea and lacinial distinctly separate. Maxillary stridulatory area with teeth. Terga of abdominal segments 3 to 7, with 2 dorsal annulets. Anal opening sometimes V- or Y-shaped, in some genera surrounded by fleshy lobes. Legs 2- to 4-segmented. Mesothoracic and metathoracic legs often with stridulatory organs; metathoracic sometimes greatly reduced in size (most Geotrupini). Spiracles cribriform or biforous.

Key to Tribes

1. Last abdominal segment obliquely flattened (Figs. 89 and 91) and with conspicuous lateral callosities (Figs. 130, 131, 133, 134, and 136). Hypopharyngeal oncyli asymmetrical (Fig. 119) 2
   Last abdominal segment rounded posteriorly (Figs. 93 and 95) and without conspicuous lateral callosities (Figs. 132 and 135). Hypopharyngeal oncyli symmetrical (Fig. 120) ....**Bolboceratini**
   page 44
2. Anal opening surrounded by flap-like dorsal anal lobe and a pair of ventral anal lobes (Figs. 130, 131, 133, and 136) ....**Geotrupini**
   page 41
   Anal opening surrounded by 6 simple radiating rays (Fig. 134) ....**Lethrini**
   page 44

TRIBE Geotrupini

Four genera belonging to this tribe are now represented in the United States, north of Mexico. These are *Geotrupes, Peltotrupes, Mycotrupes,* and *Ceratophyus. Ceratophyus,* an old-world genus, has recently (1962) become established in Santa Barbara County, California.

Larvae of Geotrupini have been described by Schiödte (1874), Paulian (1939), van Emden (1941), Ritcher (1947b), and others. Howden, in his excellent monograph of the North American Geotrupinae (1955) described the larvae of several species of *Geotrupes* and presented a key to larvae of six species of the genus. The larva of *Peltotrupes profundus* Howden was described by Howden in 1952 and the larva of *Mycotrupes gaigei* Olson and Hubbell was described by Howden in 1954. The larva of *Typhoeus* was described by van Emden in 1941. Golovianko (1936) described the larva of *Lethrus*. The larva of *Ceratophyus* is as yet unknown.

Key to Known Genera of the Tribe Geotrupini
(In part from Howden, 1955)

1. Mesothoracic and metathoracic legs similar in size, without stridulatory organs ....*Typhoeus*

    Mesothoracic legs greatly reduced in size (Figs. 89 and 110); mesothoracic and metathoracic legs with stridulatory organs 2

2. Last antennal segment greatly reduced (Fig. 99); abdomen greatly swollen (Fig. 97) ....*Peltotrupes*

    Last antennal segment at least one fourth as long as second segment (Fig. 98); abdomen moderately swollen (Fig. 89) 3

3. Endoskeletal figure of ventral anal lobe below anal opening laterally expanded with sharp, fairly truncate angles ....*Geotrupes*

    Endoskeletal figure of ventral anal lobe below anal opening lacking sharp angles, broadly rounded........
    .... *Mycotrupes*

Genus *Geotrupes* Latreille

Howden (1955) described the larvae of *Geotrupes blackburnii blackburnii* (Fab.), *G. splendidus splendidus* (Fab.), *G. egeriei* Hald., and *G. hornii* Blanchard and gave a key for separating these species, *G. blackburnii excrementi* Say and *G. ulkei* Blanchard (described by Ritcher, 1947b) and the introduced *G. stercorarius* (described by Schiödt, 1874, Spaney, 1910, etc.).

Species of this genus are only found east of the Rocky Mountains. Adults of the various species feed on either dung or fungi with one species *G. egeriei* feeding on both materials (Howden, 1955). The same writer reported that larvae developed in brood cells, at the end of burrows in the soil which had been provisioned by the adults with dung or dead leaves.

Geotrupes blackburnii excrementi Say, Third-Stage
Larva
(Figs. 89, 98, 108 to 110, 114, 119, 126, and 131)

This description is based on the following material and reprinted from my 1947 paper:

Cast skin of third-stage larva, found in pupal cell with adult. Adult and associated skin dug from pasture soil, November 30, 1942, near Lexington, Kentucky, by P. O. Ritcher. Adult identified by Chapin.

Five third-stage larvae dug June 24, 1943, at Lexington, Kentucky, by P. O. Ritcher from the soil of a cage in which 20 adults were put March 27, 1943, and supplied with fresh cow manure. No. 43-15A.

One third-stage larva dug June 7, 1944, at Lexington, Kentucky, by P. O. Ritcher from soil of a cage in which adults were put May 8, 1944, and supplied with fresh cow manure. No. 44-7K.

Seven third-stage larvae dug in June 1945, at Lexington, Kentucky, by P. O. Ritcher from soil of a cage in which adults were put in May 1945, and supplied with fresh cow manure. No. 45-3J.

Maximum width of head capsule 3.9 to 4.6 mm. Surface of cranium yellow-brown, shining, and vaguely wrinkled. Frons on each side with 1 posterior frontal seta or none, 3 to 5 setae in each anterior angle, a single (rarely 2) exterior frontal seta, and 1 or 2 anterior frontal setae (Fig. 98). Clypeo-frontal suture absent. Labrum trilobed, wider than long; clypeus somewhat asymmetrical. Antenna fairly long, 3-segmented. First antennal segment the longest, cylindrical. Second antennal segment clavate. Third antennal segment cylindrical, much reduced in diameter. Penultimate segment distally with a single hemispherical sense organ. Mandibles (Figs. 108 and 109) with nearly symmetrical scissorial areas but asymmetrical molar areas. Each scissorial area with a rather narrow, blade-like portion separated from a posterior tooth-like portion by the scissorial notch. Inner margin of mandible between scissorial and molar areas with a prominent bifurcate process. Grinding surface of left molar area strongly concave, overhung proximally by an acia; grinding surface of right molar area rather flat. Maxillary stridulatory area (Fig. 119) consisting of a sparsely set row of 5 to 7 conical teeth on each stipes and a similar row of 3 to 5 teeth along the posterior margin of the palpifer. Hypopharynx with 2 well-developed, strongly asymmetrical oncyli. Glossa emarginate. Epipharynx (Fig. 126) very similar to that of scarabaeine larvae with tormae united mesally, posterior and anterior epitormae present, and phobae surrounding pedium. Pedium covered with spicules. Each chaetoparia with 12 or more chaetae. Respiratory plates of spiracles (Fig. 89) crescent-shaped with concave margins facing ventrally or slightly cephalad. Spiracles cribriform. Respiratory plates with a great many small "holes" arranged in a large number of definite transverse rows. Spiracles of eighth abdominal segment much smaller than those of abdominal segments 1 to 7 inclusive.

Body (Fig. 89) swollen posteriorly with swollen anal lobes protruding on each side of the last abdominal segment. Dorsa of abdominal segments 1 to 8 each with 2 rather vaguely defined dorsal annulets. Anterior annulet of first abdominal segment bare; posterior annulet with a narrow, transverse row of setae most of which are long. Anterior annulet of second abdominal segment with a small, sparsely set patch of short setae; posterior annulet with a transverse band of short setae fringed caudally with long setae. Anterior annulets of abdominal segments 3 to 5 inclusive, each with a triangular patch of short setae; posterior annulets each bare mesally and with a transverse patch of setae on each side.

Dorsal annulets of abdominal segments 6 to 8 inclusive each with a transverse patch of rather long setae. Dorsum of abdominal segment 9 with a transverse patch of setae. Last abdominal segment short, obliquely flattened caudally, and with protruding lateral and ventral lobes (Fig. 131). Defining sclerotized lines of ventral anal lobes not meeting at any point. Sclerotized lines of endoskeletal figure converging toward anal opening. Endoskeletal figure narrowed toward its middle by a process from each ventral anal lobe forming a triangular area. Ventrad, triangular area of endoskeletal figure constricted into narrow stalk. Venter of last abdominal segment, anterior to ventral anal lobes, bare except for 1 to 3 setae on each side. Legs 3-segmented (Figs. 89 and 110). Prothoracic and mesothoracic legs rather long; metathoracic legs much reduced in size (Figs. 89 and 110). Legs with well-developed stridulatory organs, those on mesothoracic legs consisting of a striated area on outer apical portion of each coxa (Fig. 110), those on metathoracic legs consisting of a sparsely set single row of 9 or 10 teeth on lower, inner surface of fused trochanter-femur and a single small tooth on tibio-tarsus (Fig. 114). All legs with minute claws.

Geotrupes blackburnii excrementi Say occurs from Michigan to West Virginia and westward to Kansas and Texas according to Howden (1955). It is fairly common at Lexington, Kentucky, especially in the early spring when it can be collected in numbers from soil beneath fresh cow dung. In March 1943, the writer dug over 20 adults from burrows in soil beneath the contents of a cat box. The species is also attracted to lights. The adult is oval, shining black, moderately convex, 13 to 17 mm long and 8 to 10 mm wide. Adults construct winding vertical burrows in the soil and pack the lower end of each with an elongate wad of dung in which a single larva develops. Winter is passed in the adult stage.

Geotrupes ulkei Blanch., Third-Stage Larva
(Fig. 130)

This description is based on the following material and reprinted from my 1947b paper:

Two third-stage larvae (and one second-stage larva) associated with 2 adults. Material collected July 7, 1934, at Cold Spring, Monte Sano, Madison County, Alabama, by W. B. Jones. Loaned by the United States National Museum.

Maximum width of head capsule 2.45 to 2.52 mm. Surface of cranium straw-colored, shiny. Frons on each side with 1 posterior frontal seta, 1 or 2 setae in each anterior angle, 1 exterior frontal seta, and 1 or 2 anterior-frontal setae. Antenna 3-segmented with second segment slightly longer than first. Mandibles and maxillae practically identical with *G. blackburnii*, except that maxillary stridulatory area of *ulkei* consists of a sparsely set row of 5 or 6 conical teeth on each stipes but with only 1 or 2 teeth on each palifer. Epipharynx similar to *G. blackburnii* but each chaetoparia with only 5 or 6 chaetae.

Dorsal annulets of abdominal segments 6 to 8 inclusive each with a short, anterior patch of moderately long setae and a long, caudal, transverse band of long setae. Last abdominal segment (Fig. 130) with defining sclerotized lines of ventral anal lobes not meeting at any point. Endoskeletal figure oblong in general outline; posteriorly, sharply narrowed for less than one third of length.

Geotrupes ulkei Blanch. is an uncommon species occurring in Virginia (Blanchard, 1888-90) and in Alabama (Loding, 1935). The adult is oval, rather strongly convex, shining dark brown with metallic luster, ranging from 11 to 12 mm in length and 6.5 to 7 mm in width. According to Blanchard, the adults are found in fungi. Loding and Jones in June, 1934, collected a number of adults under leaves, on level ground, at the side of mountain paths on Monte Sano, Madison County, Alabama. Each adult was found in or near a small cylindrical hole the size of a lead pencil and 1½ to 2 inches deep, containing leaf frass in the bottom (Loding, 1935). The following July, Jones visited the same site and collected several adults and the 3 larvae described in this paper. Two of the larvae were found in one burrow and one in another. The food appeared to be decomposing leaves.

Genus *Peltotrupes* Blanchard
(Figs. 97, 99, 125, and 136)

The larva of *Peltotrupes youngi* Howden was described by Howden (1952) under the name of *Geotrupes profundus* Howden. According to Howden (1955), the larva of *P. youngi* resembles larvae of *Geotrupes* in many respects but may be distinguished by the following characters: last (third) antennal segment greatly reduced (Fig. 99); glossa not emarginate, epipharynx with 25 or more chatae on each chaetoparia (Fig. 125);

abdomen greatly swollen (Fig. 97). The endoskeletal figure of the anal lobes is also quite distinctive (Fig. 136).

Genus *Mycotrupes* Le Conte

This genus was revised by Olson and Hubbell (in Olson, Hubbell, and Howden, 1954). The known species occur in Florida, South Carolina, and Georgia. According to Howden (1955), the larva of *Mycotrupes gaigei* Olson and Hubbell can be separated from larvae of *Geotrupes* by differences in the shape of the endoskeletal figure of the ventral anal lobe.

Genus *Typhoeus* Leach
Typhoeus typhoeus L., Third-Stage Larva[11]
(Fig. 133)

This description is taken from van Emden (1941). Larva were not seen.

Antenna 3-segmented, first antennal segment at least as long as second, much longer than third. Penultimate antennal segment with a sensorial appendage at inner side of apex. Mandibles and epipharynx exactly like that of *Geotrupes*.[12] Anal lobes swollen much as in *Geotrupes*. Sclerotized line of anal segment with a deep indentation on dorsoexterior part, near base of lateral callus (Fig. 133). Endoskeletal figure of ventral anal lobe broad at anus, suddenly narrowed into a stalk which reaches the ventral edge; the two defining lines do not meet at any point and follow the ventral edge on each side. Legs normal, without stridulatory organs.

TRIBE Lethrini
Genus *Lethrus* Scop.
Lethrus apterus Laxm., Third-Stage Larva[12]
(Figs. 91 and 134)

This description is taken from Golovianko (1936). Larva were not seen.

Antenna 3-segmented. Labrum trilobed as in *Geotrupes*. Body almost naked. Anal lobes swollen. Last abdominal segment (Fig. 134) slantingly flattened, with round anal opening surrounded by 6 simple, radiating rays. Legs 3-segmented (Fig. 91), almost equally short, conical, each ending with a straight claw. Legs without stridulatory organs. Length of larva up to 40 mm.

---

[11] Extra limital, occurs in Europe.

[12] According to correspondence with van Emden who very kindly examined larvae of *Typhoeus* in the British Museum collection.

Adults of *Lethrus apterus* Laxm. cut the young shoots of plants, especially of grape, and carry them into their burrows (Boucomont, 1902).

TRIBE Bolboceratini

Larvae of Bolboceratini were unknown until the description of larvae of *Bolboceras* (*Odontaeus*), *Bolbocerosoma*, and *Eucanthus* (Ritcher, 1947b). Howden (1955) described the larvae of two additional species of *Bolboceras* (*Odontaeus*), of one additional species of *Bolbocerosoma*, and of a new subspecies of *Eucanthus*.

Key to Genera of the Tribe Bolbocerini

1. Legs 2-segmented (Figs. 111 and 117). Mesothoracic and metathoracic legs with stridulatory organs. Claws absent. Penultimate antennal segment with a single distal sense organ. Last abdominal segment with a single, unpaired lower anal lobe (Fig. 132).. *Bolboceras* (*Odontaeus*)

   Legs 3- or 4-segmented (Figs. 93, 95, 113, and 115), without stridulatory organs. Claws present. Penultimate antennal segment with 2 or more distal sense organs (Figs. 101 and 102). Last abdominal segment with paired lower anal lobes (Fig. 135) 2
2. Legs short, 3-segmented (Figs. 93 and 113). First antennal segment very short (Fig. 102); penultimate segment distally with 4 or 5 conical sense organs. Mandibles with a toothlike process distad of the molar area (Figs. 105 and 106). Glossa of labium not emarginate (Fig. 116). Venter of last abdominal segment with spinose raster ....*Eucanthus*

   Legs well developed, 4-segmented (Figs. 95 and 115). First two segments of antenna nearly equal in length (Fig. 101), penultimate segment distally with 2 conical sense organs. Mandibles without a toothlike process distad of the molar area (Fig. 104). Glossa emarginate (Fig. 120). Venter of last abdominal segment almost bare ....*Bolbocerosoma*

Genus *Eucanthus* Westwood
Eucanthus lazarus (Fab.), Third-Stage Larva
(Figs. 92, 93, 102, 105, 106, 113, 116, and 129)

This description is based on the following material and reprinted from my 1947b bulletin:

One third-stage larva associated with 2 adults reared from 2 other larvae. The 3 larvae were collected during the summer of 1942, near Fayetteville, Arkansas, by Milton W. Sanderson, from the soil of a vineyard.

Maximum width of head capsule 2.3 mm. Surface of cranium light yellow-brown, faintly reticulate. Frons on each side with a pair of posterior frontal setae, a single seta in each anterior angle, and a single exterior frontal seta (Fig. 102). Anterior frontal setae and clypeo-frontal suture absent. Labrum trilobed, wider than long. Antenna 3-segmented, with the first segment very short but of the same diameter as the second, which is cylindrical and about 4 times as long. Third segment cylindrical, much smaller in diameter than second segment. Penultimate segment distally with 4 or 5 conical sense organs. Distal half of last antennal segment covered with tactile points.

Mandibles (Figs. 105 and 106) nearly symmetrical, each with a blade-like scissorial area. Just distad of the molar area is a tooth-like process. Maxillary stridulatory area (Fig. 116) with a patch of 10 to 11 sharp, conical teeth on each side. Hypopharynx with 2 symmetrical bulbous oncyli covered with cilia. Glossa not emarginate. Tormae of epipharynx (Fig. 129) not united mesally, each with a prominent pternotorma. Haptolachus with 3 sensilla on each side, mesally with a dense phoba.

Spiracles not cribriform, somewhat resembling the "biforous" type (Fig. 92). Thoracic spiracles slightly smaller than abdominal spiracles.

Body (Fig. 93) not humped or conspicuously swollen. Dorsa of abdominal segments 1 to 8 inclusive each with 2 annulets. Each anterior annulet with a short, transverse, rather sparsely set patch of short setae. Each posterior annulet with a long, transverse, rather dense covering of longer setae. Last abdominal segment with a pair of ventral lobes and an unpaired dorsal anal lobe; lobes not fleshy. Anal opening V-shaped. Raster, on each side, consisting of a large, rather sparsely set patch of fairly short, stiff, laterad-pointing setae. Legs (Figs. 93 and 113) short, similar in size, 3-segmented, each with a sharp terminal claw. Legs without stridulatory organs.

Eucanthus lazarus (Fab.) is widely distributed over the United States, east of the Rocky Mountains. It is fairly common in Illinois, where it is often taken at lights, but seems quite rare in Kentucky as only a single specimen has been collected, taken April 28, 1946, at Sand Gap. The adult is oval, reddish-brown, with a bifurcate process on the head and ranges from 9 to 11 mm long and 5.5 to 6 mm wide. Brown (1928) reports collecting adults of *E. lazarus* near Stillwater, Oklahoma, from burrows along country lanes or in pasture land in company with adults of *Bolboceras* and *Bolbocerosoma*. Sim found adults burrowing in golf courses and into the sides of deep wheel-ruts in old roads (Wallis, 1928).

In 1955 Howden described a new subspecies, *Eucanthus lazarus subtropicus*, which occurs in the southeastern states. Manee's (1908) observations at Southern Pines, North Carolina, probably refer to this subspecies.

Genus *Bolbocerosoma* Schaeffer
Bolbocerosoma sp., (probably *tumefactum* [Beauv.])
Third-Stage Larva
(Figs. 94, 95, 101, 104, 112, 115, 120, 127, and 135)

This description is based on the following material and reprinted from my 1947b paper:

One third-stage larva collected October-November 1941, near Dover, Delaware, by J. M. Amos, from soil. (Loaned by the U. S. National Museum.)

Maximum width of head capsule 3.2 mm. Surface of cranium straw colored. Frons on each side with a widely separated pair of posterior frontal setae, 1 to 3 setae in each anterior angle, 2 exterior frontal setae, and 3 to 4 anterior frontal setae (Fig. 101). Clypeo-frontal suture absent. Labrum entire, broadly rounded, wider than long. Antenna 3-segmented, the first 2 segments nearly equal in length. Third segment cylindrical, much smaller in diameter than first two. Penultimate segment distally with a pair of conical sense organs. Mandibles (Fig. 104) nearly symmetrical; each with a scissorial area consisting of a distal blade-like portion separated from a blunt proximal tooth-like portion by the scissorial notch. Region distad of the molar area without a distinct tooth-like process. Maxillary stridulatory area (Fig. 120) with a band of 20 to 25 sharp conical teeth arranged in 2 or 3 irregular rows. Hypopharynx symmetrical with 2 bulbous oncyli. Glossa emarginate. Tormae of epipharynx (Fig. 127) not united mesally, each with a pternotorma in the curve of which are found 5 to 7 sensilla. Haptolachus mesally with 2 dense, longitudinal mesophobae, between whose proximal portions are 4 macrosensilla.

Spiracles not cribriform (Fig. 94). Respiratory plates with a number of prominent trabeculae. Thoracic spiracles much larger than abdominal spiracles.

Body (Fig. 95) not humped or conspicuously swollen. Dorsa of abdominal segments 1 to 8 inclusive, each with 2 annulets. Each anterior annulet (prescutum) is short and diamond-shaped when viewed from above; each posterior annulet (scutum) is broad and extends from the spiracular area of one side across the dorsum to the other. Prescuta of abdominal segments 3 to 5 inclusive and scuta of abdominal segments 1 to 8 inclusive are densely covered with setae. Last abdominal segment (Figs. 95 and 135) with a pair of rather fleshy oval, ventral anal lobes and an unpaired, rather fleshy, dorsal anal lobe. Anal opening Y-shaped. Venter of last abdominal segment bare except for a few scattered setae on each side. Legs (Figs. 95 and 115) well developed, all three pairs similar in size and shape, 4-segmented. Claws on mesothoracic and metathoracic legs falcate and sharp, those on prothoracic legs very short. Legs without stridulatory organs.

Bolbocerosoma tumefactum (Beauv.) occurs in Connecticut, New Jersey, Maryland, Virginia, North Carolina, and Alabama (Dawson and McColloch, 1924). A single male was collected June 29, 1893, at Lexington, Kentucky. According to the same writers adults range from 8 to 11 mm in length and from 5.7 to 8 mm in width. Adults are oval, strongly convex, orange-brown and black, with the black distal spot on each elytron not meeting the sutural stripe. According to Sim (1930) this beetle is often reported as a pest of golf courses from its habit of throwing up mounds of sand at the mouths of its burrows on golf greens and fairways. Sim also found the species frequenting old roadways, near Rancocas, New Jersey, that were not too well shaded, and between August 21 and October 3, 1927, he collected 75 adults.

Genus *Bolboceras* (*Odontaeus*) Kirby
(Figs. 90, 96, 100, 103, 107, 111, 117, 118, 121 to 124, 128, and 132)

Howden (1955) characterized larvae of this genus as follows: "Antennae with one conical sense organ on penultimate segment; first segment longer than second, which is longer than the third; body not humped; anal opening transverse with lower anal lobe unpaired; prothoracic legs smaller than others, which are similar in size. Legs 2-segmented, lacking claws."

The larva of *Bolboceras simi* (Wallis) was first described by Ritcher (1947b) and the description is reprinted in the following pages. Howden (1955) described the larvae of *B. liebecki* (Wallis) and *B. darlingtoni* (Wallis) and presented a key for separating larvae of the three known species. A fourth species, *B. obesus* (Lec.) is described in this monograph.

Key to Larvae of *Bolboceras*

1. Prothorax with an anterior process on each side (Fig. 122). Epipharynx without phobae on anterior margin of pedium (Fig. 118) ....*B. obesus*
   Prothorax without anterior processes (Fig. 123). Pedium of epipharynx bordered anteriorly with phobae 2

2. Glossa not emarginate ....*B. simi*
   Glossa emarginate 3

3. Pedium of epipharynx with a large, prominent anterior phoba (Fig. 107) ....*B. darlingtoni*
   Pedium of epipharynx with a small anterior phoba..*B. liebecki*

Bolboceras simi (Wallis), Third-Stage Larva
(Figs. 90, 100, 103, 111, 117, 128, and 132)

This description is based on the following material, and reprinted from my 1947b paper:

Two third-stage larvae. Larvae and associated adults dug in September 1927, at Riverside, New Jersey, by R. J. Sim. One larva pupated November 10, 1927. Loaned by the United States National Museum.

Maximum width of head capsule 2.1 to 2.2 mm. Surface of cranium straw colored, faintly reticulate. Frons on each side with about 4 posterior frontal setae, 2 setae in each anterior angle, and 1 to 3 exterior frontal setae (Fig. 100). Anterior frontal setae absent. Clypeofrontal suture indistinct. Labrum trilobed, wider than long. Antenna 3-segmented, the segments subequal in length. Third antennal segment obovate, smaller in diameter than the first 2 segments. Penultimate segment with a single, distal, subconical sense organ. Distal half of last antennal segment covered with tactile points. Mandibles (Fig. 103) nearly symmetrical, each with a scissorial area consisting of a distal bladelike portion separated from a blunt, proximal toothlike portion by the scissorial notch. Region distad of the molar area without a tooth-like process. Maxillary stridulatory area with a small group of 7 to 8 sharp teeth. Hypopharynx symmetrical with 2 bulbous oncyli. Glossa not emarginate. Tormae of epipharynx (Fig. 128) united mesally, having a prominent posterior epitorma and a smaller anterior epitorma. Pternotormae absent. Pedium surrounded by phobae and covered with spicules.

Spiracles not cribriform (Fig. 90). Thoracic spiracle only slightly larger than abdominal spiracles.

Body not humped; thorax and first 4 abdominal segments rather swollen. Dorsa of abdominal segments 1 to 8 inclusive with 2 annulets as in *Bolbocerosoma* and *Eucanthus*. Each annulet on dorsa of abdominal segments 2 to 8 with a rather dense covering of setae. Last abdominal segment (Fig. 132) with a single, unpaired ventral anal lobe and an unpaired dorsal anal lobe; lobes not fleshy. Anal opening transverse, faintly V-shaped. Venter of last abdominal segment with a sparse covering of about 15 setae anterior to the lower anal lip. Legs (Figs. 111 and 117) 2-segmented, the prothoracic legs noticeably smaller than the mesothoracic and metathoracic legs, which are similar in size. Claws entirely absent. Mesothoracic and metathoracic legs with stridulatory organs, those on the mesothoracic leg consisting of a file-like area on the coxa (Fig. 111) and those on the metathoracic leg consisting of a sparse row of 4 sclerotized teeth on the fused trochanter-femur-tibiotarsus (Fig. 117).

Bolboceras simi (Wallis) occurs from Michigan and Indiana eastward into Connecticut, Massachusetts, and New Jersey (Wallis, 1928). According to the same writer, adults are oblong-oval, dark yellowish-brown, with a movable horn on the clypeus. Adults range from 5.18 to 7.49 mm long and from 3.29 to 4.83 mm wide.

R. J. Sim, after whom the species is named, dug adults from vertical burrows in the greens and fairways of golf courses in New Jersey. The burrows vary from sevaral inches to more than a foot in depth and are marked at the surface by a pile of sand "sausages" the diameter of the hole. He also collected this species in a pine-oak bush lot near Riverside and in and near an old road at Arney Mountain, New Jersey (Sim, 1930). Sim thinks the female oviposits in a mass of humus packed into an elongate mass at the lower end of the burrow (Wallis, 1928).

Bolboceras obesus (Lec.), Third-Stage Larva (Figs. 96, 118, 122, and 124)

This description is based on the following material:

Two third-stage larvae dug May 15, 1955, at Mill Valley, California, under shrub and grass roots, by Hugh B. Leech.

Maximum width of head capsule 2.96 to 3.07 mm. Surface of cranium whitish, shining, feebly rugulose. Frons, on each side, with an irregular patch of 6 to 9 setae and 1 or 2 anterior angle setae. Epicranial setae 13 to 16 on each side. Labrum trilobed, clypeo-frontal suture absent. Antenna 3-segmented, last segment reduced in size. Penultimate segment of antenna with a single conical, apical sensory appendage. Mandibles nearly symmetrical, molar areas strongly concave on inner face. Epipharynx (Fig. 118) trilobed; haptomeral region feebly raised and with 12 sensilla. Chaetoparia consisting of 14 to 19 setae on each side. Pedium posteriorly, on each side, with a flat process and a spiculate phoba; surface of pedium not spiculate. Tormae symmetrical, united medially. Pternotormae poorly developed. Anterior epitorma very small; posterior epitorma large, triangular in shape. Maxilla (Fig. 124) with galea and lacinia distinctly separate. Lacinia with 2 apical unci fused at their bases. Maxillary stridulatory area with a patch of about 13 to 23 teeth on the stipes; also with 5 to 9 similar teeth on the first segment of the maxillary palpus.

Larva stout-bodied (Fig. 96) with abdominal segments 6 to 10 reduced in size. Prothorax, on each side, with an anterior process (Fig. 122), absent in other known species of the genus. Anterior margin of dorsum of prothorax distinctly biarcuate. Abdominal segments 1 to 8 each with 2 dorsal folds. Each scutum on segments 2 to 8 with a short, transverse patch of fine, slightly curved setae; each scutellum on segments 1 to 8 also with a long, transverse band of fine, slightly curved setae. Venter of last abdominal segment with a sparsely set covering of about 37 to 45 slender setae. Lower anal lobe entire. Anal slit transverse, slightly curved. Anus on dorsal surface of last abdominal segment. Legs similar to those of *B. simi* (Wallis). Coxa of each mesothoracic leg with a stridulatory "file" having 8 or 9 widely separated striae. Metathoracic leg with a sparsely set row of 3 or 4 teeth which appose the "file." Claws absent.

Bolboceras (*Odontaeus*) *obesus* Lec. occurs in California, Oregon, and Washington and is the only species of the genus occurring in these states. It is sometimes fairly common; 25 specimens are present in the Entomology Department collection at Oregon State University. Adult males range from 8.5 to 12 mm in length and from 6.5 to 8 mm in width. Females range from 8.5 to 12 mm in length and 5 to 8 mm in width. Adults are dark reddish-brown to black in color, strongly convex, and have elytra with prominent punctate striae. The head of the male has a long, slender, caudally curved horn which sometimes extends almost to the scutellum. The pronotum of the male, on each side, has a longitudinal ridge and posteriorly has a median, raised, U-shaped protuberance.

Little is known about the biology of this species. Linsley and Michener (1943) dug several hundred pupae and a few larvae from sandy soil of an old road bed in the midst of a manzanita chute, near Mt. Lassen, California, but their material has been lost. On May 19, 1960, the writer and David Fellin observed several "pushups" in a wooded area near Trail, Oregon, and dug two pairs of adults and a third female from the winding burrows beneath. Specimens are occasionally taken at light.

Plate VIII

FIGURE 89.   *Geotrupes blackburnii excrementi* Say. Third-stage larva, left lateral view (setation omitted).

FIGURE 90.   *Bolboceras simi* (Wallis). Thoracic spiracle.

FIGURE 91.   *Lethrus apterus* Laxm. (after Golovianko). Third-stage larva, left lateral view (setation omitted).

FIGURE 92.   *Eucanthus lazarus* (Fab.). Thoracic spiracle.

FIGURE 93.   *Eucanthus lazarus* (Fab.). Third-stage larva, left lateral view (setation omitted).

FIGURE 94.   *Bolbocerosoma* (prob. *tumefactum* [Beauv.]). Thoracic spiracle.

FIGURE 95.   *Bolbocerosoma* (prob. *tumefactum* [Beauv.]). Third-stage larva, left lateral view (setation omitted).

FIGURE 96.   *Bolboceras obesus* (Le Conte). Third-stage larva, left lateral view.

FIGURE 97.   *Peltotrupes youngi* Howden. Third-stage larva, left lateral view.

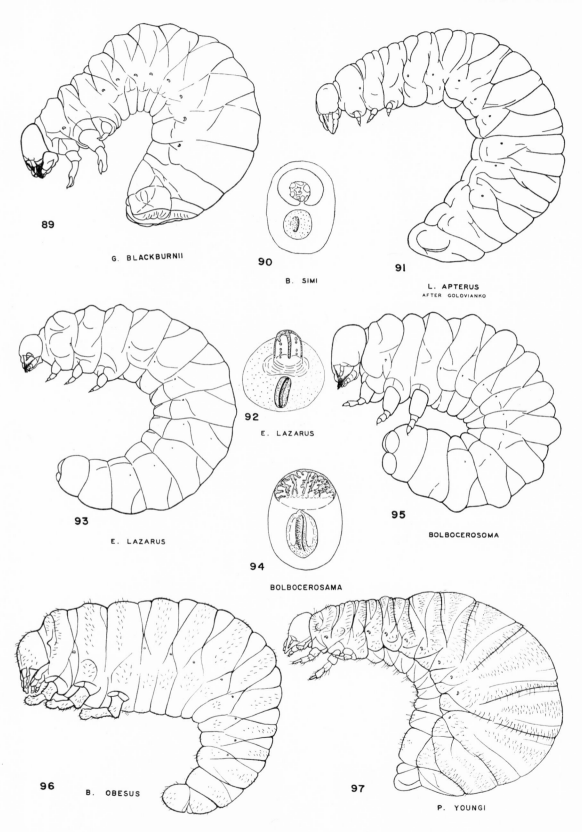

89

G. BLACKBURNII

90

B. SIMI

91

L. APTERUS

AFTER GOLOVIANKO

93

E. LAZARUS

92

E. LAZARUS

95

BOLBOCEROSOMA

94

BOLBOCEROSAMA

96   B. OBESUS

97

P. YOUNGI

Plate IX

FIGURE 98.   *Geotrupes blackburnii excrementi* Say. Head.

FIGURE 99.   *Peltotrupes youngii* Howden. Head.

FIGURE 100.   *Bolboceras simi* (Wallis). Head.

FIGURE 101.   *Bolbocerosoma* (prob. *tumefactum* [Beauv.]). Head.

FIGURE 102.   *Eucanthus lazarus* (Fab.). Head.

FIGURE 103.   *Bolboceras simi* (Wallis). Left mandible, dorsal view.

FIGURE 104.   *Bolbocerosoma* (prob. *tumefactum* [Beauv.]). Left mandible, dorsal view.

FIGURE 105.   *Eucanthus lazarus* (Fab.). Left mandible, dorsal view.

FIGURE 106.   *Eucanthus lazarus* (Fab.). Right mandible, dorsal view.

FIGURE 107.   *Bolboceras darlingtoni* (Wallis). Epipharynx.

FIGURE 108.   *Geotrupes blackburnii excrementi* Say. Left mandible, dorsal view.

FIGURE 109.   *Geotrupes blackburnii excrementi* Say. Right mandible, dorsal view.

Symbols Used

A—Antenna
AA—Setae of anterior angle of
frons
AF—Anterior frontal setae
BR—Brustia
DMS—Dorsomolar setae
E—Epicranium

EFS—Exterior frontal setae
ES—Epicranial suture
F—Frons
FS—Frontal suture
L—Labrum
MO—Molar region
PA—Preartis

PC—Preclypeus
PFS—Posterior frontal setae
PSC—Postclypeus
PTA—Postartis
SA—Scissorial area
VP—Ventral process

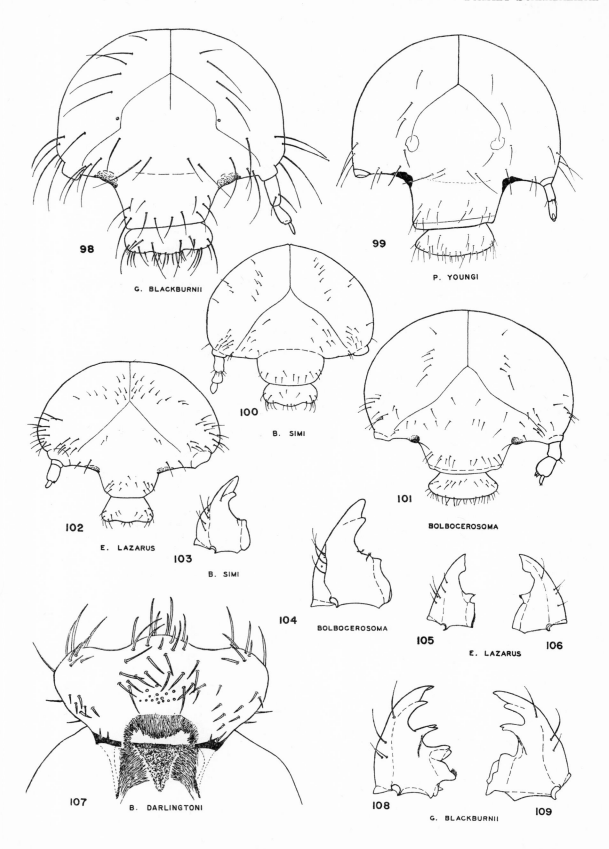

98   G. BLACKBURNII

99   P. YOUNGI

100   B. SIMI

101   BOLBOCEROSOMA

102   E. LAZARUS

103   B. SIMI

104   BOLBOCEROSOMA

105   106   E. LAZARUS

107   B. DARLINGTONI

108   109   G. BLACKBURNII

Plate X

FIGURE 110. *Geotrupes blackburnii excrementi* Say. Left metathoracic and mesothoracic legs, lateral view.

FIGURE 111. *Bolboceras simi* (Wallis). Left mesothoracic leg, lateral view.

FIGURE 112. *Bolbocerosoma* (prob. *tumefactum* [Beauv.]). Third abdominal segment. Left, lateral view.

FIGURE 113. *Eucanthus lazarus* (Fab.). Left metathoracic leg, lateral view.

FIGURE 114. *Geotrupes blackburnii excrementi* Say. Left metathoracic leg.

FIGURE 115. *Bolbocerosoma* (prob. *tumefactum* [Beauv.]). Left metathoracic leg, lateral view.

FIGURE 116. *Eucanthus lazarus* (Fab.). Left maxilla, labium and hypopharynx.

FIGURE 117. *Bolboceras simi* (Wallis). Left metathoracic leg.

FIGURE 118. *Bolboceras obesus* (LeConte). Epipharynx.

FIGURE 119. *Geotrupes blackburnii excrementi* Say. Left maxilla, labium and hypopharynx.

FIGURE 120. *Bolbocerosoma* (prob. *tumafactum* [Beauv.]). Right maxilla, labium and hypopharynx.

FIGURE 121. *Bolboceras darlingtoni* (Wallis). Head.

FIGURE 122. *Bolboceras obesus* (LeConte). Head and prothorax, dorsal view.

FIGURE 123. *Bolboceras darlingtoni* (Wallis). Head and prothorax, dorsal view.

FIGURE 124. *Bolboceras obesus* (LeConte). Hypopharynx and right maxilla, dorsal view.

Symbols Used

AP—Anterior process	H—Head	PRS—Prescutum
CL—Claw	LP—Labial palpus	SCU—Scutum
CPA—Chaetoparia	LPH—Laeophoba	SD—Stridulatory area
CX—Coxa	O—Oncylus	SP—Spiracle
DPH—Dexiophoba	PLL—Pleural lobe	TRFE—Trochanter-femur
EUS—Eusternum	PE—Pedal area	TT—Tibiotarsus
GL—Glossa	PRO—Prothorax	UN—Uncus

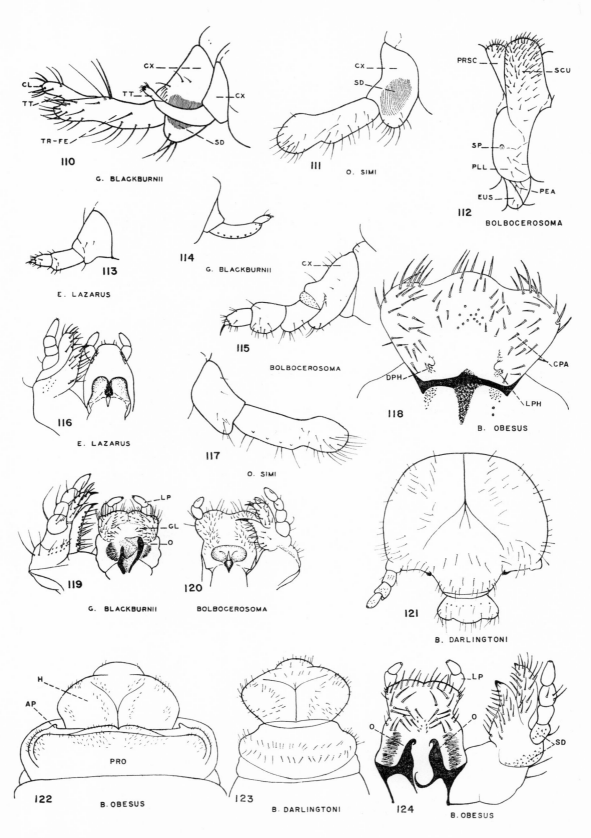

110 G. BLACKBURNII

111 O. SIMI

112 BOLBOCEROSOMA

113 E. LAZARUS

114 G. BLACKBURNII

115 BOLBOCEROSOMA

116 E. LAZARUS

117 O. SIMI

118 B. OBESUS

119 G. BLACKBURNII

120 BOLBOCEROSOMA

121 B. DARLINGTONI

122 B. OBESUS

123 B. DARLINGTONI

124 B. OBESUS

Plate XI

Figure 125.   *Peltotrupes youngi* Howden. Epipharynx.

Figure 126.   *Geotrupes blackburnii excrementi* Say. Epipharynx.

Figure 127.   *Bolbocerosoma* (prob. *tumefactum* [Beauv.]). Epipharynx.

Figure 128.   *Bolboceras simi* (Wallis). Epipharynx.

Figure 129.   *Eucanthus lazarus* (Fab.). Epipharynx.

Figure 130.   *Geotrupes ulkei* Blanch. Caudal view of last abdominal segment.

Figure 131.   *Geotrupes blackburnii excrementi* Say. Caudal view of last abdominal segment.

Figure 132.   *Bolboceras simi* (Wallis). Caudal view of last abdominal segment.

Figure 133.   *Typhoeus typhoeus* L. Caudal view of last abdominal segment (after van Emden).

Figure 134.   *Lethrus apterus* Laxm. Caudal view of last abdominal segment (after Golovianko).

Figure 135.   *Bolbocerosoma* (prob. *tumefactum* [Beauv.]). Caudal view of last abdominal segment.

Figure 136.   *Peltotrupes youngi* Howden. Caudal view of last abdominal segment.

Symbols Used

ASL—Anal slit
CPA—Chaetoparia
DAL—Dorsal anal lobe
DX—Dexiotorma
ESF—Endoskeletal figure

ETA—Anterior epitorma
ETP—Posterior epitorma
LT—Laeotorma
MPH—Mesophoba

PE—Pedium
PH—Phoba
PTT—Pternotorma
VAL—Ventral anal lip

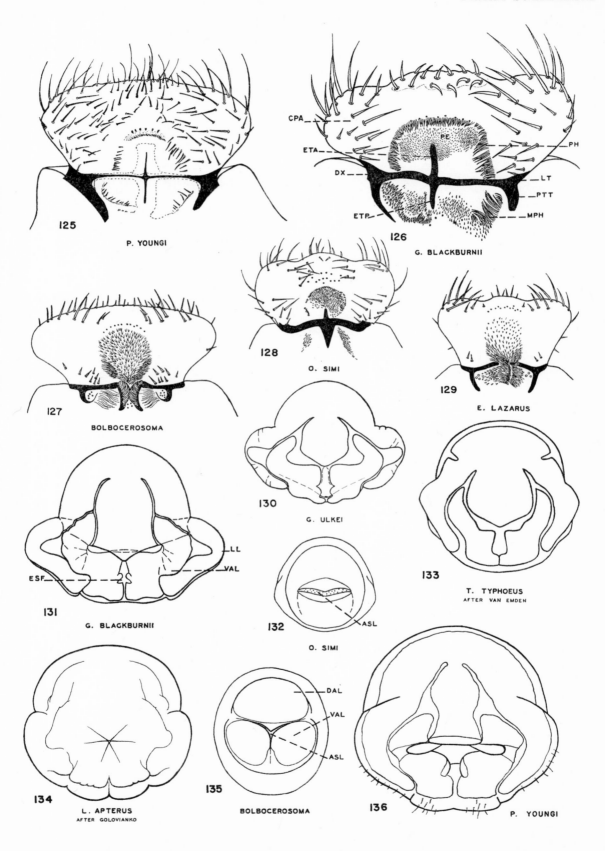

125 P. YOUNGI

126 G. BLACKBURNII

CPA
ETA
DX
ETP
PE
PH
LT
PTT
MPH

127 BOLBOCEROSOMA

128 O. SIMI

129 E. LAZARUS

130 G. ULKEI

131 G. BLACKBURNII

LL
VAL
ESF

132 O. SIMI

ASL

133 T. TYPHOEUS
AFTER VAN EMDEN

134 L. APTERUS
AFTER GOLOVIANKO

135 BOLBOCEROSOMA

DAL
VAL
ASL

136 P. YOUNGI

Subfamily Pleocominae

Leng (1920) and Cazier (1953) have included two genera, *Pleocoma* and *Acoma*, in the subfamily Pleocominae. Howden (1958) states that the placement of *Acoma* is likely to remain in doubt until females are discovered. The exact systematic position of the genus *Pleocoma* also remains in some doubt, but a number of workers have felt that it belongs in a subfamily closely related to the geotrupids (Horn, 1883 and 1888; Davis, 1934; and Ritcher, 1947a).

Paulian (1941), on the basis of adult structures, concluded that *Pleocoma* is a geotrupid genus and Ritcher (1947b) supported that view, based on studies of larvae of several genera of geotrupid larvae. The finding that *Pleocoma* larvae have a large number of moults (8 to 13), unlike all other Scarabaeidae which have only 3 (Ellertson and Ritcher, 1959), leads me to believe that the genus is less closely related to the Geotrupinae than I had formerly supposed. Also, I find that female *Pleocoma* possess 18 ovarioles on each side, while, according to Robertson (1961), Geotrupinae possess 1 to 6.

The genus *Pleocoma* contains 29 described species, all confined to western North America. In distribution they range from northern Mexico (Baja, California) through California and Oregon into southern Washington (Linsley, 1938). Two species have been described from Utah (locality unknown) and one from southern Alaska.

The adults are called rain beetles because their flight usually follows fall rains. The winged males emerge from the soil and seek out the burrows of wingless females where mating occurs. For some species, the heaviest flights occur before or just after dawn. Eggs are deposited from 11 to 30 inches deep in the soil, in spiral fashion, in a prepared core of soil, the following May or June (Ritcher and Beer, 1956 and Ellertson and Ritcher, 1959). The larvae moult once a year and live for 8 to 12 years before pupating (Ellertson and Ritcher, 1959). They feed on live roots of forest trees (Stein, 1963) or of grasses (a few species only) (Ritcher, 1947a), and several species also damage the roots of pear, apple, and cherry trees (Ellertson and Ritcher, 1959; Ellertson, 1956). The adults do not feed.

Pleocoma larvae have been described by Osten Sacken (1874-1876), Ritcher (1947a and 1947b), and Hayes and Peh-I Chang (1947) and included in the key of Böving and Craighead (1930). Larvae of *Pleocoma australis* Fall, *P. badia* Fall, *P. carinata* Linsley, *P. crinita* Linsley, *P. dubitalis* Davis, *P. edwardsii* Lec., *P. fimbriata* Lec., *P. minor* Linsley, *P. oregonensis* Leach, *P. simi* Davis, and *P. hirticollis vandykei* Linsley, have been studied by the writer. Except for the characters given in the tentative key, no good morphological characters have been found for separating them.

The subfamily Pleocominae (genus *Pleocoma*) may be characterized as follows: Haptomerum of epipharynx with a longitudinal group of heli. Epipharynx with plegmatia and with prominent chaetoparia and acanthoparia. Hypopharynx without oncyli. Antenna 3-segmented with apical segment greatly reduced in size. Second antennal segment with an apical sensory spot. Legs 4-segmented, each with an apical claw bearing 2 basal setae. Trochanters and femora of mesothoracic and metathoracic legs with fossorial setae. Mesothoracic legs and metathoracic legs with stridulatory organs. Spiracles with concavities of respiratory plates facing ventrally. Terga of abdominal segments 3 to 7, inclusive, each with 4 dorsal annulets. Anal opening V or Y-shaped, not surrounded by fleshy lobes.

Genus *Pleocoma* Le Conte

Tentative Key to Species of *Pleocoma* Based on Characters of the Larvae

1. Maxilla with uncus of lacinia having a distinct basal tooth 2
   Maxilla with uncus of lacinia having a more or less distinct shoulder at base but no tooth 4

2. Left mandible with 25 or more dorsomolar setae (DMS) ....*P. hirticollis vandykei*
   Left mandible with 15 or fewer dorsomolar setae (DMS) 3

3. Trochanter of metathoracic leg with 6 or 7 prominent stridulatory teeth ....*P. crinita*
   Trochanter of metathoracic leg with 9 or more prominent stridulatory teeth ....*P. oregonensis*

4. Trochanter of metathoracic leg with 10 or more prominent stridulatory teeth ....*P. simi*
   Trochanter of metathoracic leg with 9 or fewer prominent stridulatory teeth ....
   ....*P. dubitalis, P. minor, P. edwardsi,* and *P. carinata*

Pleocoma hirticollis vandykei Linsley, Late-Stage Larva
(Figs. 137 to 146)

This description is based on the following material and reprinted from Ritcher (1947b):

Nine late-stage larvae and a cast skin of a larva found associated with a dead male in its pupal cell. These specimens, together with about 30 other larvae, were dug from pasture soil, March 8, 1946, near Patterson Pass, California, by R. F. Smith, J. W. McSwain, several of Linsley's graduate students, and P. O. Ritcher.

Larva typically scarabaeiform with whitish body and light yellow-brown head. Length of mature larva ranging from about 45 to 50 mm. Maximum width of cranium ranging from 6.5 to 9 mm with a mean of 8 mm. Surface shining and generally smooth with a series of fine longitudinal striae on each side of the epicranial suture (ES). Frons bearing on each side an irregular, transverse row of 5 to 10 posterior frontal setae, 1 large seta in each anterior angle (rarely with an additional small seta), a single large exterior frontal seta, and often with 1 or 2 small setae, and a single large anterior frontal seta plus 2 to 4 small setae (Fig. 138). Epicranium with 2 large dorsoepicranial setae on each side of the epicranial suture. Clypeo-frontal suture present. Labrum slightly wider than long, symmetrical, and apically trilobed.

Antenna (Fig. 138) almost as long as cranium, fairly slender, 3-jointed, and borne on a cylindrical basal piece fused to the epicranium. First segment as long as second and third segments together; third segment very small, about one third as long and half as wide as second segment. First and second segments bearing numerous setae. End of second antennal segment, below juncture with apical segment, with a small oval sensory spot. Apical segment without sensory spots; apex with 2 to 4 olfactory pegs.

Mandibles (Figs. 140 and 141) shorter than cranium, approximately symmetrical, subtriangular in outline, each with a strong, bladelike scissorial area (SA) and a rather small, molar area (MO). Scissorial area blackish with a slightly concave or nearly straight cutting edge. Molar areas also blackish, those of left and right mandibles similar in size but somewhat asymmetrical. Molar area of left mandible with 2 dorsal and 4 ventral lobes bounding a central concavity. Molar area of right mandible with a transverse apical lobe and a curved median lobe, the latter surrounding a small, longitudinal proximal lobe. At the base of each molar area is a dense brush of setae or brustia (BR).

Epipharynx (Fig. 137) trilobed apically. Apical lobe, or corypha (CO), set with coarse setae and bounded on each side by a clithrum (CL). Plegmatia present. Each plegmatium consisting of about 11 to 15 semicircular, sclerotized plegmata, each surrounding the base of a course acanthoparial seta. Chaetoparia (CPA) large, separated from the acanthoparia by a narrow gymnoparia (GP). Each chaetoparia consisting of a dense patch of sharp setae not interspersed with sensilla. Chaetae stoutest toward the pedium (PE). Haptomerum (H) bearing a rather sparsely set, semicircular group of 9 to 12 large, stout heli. Tormae symmetrical, not branched, not meeting mesally. Haptolachus incomplete, nesia absent. Caudomesad of the inner end of each torma is a longitudinal curved phoba (PH). The area laterad of each phoba has 20 to 30 crepidial punctures (CP). Four macrosensilla (MS) are found between the caudal ends of the phobae. Posterior to the phobae is the curved, transverse crepis (CR).

Respiratory plates of spiracles kidney-shaped, with their concavities facing ventrally (Fig. 146). Spiracles cribriform but "holes" of respiratory plate rather opaque, not arranged in definite rows. Thoracic spiracles considerably larger than the abdominal spiracles which are alike in size.

First and second abdominal segments each with 3 dorsal areas. Each prescutum with a transverse band of short, stout setae; that on the first abdominal segment 2 or 3 rows wide. Each scutum with a long, transverse band of setae consisting anteriorly of about 5 irregular rows of short, stout setae and posteriorly of a single sparsely set row of fairly long, slender setae. Scutellum of first abdominal segment with a short, sparsely set transverse patch of short, stout setae on each side and a bare, mid-dorsal area. Scutellum of second abdominal segment with a long, narrow, transverse, single or double row of short, stout setae. Abdominal segments 3 to 7 inclusive each with 4 dorsal annulets, a prescutum, scutum, scutellum, and postscutellum (Fig. 146). Each prescutum bears a transverse patch of about 5 to 7 irregular rows of short, stout setae. Each scutum with a short, transverse, irregular, double row of short, stout setae. Each scutellum has a long, transverse band of short, stout setae posterior to which is a single sparsely set row of rather long, slender setae. Each postscutellum, except that of abdominal segment 7 which is bare, with a long, transverse, irregular double or triple row of short, stout setae.

Dorsa of abdominal segments 8 and 9 not divided into distinct annulets. Each dorsum anteriorly with scattered slender setae and an occasional short, stout seta, posteriorly with a sparsely set, transverse row of long, slender setae. Dorsum of tenth abdominal segment with a transverse, sinuate dorsal impressed line. Region between dorsal impressed line and anterior margin of abdominal segment 10 set with scattered short, stout setae. Dorsal area, between dorsal anal lip (DAL, Fig. 145) and dorsal impressed line, clothed with short, stout setae interspersed with several long, slender setae.

Anal slit (Fig. 145) Y-shaped, basal cleft short. Anus bordered dorsally by the triangular dorsal anal lip (DAL) and ventrally by the mesally cleft ventral anal lip (VAL). Both anal lips covered with short, stout setae.

Raster (Fig. 145) consisting of a simple teges (T) in the form of a narrow, transverse band of 50 to 70 narrow, short, stout, caudally directed setae, located cephalad of the lower anal lip. Cephalad of the teges, especially toward each side, are several long, slender setae. Anterior to these setae and the teges, the venter of the tenth abdominal segment is bare.

Legs well developed with the prothoracic shorter than the metathoracic pair. Each leg 4-jointed, consisting of a fairly long, stout, subcylindrical coxa (CX), a slightly shorter trochanter (TR), a short femur (FE), and a short tibiotarsus (TT) which bears a terminal claw. Mesothoracic and metathoracic legs with stridulatory organs consisting of a finely striated area on the posterior surface of the mesocoxa and a V-shaped row of about 12 sclerotized teeth, each tooth at the base of a seta, on the anterior surface of the metatrochanter (Fig. 143). Anterior ventral surface of mesothoracic and metathoracic trochanters and ventral surface of mesothoracic and metathoracic femora densely set with stout, spine-like setae undoubtedly useful in burrowing through hard soil. Claws simple, each consisting of a straw-colored base and a dark, slender, sharp distal portion. Base of each claw with 2 setae. Phothoracic claws much longer than mesothoracic and metathoracic claws.

Pleocoma hirticollis vandykei Linsley occurs in Sonoma, Contra Costa, and Alameda counties, California, and is very abundant in a small pasture area near Patterson Pass in open, hill grassland in the latter county. Adult males are oval, black, and range from 20 to 24.5 mm in length; females are oval, brown, and range from 24.5 to 29 mm in length (Linsley, 1938, 1941, and 1945). The males fly in the fall after rain of half an inch or more and seek out the females which remain in their open burrows. On October 30, 1944, Smith and Potts (1945) collected 37 males, and on the following November 4 they collected 78 males from the same spot. Seventeen females were dug from open burrows, most of them at depths of 3 to 10 inches and accompanied by cast pupal or larval skins. Four larvae were taken. While visiting at the University of California, in Berkeley, early in 1946, the writer was shown 8 larvae of the same species collected February 26, 1946, by Smith and McSwain from the same spot. A second trip to Patterson Pass, March 8, 1946, by Smith, McSwain, several of Linsley's graduate students, and the writer, yielded over 40 additional larvae (Ritcher, 1947). They were found from 1 to 3 inches deep in the pasture soil, feeding on grass roots. A dead male was found in its pupal cell at a depth of 5 inches.

Plate XII

Pleocoma hirticollis vandykei Linsley

FIGURE 137. Epipharynx.

FIGURE 138. Head.

FIGURE 139. Labium and hypopharynx, dorsal view.

FIGURE 140. Left mandible, ventral view.

FIGURE 141. Left mandible, dorsal view.

FIGURE 142. Right maxilla and labium, ventral view.

FIGURE 143. Right metathoracic leg.

FIGURE 144. Left maxilla, dorsal view.

FIGURE 145. Caudal view of last abdominal segment.

FIGURE 146. Third abdominal segment, left lateral view.

Symbols Used

A—Antenna
AA—Setae of anterior angle of frons
AF—Anterior frontal setae
ASL—Anal slit
BR—Brustia
CAR—Cardo
CL—Claw
CLI—Clithrum
CO—Corypha
CP—Crepidal punctures
CPA—Chaetoparia
CR—Crepis
CX—Coxa
DAL—Dorsal anal lobe
DMS—Dorsomolar setae
DX—Dexiotorma
E—Epicranium
EFS—Exterior frontal setae
ES—Epicranial suture
EUS—Eusternum
F—Frons

FE—Femur
FOS—Fossorial setae
FS—Frontal suture
G—Galea
GL—Glossa
GP—Gymnoparia
H—Haptomerum
HE—Helus
HL—Haptolachus
HP—Hypopharynx
L—Labrum
LA—Lacinia
LAL—Lower anal lobe
LL—Lateral lobe
LP—Labial palpus
LT—Laeotorma
MO—Molar area
MP—Maxillary palpus
PA—Preartis
PC—Preclypeus
PE—Pedium
PEA—Pedal area

PFS—Posterior frontal setae
PH—Phoba
PL—Plegmum
PLL—Pleural lobe
PMP—Postmentum
PRM—Prementum
PRSC—Prescutum
PSC—Postclypeus
PSCL—Postscutellum
PTA—Postartis
SA—Scissorial area
SC—Scutum
SCL—Scutellum
SD—Stridulatory area
SP—Spiracle
SPA—Spiracle area
ST—Stipes
T—Teges
TR—Trochanter
TT—Tibiotarsus
UN—Uncus
VP—Ventral process

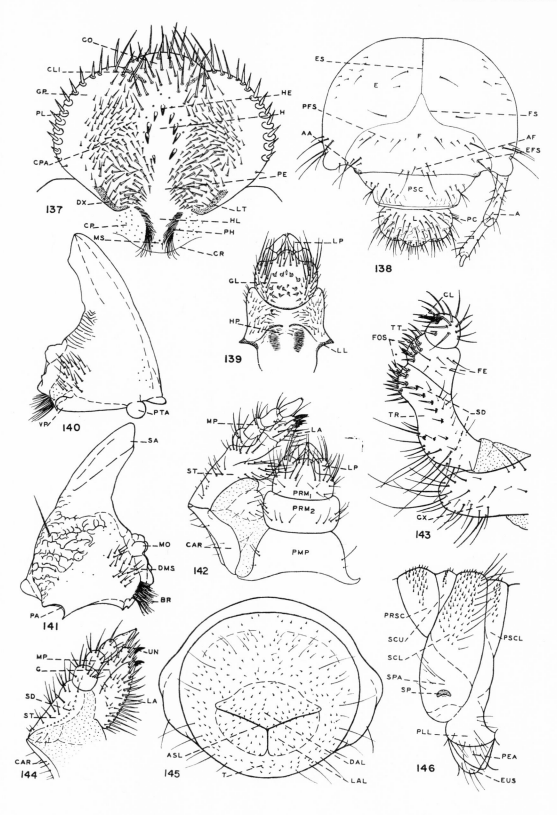

Subfamily Glaphyrinae

This small, odd subfamily is represented in North America only by the genus *Lichnanthe* (*Amphicoma*). A few representatives of the subfamily, all belonging to other genera, are found in Peru (one species) and Chile (nine species), but none are known from Mexico or Central America (Blackwelder, 1944).

Leng (1920) lists seven species of *Lichnanthe* as occurring in the United States. According to Van Dyke (1928), there are only four valid species in this country, two eastern, *L. lupina* Lec. and *L. vulpina* (Hentz) and two western, *L. ursina* Lec. and *L. rathvoni* Lec. The western species each have several color forms which were formerly considered separate species.

The hairy adults are diurnal and ressemble bumble bees when in flight. They are not known to feed. In Oregon, the bluish "hairy" larvae of *L. rathvoni* are sometimes common in sandy areas, especially along streams. In Massachusetts, larvae of *L. vulpina* (Hentz) are serious pests in old cranberry bogs (Franklin, 1950) where they destroy the fibrous roots.

Hayes (1929) described and figured the epipharynx of *Lichnanthe* (*Amphicoma*) sp. and figured the venter of the last abdominal segment. His material came from Massachusetts, so it probably represents *L. vulpina*. I have four Massachusetts larvae in my collection which are part of the same lot. I am unable to find any characters for separating them from larvae of *L. rathvoni*.

Lichnanthe rathvoni Lec., Third-Stage Larva
(Figs. 147 to 155)

This description is based on the following material:

Three third-stage larvae and cast skins of 9 third-stage larvae reared to the adult stage. Larvae collected April 27, 1954, from sandy loam on Kiger Island, four miles southeast of Corvallis, Oregon, by Vincent Roth, Frank Beer, and P. O. Ritcher.

Eight third-stage larvae collected from sandy loam near the Willamette River, Corvallis, Oregon, April 29, 1961, by P. O. Ritcher.

Maximum width of head capsule 3.6 to 4.3 mm. Surface of cranium pitted, reddish-brown, with a rather uniform covering of slender setae which are quite long on the epicranium. Frons rather flat and with a conspicuous, median, circular depression (Fig. 147); surface of frons coarsely pitted, each pit with a single, short seta. Eye spots absent. Labrum trilobed, symmetrical. Epipharynx (Fig. 149) with clithra; epizygum, zygum, proplegmatia, and plegmatia absent. Chaetoparia well developed with numerous long chaetae.

Phobae absent. Sclerotized plate (nesium externum) poorly developed, often absent; sensory cone (nesium internum) present, well developed. Tormae strongly asymmetrical with dexioterma extending across median line. Labial palpi 2-segmented, unlike those of any other scarabaeid; distal segment much reduced in size and projecting mesally. Mandible (Figs. 150 and 151) with a small blade-like portion anterior to the scissorial notch and a single tooth posterior to the same notch. Molar area strongly developed. Lateral face of mandible heavily wrinkled, scrobis absent. Maxilla (Fig. 155) with galea and lacinia distinctly separate. Lacinia with 3 unci arranged in a longitudinal row. Stipes with a row of 10 to 13 conical stridulatory teeth. Antenna (Fig. 148) 4-segmented; last segment very small, reduced in diameter, hardly as long as apical projection of third antennal segment. Third antennal segment with numerous small sensory pits.

Spiracles cribriform; those on the prothorax with the emarginations of the respiratory plates facing caudoventrad and those on the abdominal segments with the emarginations of their respiratory plates facing cephaloventrad. Respiratory plates reniform, not surrounding the bullae. Spiracles of abdominal segments 1 to 5 progressively smaller in size, those of abdominal segments 6 to 8 similar in size to three of abdominal segment 5.

Thorax and abdomen covered with numerous soft, hair-like setae. Abdominal segments 1 to 8 each with 3 dorsal annulets. Setation of venter of last abdominal segment (Fig. 153) similar to that of rest of body; no separate lower and upper anal lobes. Anal slit transverse, located caudally on the dorsum of the last abdominal segment. Legs 4-segmented, without stridulatory organs. Claws (Fig. 154) very long and curved, each with 4 prominent basal setae.

Lichnanthe rathvoni Lec. is a common species in sandy areas along streams in the Willamette Valley of Oregon. According to Leng (1920) it has been found in Nevada, California, and Washington. The adults are hairy, day-flying scarabaeids with three color forms in Oregon, one all black, which is least common, a second with yellow and black bands of hairs on the abdomen, and a third with orange and black bands of hairs on the abdomen. In size the beetles range from 11 to 14.5 mm in length and from 5.5 to 8.5 mm in width, with females averaging slightly larger than males.

The bluish larvae are common in sand drifts along streams under willows where they feed on the thin layers of buried decaying leaves and other plant debris, usually at depths of 8 to 15 inches. More mature third-stage larvae are yellow in color. The adults, which do not feed, begin emerging in early July. Eggs are laid that same month and hatch in August. Overwintering larvae are of all three instars. Rearing studies indicate that the length of life cycle is usually 3 years with some individuals requiring 4 years. Pupation occurs in May at a depth of 3 to 10 inches, with an average depth of 6.4 inches. The pupal stage at prevailing soil temperatures of 15° C lasts from 30 to 33 days.

Plate XIII

Lichnanthe rathvoni Lec.

FIGURE 147.   Head, cephalic view.

FIGURE 148.   Right antenna, side view.

FIGURE 149.   Epipharynx.

FIGURE 150.   Right mandible, dorsal view.

FIGURE 151.   Right mandible, ventral view.

FIGURE 152.   Hypopharynx and labium.

FIGURE 153.   Venter of ninth and tenth abdominal segments.

FIGURE 154.   Claw of metathoracic leg.

FIGURE 155.   Right maxilla, dorsal view.

Symbols Used

CPA—Chaetoparia   LP—Laeophoba   PE—Pedium
DX—Dexiotorma   LT—Laeotorma   SC—Sense cone

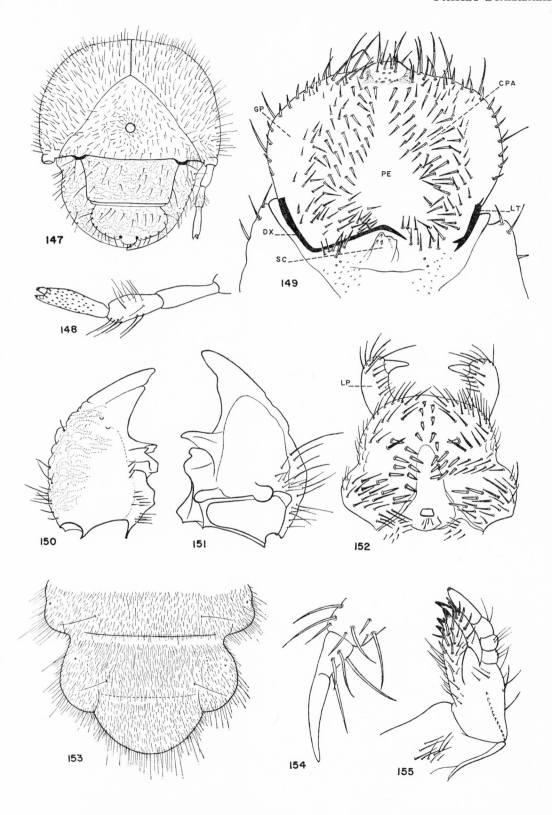

Subfamily Acanthocerinae

Only two genera of this small subfamily occur in the United States. These are *Cloeotus* (two species) and *Acanthocerus* (one species). According to Hamilton (1887) and Lugger (1888), *Cloeotus aphodioides* Ill. is found under the bark of dead trees. Lugger, who frequently bred the beetles from their eggs, states that he found all stages *in situ* under bark. Several oriental species of the genus *Haroldius* are said to be myrmecophilous (Wassman, 1918, and Pereira, 1954).

Based on larval characters, the Acanthocerinae are closely related to the Hybosorinae. Both have rather similar epipharynges and spatulate setae on their rasters. Also species of both subfamilies have stridulatory organs on the legs.

Larvae of Acanthocerinae (*Cloeotus*) have been described in part by Böving and Craighead (1930) in their key. The subfamily may be characterized as follows: Anterior margin of labrum strongly serrate; epipharynx with a dextral beak-like process. Maxilla with separate galea and lacinia. Maxillary stridulatory area with a row of conical teeth. Antenna 4-segmented; last segment not reduced in diameter and with a dorsal sensory spot. Thoracic spiracle with concavity of respiratory plate facing posteriorly; abdominal spiracles with concavities of respiratory plates facing anteriorly. Dorsa of abdominal segments 2 to 5 plicate. Raster with a transverse palidia of spatulate setae. Legs 4-segmented, with well-developed claws. Stridulatory organ present on mesothoracic and metathoracic legs.

Genus *Cloeotus* Germar
Cloeotus aphodioides (Ill.), Third-Stage Larva
(Figs. 156 to 161, 163, and 164)

This description is based on the following material:

Six third-stage larvae and 1 cast larval skin, associated with 2 adults, collected July 26, 1949, at College Park, Maryland, under bark of a standing red or black oak, by H. S. Barber and G. B. Vogt.

Maximum width of head capsule 1.7 to 1.8 mm. Surface of cranium smooth, yellow-brown. Frons, on each side, with 1 anterior frontal seta, 1 seta in the anterior frontal angle, 1 exterior frontal seta, and 1 posterior frontal seta (Fig. 156). Dorsoepicranial setae, 2 on each side. Clypeus produced at the anterior lateral angles. Labrum somewhat asymmetrical and with a number of gibbosities; anterior margin strongly serrate.[13] Epipharynx (Fig. 157) with a dextral beak-like process; paria without chaetae. Left paria with a laeophoba of about 7 filaments. Mandibles (Figs. 159 and 160) each with a 2-toothed scissorial area.

Maxilla (Fig. 158) with separate galea and lacinia. Maxillary stridulatory area consisting of a slightly irregular row of 10 to 13 conical teeth. Antenna 4-segmented with the second segment twice as long as the first and equal in length to the combined third and fourth segments. Third antennal segment with an apical sensory appendage. Last segment not reduced in diameter and with 1 dorsal sensory spot which extends around to the ventral surface (Fig. 156).

Spiracles cribriform. Thoracic spiracle (Fig. 164) with concavity of respiratory plate facing posteriorly; abdominal spiracles smaller in size and with the concavities of their respiratory plates facing anteriorly. Dorsa of abdominal segments 2 to 5 with three dorsal folds, each with a sparse covering of short setae; each scutellum also with a few long setae. Venter of last abdominal segment (Fig. 163) with a sparsely set, transverse, slightly curved palidia of 9 to 11 spatulate setae. Anal slit transverse, located on dorsal surface of last abdominal segment. Legs 4-segmented with well-developed claws. Stridulatory organs present on mesothoracic and metathoracic legs, consisting of scattered, minute, truncate carina on the femora of the metathoracic legs (Fig. 161) and a granular area on each coxa of the mesothoracic legs (discernible only with a compound microscope).

---

[13] Also serrate in *Philarmostes* (see figures 162 and 165).

Plate XIV

FIGURE 156. *Cloeotus aphodioides* (Ill.). Head.

FIGURE 157. *Cloeotus aphodioides* (Ill.). Epipharynx.

FIGURE 158. *Cloeotus aphodioides* (Ill.). Left maxilla, dorsal view.

FIGURE 159. *Cloeotus aphodioides* (Ill.). Left mandible, dorsal view.

FIGURE 160. *Cloeotus aphodioides* (Ill.). Right mandible, dorsal view.

FIGURE 161. *Cloeotus aphodioides* (Ill.). Portion of stridulatory area on metathoracic leg (highly magnified).

FIGURE 162. *Philarmostes* sp. Epipharynx.

FIGURE 163. *Cloeotus aphodioides* (Ill.). Venter of last abdominal segment.

FIGURE 164. *Cloeotus aphodioides* (Ill.). Thoracic spiracle.

FIGURE 165. *Philarmostes* sp. Head.

Symbols Used

AA—Seta of anterior frontal angle
AFS—Anterior frontal seta
E—Epicranium
EFS—Exterior frontal seta
DX—Dexiophoba
EFS—Exterior frontal seta

F—Frons
G—Galea
H—Haptomerum
LA—Lacinia
LPH—Laeophoba
LT—Laeotorma

PLA—Palidium
PFS—Posterior frontal seta
PSC—Preclypeus
PTA—Pternotorma
SD—Stridulatory teeth
UN—Uncus

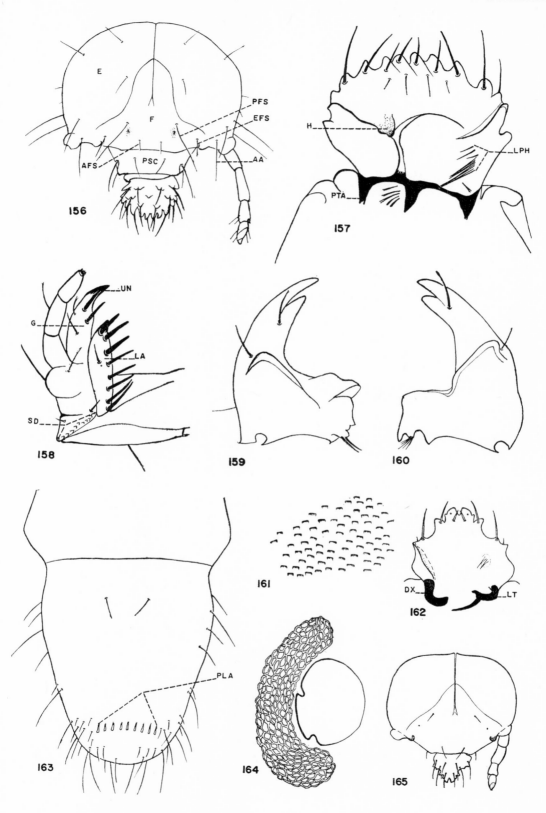

Subfamily Troginae

The subfamily Troginae,[14] in North America, includes only two genera, *Glaresis* and *Trox*. The latter genus has been revised by Mrs. Vaurie (1955). In her monograph she discusses the taxonomy and distribution of 41 North American species of *Trox* and gives information on the biology where known. In general, the grayish, often mud-covered adults are found on or under dry animal and bird carcasses or in the nests of birds or burrows of small mammals. They may also be found on the faeces of carnivorous animals and are often attracted to light.

Trox larvae occur at shallow depths in the soil beneath carcasses where the adults have been feeding. Larvae of some species occur in birds' nests in situations where there is enough moisture (Robinson, 1941). In Africa and South America, *Trox* larvae have been found feeding on locust eggs (Denier, 1936; Hayward, 1936; and van Emden, 1948).

Only the larvae of several species of *Trox* have been previously described; larvae of *Glaresis* are unknown. Böving and Craighead (1930) characterized the genus *Trox* in a key and were apparently the first to mention that some species possess cribriform spiracles while others have biforous spiracles. Sim (1934) briefly described the larva of *Trox suberosus* (Fab.). Van Emden (1941) characterized the genus and gave a key for separating larvae of *T. hispidus* Pont and *T. scaber* (L.). In 1948 van Emden described in detail the larva of *T. procerus*. Panin (1957) described the larva of *T. sabulosus* L.

[14] This subfamily is discussed only briefly since Charles Baker, one of my graduate students, has studies under way on the larval taxonomy and biology.

Preliminary studies indicate that, based on larval characters, the Omorgus group could be raised to generic level as proposed by Erickson (1847) and later by Le Conte (1854). Mrs. Vaurie (1955) has not recognized the group even as a subgenus because the characters separating adults of the Omorgus species in North America do not always apply to species in that group from other continents.

Larvae of Troginae (*Trox*) may be characterized as follows: Labrum bilobed. Epipharynx with an oval pedium often surrounded by phobae; heli absent, tormae united. Maxilla with galea and lacinia distinctly separate; maxillary stridulatory area with a row of teeth or a patch of minute teeth. Antenna 3-segmented; apical segment much reduced. Dorsa of abdominal segments

Key to Larvae of *Trox*

1. Spiracles cribriform. Maxilla with a stridulatory area consisting of a single row of teeth (Fig. 171).......... .... *Omorgus* group

   Spiracles biforous. Maxilla with stridulatory area consisting of a patch of minute teeth (Figs. 170 and 172) .... Other groups

1 to 6 with 3 folds, each bearing 1 or more transverse rows of short, stiff setae. Spiracles cribriform or biforous; if cribriform, with concavities of respiratory plates facing cephaloventrally. Last abdominal segment with bare, fleshy lobes surrounding anus. Legs 4-segmented, well developed, each with a long, curved claw bearing 2 setae at its base.

Plate XV

FIGURE 166.   *Trox spinulosus dentibius* Rob. Head.

FIGURE 167.   *Trox suberosus* Fab. Head.

FIGURE 168.   *Trox spinulosus dentibius* Rob. Epipharynx.

FIGURE 169.   *Trox suberosus* Fab. Epipharynx.

FIGURE 170.   *Trox unistriatus* Beauv. Left maxilla and labium, dorsal view.

FIGURE 171.   *Trox suberosus* Fab. Left maxilla, dorsal view.

FIGURE 172.   *Trox scaber* (L.). Right maxilla, dorsal view.

FIGURE 173.   *Trox spinulosus dentibius* Rob. Left mandible, dorsal view.

FIGURE 174.   *Trox spinulosus dentibius* Rob. Left mandible, ventral view.

FIGURE 175.   *Trox unistriatus* Beauv. Venter of last abdominal segment.

Symbols Used

A—Antenna
AA—Setae of anterior frontal angle
ACP—Acanthoparia
AFS—Anterior frontal setae
CL—Clithrum
CPA—Chaetoparia
DX—Dexiophoba

EFS—Exterior frontal seta
F—Frons
GP—Gymnoparia
L—Labrum
LT—Laeotorma
M—Mandible
PC—Postclypeus
PE—Pedium

PFS—Posterior frontal setae
PH—Phoba
PSC—Preclypeus
PTA—Pternotorma
SC—Sense cone
SD—Maxillary stridulatory area
SP—Sclerotized plate
UN—Uncus

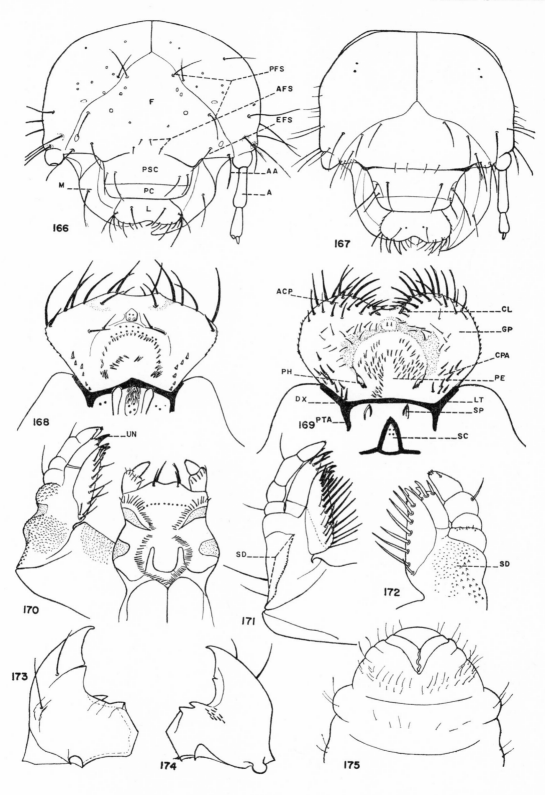

Subfamily Melolonthinae

The scarabaeid subfamily Melolonthinae contains a great many species of small to medium sized leaf-feeding beetles, most of which are nocturnal in habits and dull in color. Species of such common genera as *Phyllophaga, Serica,* and *Diplotaxis* are usually either brown, reddish-brown, or blackish in color. A few species in other genera of the subfamily are brightly colored. Among these are several white-striped species of *Polyphylla* found in the western states and several blue or green species of *Dichelonyx* peculiar to California and Oregon.

This subfamily is world-wide in distribution, and many of the most injurious species of the family Scarabaeidae belong to this group. Adults of most genera in the subfamily feed extensively on the foliage of trees and shrubs but adults of a few genera attack flowers (*Hoplia* and *Macrodactylus*) or fruit (*Macrodactylus*). Larvae of species belonging to this subfamily, known universally as white grubs, are subterranean feeders on the roots and underground stems of living plants.

In eastern United States much of the damage done by this group is caused by various species of the genus *Phyllophaga* which is represented by some 150 species. Adult *Phyllophaga*, commonly called May or June beetles, are often so abundant in the north-central states that oak, hickory, and walnut trees are completely defoliated. *Phyllophaga* grubs cause extensive damage to the roots of grasses, corn, wheat, strawberries, and tree seedlings. They also injure potatoes and vegetable crops. Other genera whose larvae are of economic importance are *Polyphylla, Serica, Diplotaxis,* and the introduced genera *Maladera* (*Autoserica*) and *Amphimallon*.

Larvae of many species and genera of the subfamily Melolonthinae have been described by workers here and abroad, but many still await description. Keys to larvae of the known genera or known tribes have been presented by Hayes (1929), Böving and Craighead (1931), Gardner (1935), van Emden (1941), and others. In the United States Böving (1936, 1942a-c, and 1945) was most active in describing larvae of the group. In 1949 the writer published a bulletin describing the larvae of many Melolonthinae and giving keys to tribes, genera, and species. Much of this material, with some corrections and additions, is reprinted in the following pages.

Larvae of the subfamily Melolonthinae may be characterized as follows: Mandible with ventral stridulatory area absent, indistinct or consisting of a patch of minute granules. Scissorial area of mandible with a distal blade-like portion which is separated from a proximal tooth by the scissorial notch. Galea and lacinia of maxilla fused proximally but separated distally; rarely, galea and lacinia separated but tightly fitted together. Lacinia with a row of 3 unci which is usually longitudinal. Maxillary stridulatory area without anterior process. Antenna 4-segmented. Last antennal segment with a single, elliptical, dorsal sensory spot. Anal opening usually angulate or Y-shaped. Lower anal lip usually with a sagittal cleft or groove. Legs well developed, 4 segmented. Each claw bearing 2 setae.

Key to Tribes of the Subfamily Melolonthinae

1. Raster with a curved, transverse row of prominent setae anterior to the ventral anal lobes (Fig. 209) or with a prominent oblique row or rows of mustache-like setae, on each side (Fig. 210) 2
Raster with longitudinal palidia (Figs. 214, 217-221, and 225-227) or with palidia absent (Figs. 215, 216, and 223) 3

2. Cardo, maxillary articulating membrane, and many other parts of the body bearing numerous black dots (Figs. 186 and 187). Dorsal anal lobe much smaller than the ventral anal lobes. Raster with a curved, transverse, comb-like palidium (Figs. 209 and 212).. .... tribe **Sericini page 76**
Black dots absent. Anal lobes similar in size. Raster with a prominent oblique row or rows of mustache-like setae, on each side, anterior to the ventral anal lobes (Fig. 210) .... tribe **Diplotaxini page 79**

3. Raster lyre-shaped and without palidia (Figs. 215 and 216) .... tribe **Dichelonycini page 80**
Raster with or without palidia, if lyre-shaped then with palidia 4

4. Haptomerum of epipharynx with from 1 to 4 heli (Figs. 193, 194, 197, and 202) 5
Haptomerum of epipharynx with 5 or more heli (Figs. 190-192, 205, 206, and 235) 6

5. Haptomerum of epipharynx with 4 heli (Fig. 193). All legs with well-developed claws. Raster with a pair of short, longitudinal palidia (Fig. 218).......... tribe **Macrodactylini page 81**
Haptomerum of epipharynx with a single helus (Fig. 194). Claws of mesothoracic and metathoracic legs reduced in size. Raster without palidia (Fig. 223).... .... tribe **Hopliini page 81**

6. Raster, on each side of the septula, with a pair of curved palidia, the outer palidium of each pair with pali directed laterad (Fig. 221) .... tribe **Pachydemini** page 82

   Raster without palidia or with a single palidium on each side of the septula (Figs. 214, 217, 219, 220, 225 to 227, and 234). Palidium, if polystichous, with all pali directed mesad (Fig. 227) 7

7. Respiratory plates of spiracles constricted (surrounding bullae) .... most **Melolonthini** page 83

   Respiratory plates of spiracles not constricted (not surrounding bullae) 8

8. Haptomerum of epipharynx with 5 or 6 heli (Fig. 190) .... tribe **Plectrini** page 82

   Haptomerum of epipharynx with 8 or more heli (Fig. 235) .... a few **Melolonthini** page 83

TRIBE Sericini

This tribe is represented in the United States by the native genus *Serica* containing about 75 species and one introduced species of the genus *Maladera* (*Autoserica*).

Larvae of this tribe may be characterized as follows: Head with few setae; anterior and exterior frontal setae absent, posterior frontal setae and setae in anterior angle of frons consisting of a single seta on each side. Eye spots usually present, black-pigmented. Labrum symmetrical, distally trilobed. Epipharynx (Fig. 197) with a well-developed epizygum. Haptomerum with 3 or 4 heli. Plegmatia present; proplegmatia absent. Maxilla with a maxillary stridulatory area consisting of a single row of peg-like teeth. Groups of dark granules present on the cardo and articulating membrane of the maxilla (Figs. 186 and 187), on the prothoracic shield, below the spiracles, on the coxae of the legs, on the sclerites adjacent to the coxae, and elsewhere on the body. Thoracic spiracles much larger than abdominal spiracles; arms of spiracular, respiratory plates not constricted. Raster (Figs. 209 and 212) with a single, transverse, slightly curved, comb-like row (palidium) of flattened pali just anterior to the ventral anal lobes. Anal slit Y-shaped, with the stem of the Y very much longer than the arms of the Y. Anal lobes densely setose. Dorsal anal lobe much smaller than ventral anal lobes. Teges consisting of a small to large number of hamate setae which are more or less separated anteriorly by a median bare area.

Key to Species of Known Larvae of Tribe Sericini

1. Raster, anterior to palidium, with many fine, slender setae, intermixed with the hamate tegillar setae........ .... *Serica perigonia* Dawson
   Raster, anterior to palidium, with few or no fine, slender setae intermixed with the hamate tegillar setae........... 2

2. Raster with a teges covering at least the posterior half of the area between the palidium and the anterior margin of the venter of the last abdominal segment (Fig. 209). Teges set with 75 or more setae............. 3
   Raster with a teges covering less than the posterior half of the area between the palidium and the anterior margin of the venter of the last abdominal segment (Fig. 212). Teges set with fewer than 50 hamate setae 6

3. Maxilla with stipes bulbous posteriorly and projecting beyond the cardo (Fig. 186) .... *Maladera castanea* Arrow
   Maxilla with base of stipes not projecting beyond the cardo (Fig. 187) 4

4. Dorsal surface of eighth and ninth abdominal segments each with a rather uniform covering of short fine setae between the anterior and posterior transverse bands of short and long setae. Short fine setae also present laterally between the teges and barbulae on the venter of the last abdominal segment............. .... *Serica intermixta* Blatchley
   Short fine setae absent on dorsal surface of eighth and ninth abdominal segments between the anterior and posterior transverse bands of setae. Short fine setae absent in region between teges and barbulae 5

5. Palidium set with fairly short pali. Teges with a patch of about 45 fairly long, hamate setae, on each side, which are only slightly shorter than the pali .... *Serica vespertina* (Gyll.)
   Palidium set with fairly long pali. Teges with a patch of 60 or more short setae, on each side..*Serica sericea* (Ill.)

6. Maxilla with stipes not bulbous posteriorly (Fig. 187). Maxillary stridulatory area with posterior teeth sparsely set (separated by a distance at least 3 times the width of a tooth at its base) (Fig. 187) 7
   Maxilla with stipes bulbous posteriorly. Maxillary stridulatory area with posterior teeth rather closely set (separated by a distance of twice the width of a tooth at its base, or less) 8

7. Eye spots inconspicuous or absent........*Serica ligulata* Dawson
   With distinct, black-pigmented eye spots......................... .... *Serica anthracina* Lec.

8. Claw of metathoracic leg only slightly shorter than claws of prothoracic and mesothoracic legs .... *Serica peregrina* Chapin
   Claw of metathoracic leg much shorter than claws of prothoracic and mesothoracic legs 9

9. Claw of metathoracic leg minute............. *Serica solita* Dawson
   Claw of metathoracic leg short but not minute 10

10. Eye spots present, black-pigmented..........*Serica curvata* Lec.
    Eye spots indistinct...............................*Serica falli* Dawson

Genus *Maladera* Muls.
Maladera castanea (Arrow), Third-Stage Larva
(Figs. 181, 186, and 209)

This description is based on the following material and reprinted from my 1949a bulletin:

Ten third-stage larvae dug from soil at Green Farms, Connecticut, November 24, 1941, by J. Peter Johnson.

Maximum width of head capsule ranging from 2.66 to 2.8 mm. Surface of head yellow-brown with shiny, faintly reticulated surface. Eye spots present, black-pigmented. Maxilla (Fig. 186) with greatly swollen stipes which is bulbous proximally and overhangs the cardo. Lacinia with 2 apical unci fused at their bases and a subapical uncus (Fig. 181). Maxillary stridulatory area consisting of a regular row of 17 to 20 long, nearly erect, peg-like teeth which are progressively closer set distally. Teeth absent on proximal third of the stipes. Raster (Fig. 209) with a transverse palidium of 21 to 27 stout, closely set pali which are considerably longer and stouter than the tegillar setae. Teges, on each side, with a subtriangular patch of about 50 short, curved hamate setae, the entire teges covering the caudal half of the area between the base of the palidium and the cephalic margin of the last abdominal segment. Claws of metathoracic legs very short.

Maladera (Autoserica) castanea (Arrow) is a small, chestnut brown beetle with irridescent, velvety sheen (Hawley and Hallock, 1936). Known as the Asiatic garden beetle, this introduced species was first found in New Jersey in 1921. It is now widely distributed from New York to South Carolina in states along the eastern seaboard (Metcalf, Flint, and Metcalf, 1962).

Adults are nocturnal, feeding on foliage of many kinds of plants, especially on many vegetables and flowers. The larvae damage roots of turf in lawns and roots of strawberries, vegetable seedlings, and flowers. The insect has a one year life cycle (Hawley and Hallock, 1936).

Genus *Serica* MacLeay

Adult *Serica* are small brown, blackish, or castaneous beetles, often with a bluish irridescence. Their prominent humeral angles and more robust shape serve to distinguish them from *Diplotaxis*. In the eastern United States, *Serica* are uncommon in cultivated soils but may be found occasionally in wooded areas. Adults of some species in Illinois and North Carolina are diurnal or fly on overcast days. In California, *Serica* adults are abundant locally and sometimes damage the foliage of avocado, peach, plum, prune, apple, and pear trees (Essig, 1958). The larvae feed on the roots of grasses and other plants.

Serica anthracina Lec., Third-Stage Larva

This description is based on the following material and reprinted from my 1949a bulletin:

Three third-stage larvae and cast skins of 3 third-stage larvae reared to the adult stage. Seven larvae were found close to the surface of the soil under short grasses, in lower ground in the sand hill country near Ben Lomond, California, by L. W. Saylor and P. O. Ritcher. No. 46-10B.

Two third-stage larvae, 1 prepupa, and cast skins of 3 third-stage larvae found with pupae which were later reared to the adult stage. Two larvae, 1 prepupa, and 6 pupae were found in the soil, from 4 to 6 inches deep, on a high slope at Sigmund Stern Grove, San Francisco, California, by Peter Ting, R. W. L. Potts, and P. O. Ritcher, February 26, 1948. No. 46-11A (1).

Maximum width of head capsule ranging from 2.1 to 2.31 mm. Head smooth, light yellow-brown in color. Body yellow. Eye spots present, black-pigmented. Stipes not swollen. Lacinia with 3 unci arranged in a nearly longitudinal row. Maxillary stridulatory area consisting of a single regular row of 16 to 20 teeth extending nearly to the base of the stipes. Teeth progressively closer set distally with the proximal 7 or 8 teeth separated by a distance as great as 3 times their width. Raster with transverse palidium of 16 to 23 depressed, stout pali which are about 3 times as long as their greatest width. Palidium closely set except for 4 or 5 pali on each side which are sparsely set. Teges, on each side, very sparsely set with 13 to 20 hamate setae; teges covering less than the caudal half of the area between the palidium and the cephalic margin of the last abdominal segment. Claws of metathoracic leg reduced in size.

S. anthracina Lec., the manzanita *Serica,* is a common species found in British Columbia, Washington, Oregon, and California. The adults feed on native shrubs and may invade orchards in the spring to feed on apple, prune, plum, and other fruit trees (Essig, 1926). Larvae are common in sandy areas in the Bay area of California, under grasses.

Serica curvata Lec., Third-Stage Larva

This description is based on the following material and reprinted from my 1949a bulletin:

Four third-stage larvae and the cast skin of a third-stage larva reared to the adult stage. Larvae collected from soil beneath *Artemesia,* at Oakland, California, March 4, 1946, by P. O. Ritcher. No. 46-7F.

Maximum width of head capsule ranging from 2.5 to 2.87 mm. Surface of head light yellow-brown. Body white. Eye spots present, black-pigmented. Maxilla with stipes bulbous posteriorly, slightly overhanging cardo.

Maxillary stridulatory area with a rather closely set row of 19 to 23 fairly long, peg-like teeth which are progressively closer set distally. Raster with a transverse palidium of 19 to 20 fairly long pali. Teges, on each side, sparsely set with 14 to 16 hamate setae, the entire teges covering less than the caudal half of the area between the palidium and the cephalic margin of the last abdominal segment. Claws of metathoracic leg short.

Serica falli Dawson, Third-Stage Larva

This description is based on the following material:
Eight third-stage larvae and cast skins of 7 third-stage larvae reared to the adult stage. Larvae found in soil of harvester ant nests, 8 miles east of Silver Lake, Oregon, May 16, 1957, by Robert Koontz and P. O. Ritcher.

Maximum width of head capsule ranging from 2.2 to 2.3 mm. Head straw-colored. Eye spots very inconspicuous. Base of stipes swollen but not extending laterally past the cardo. Maxillary stridulatory area with a long, rather closely set row of 19 to 21 peg-like teeth, the anterior 6 or 7 teeth smaller and closer together. Dorsa of abdominal segments 8 and 9 each with 2 sparsely set transverse patches of slender setae, each patch with several longer setae posteriorly. Raster with a single slightly curved palidium of 17 to 21 depressed, rather closely set pali. Teges, on each side, consisting of a triangular patch of 13 to 19 short, flattened setae, covering less than the caudal half of the area between the palidium and the cephalic margin of the last abdominal segment. Claws of metathoracic leg very short.
Larvae of *S. falli* are common in soil beneath harvester ant nests, in central Oregon. Pupation occurs in the spring.

Serica intermixta Blatchley, Third-Stage Larva

This description is based on the following material and reprinted from my 1949a bulletin:
Eleven third-stage larvae and 4 pupae dug from soil beneath forest leaf mold at Eastport, Michigan, August 9, 1947, by P. O. Ritcher. Pupae reared to the adult stage. Adults identified by R. W. Dawson.

Maximum width of head capsule ranging from 2.45 to 2.6 mm. Head straw-colored. Eye spots present. Stipes slightly swollen. Maxillary stridulatory area consisting of a single regular, rather closely set row of 20 to 22 peg-like teeth. Teeth progressively closer set distally. Proximal teeth separated by a distance less than 3 times the width of a tooth. Raster with a sparsely set, slightly curved, transverse, palidium of 22 to 26 rather short pali which are no longer than many of the tegillar setae. Teges rather thickly set with from 75 to more than 100 hamate setae. Entire teges subtriangular in outline and covering from one-half to two-

thirds of the area between the palidium and the cephalic margin of the last abdominal segment. Laterally, among the hamate tegillar setae and the barbulae, are many short, fine setae. Dorsal surface of abdominal segments 8 and 9 with a rather uniform covering of short, fine setae between the anterior and posterior transverse bands of long and short setae. Claws of metathoracic legs about half as long as claws of the prothoracic and mesothoracic legs.

Serica ligulata Dawson, Third-Stage Larva
(Figs. 187 and 212)

This description is based on the following material and reprinted from my 1949a bulletin:
Eleven third-stage larvae and cast skins of 5 third-stage larvae reared to the adult stage. Larvae dug from soil on a hillside beneath *Adenostoma* shrubs at Potwisha, in Sequoia National Park, California, March 25, 1946, by P. O. Ritcher. No. 46-16C.
Four third-stage larvae and cast skins of 4 third-stage larvae, 3 of which were reared to the adult stage. Ten larvae were dug from soil in low ground close to a stream at Potwisha, in Sequoia National Park, California, March 25, 1946, by P. O. Ritcher. No. 46-16B.

Maximum width of head capsule ranging from 2.1 to 2.24 mm. Head yellowish-brown in color. Eye spots absent or very inconspicuous. Stipes not swollen. Lacinia with 3 unci arranged in a nearly longitudinal row. Maxillary stridulatory area (Fig. 187) consisting of a single regular row of 15 to 18 teeth; proximal 6 to 8 teeth widely separated and fairly long and distal teeth closely set and very short. Raster (Fig. 212) with a single sparsely set, curved palidium of 12 to 18 depressed pali which are more than 4 times as long as their greatest width. Teges, on each side, sparsely set with 10 to 17 hamate setae. Teges covering less than the caudal half of the area between the base of the palidium and the cephalic margin of the last abdominal segment. Claws of metathoracic legs much reduced in size.

Serica sericea (Ill.),[15] Third-Stage Larva

This description is based on the following material and reprinted from my 1949a bulletin:
One third-stage larva from the collection of the Federal Japanese Beetle Laboratory, Moorestown, New Jersey.

Maximum width of head capsule 2.38 mm. Head yellowish-brown. Eye spots inconspicuous. Stipes not swollen. Maxillary stridulatory area with a single regu-

---

[15] Described by Sim (1934) under the name *Serica parallela* Casey which is a synonym according to Leng (1920).

lar, sparsely set row of 15 to 16 peg-like teeth, the proximal 8 teeth separated by a distance twice the width of a tooth at its base. Raster with a transverse palidium of 26 fairly long pali. Palidium closely set, except for 4 or 5 pali on each side. Teges with a patch of 60 or more rather short hamate setae, on each side. Teges covering the caudal two-thirds of the venter of the last abdominal segment. Claws of metathoracic legs much reduced in size.

Serica peregrina Chapin, Third-Stage Larva

This description is based on the following material and reprinted from my 1949a bulletin:
> Two third-stage larvae from the collection of the Federal Japanese Beetle Laboratory, Moorestown, New Jersey.

Maximum width of head capsule 2.3 mm. Eye spots present. Stipes not swollen proximally. Maxillary stridulatory area with a closely set row of 17 to 20 fairly long, peg-like teeth. Teeth slightly more closely set distally. Raster with a transverse palidium of 15 to 20 pali. Teges, on each side, sparsely set with 15 to 21 hamate setae, many of which are nearly as long as the pali. Teges covering less than the caudal half of the area between the palidium and the cephalic margin of the last abdominal segment. Claws of metathoracic legs only slightly shorter than claws of prothoracic and mesothoracic legs.

Serica solita Dawson, Third-Stage Larva

This description is based on the following material and reprinted from my 1949a bulletin:
> Twenty-eight third-stage larvae and cast skins of 4 third-stage larvae, one of which was reared to the adult stage. Larvae found close to the surface of the soil under short grasses in low ground in the sand hill country near Ben Lomond, California, by L. W. Saylor and P. O. Ritcher. No. 46-10A.

Maximum width of head capsule ranging from 2.3 to 2.45 mm. Surface of head light reddish-brown. Body white. Eye spots present. Maxilla with stipes slightly swollen proximally. Maxillary stridulatory area with a rather closely set row of 21 to 24 fairly long, peg-like teeth. Teeth progressively more closely set distally; posterior 3 or 4 teeth separated by a distance of about twice the width of a tooth at its base. Raster with a transverse palidium of 17 to 24 closely set pali which are about 4 times as long as their greatest width. Teges, on each side, sparsely set with 14 to 21 hamate setae, the entire teges covering less than the caudal half of the area between the palidium and the cephalic margin of the last abdominal segment. Claws of metathoracic legs minute.

Serica perigonia Dawson, Third-Stage Larva

This description is based on the following material:
> One third-stage larva dug from soil beneath *Adenostoma* shrub in Banning Canyon, 4 miles north of Banning, California, April 5, 1946, by P. O. Ritcher. Associated with a reared adult. No. 46-17B2.

Maximum width of head capsule 2.5 mm. Eye spots present but inconspicuous. Stipes swollen proximally but not overhanging cardo. Maxillary stridulatory area with a closely set row of 19 to 21 peg-like teeth; posterior 7 or 8 teeth separated by a distance equal to the width of a tooth at its base. Raster with a transverse palidium of 20 long, closely set pali. Teges, on each side, sparsely set with 18 long, hamate setae, extending less than half way to the cephalic margin of the last abdominal segment. Surface between palidium and cephalic margin of last abdominal segment also covered with many fine, cylindrical setae (a feature absent on all other *Serica* that I have examined).

Serica vespertina (Gyll.), Third-Stage Larva

This description is based on the following material loaned by the United States National Museum and reprinted from my 1949a bulletin:
> One third-stage larva, associated with 2 adults, collected on the shore of Pigeon River, Retreat, North Carolina, June 3, 1893. L-259.

Maximum width of head capsule 2.8. Head yellowish-brown. Eye spots inconspicuous. Stipes not swollen. Maxillary stridulatory area with a single regular, sparsely set row of 16 to 18 peg-like teeth, the teeth progressively more closely set distally. Proximal 8 teeth separated by a distance about twice the width of a tooth at its base. Raster with a transverse palidium of 23, fairly short, closely set pali. Teges with a patch of about 45 fairly long, hamate setae, on each side, which are only slightly shorter than the pali. Teges covering caudal half of the venter of the last abdominal segment. Claws of metathoracic legs much reduced in size.

TRIBE Diplotaxini
Genus *Diplotaxis* Kirby, Third-Stage Larva
(Figs. 202 and 210)

This description is based on the following material and reprinted from my 1949a bulletin:
> Four third-stage larvae collected from bluegrass pasture, behind the plow, north of Versailles, Kentucky, April 5, 1959, by P. O. Ritcher. No. 39-15CC.

Larvae of this genus may be characterized as follows: Head without eye spots. Labrum symmetrical, with a prominent, curved, transverse ridge. Epipharynx

(Fig. 202) with prominent plegmatia; proplegmatia absent. Haptomerum with 3 to 5 stout heli in a transverse row. Haptolachus without microsensilla. Maxillary stridulatory area with 10 to 16 blunt teeth. Lacinia with a longitudinal row of 3 unci; distal uncus slightly the largest.

Anal opening Y-shaped with stem of Y equal in length to the arms of the Y. Anal lobes setose. Lower anal lobes triangular in shape, almost equal in size to the dorsal anal lobe. Raster (Fig. 210) posteriorly with an oblique palidium of long mustache-like setae, on each side, just anterior to each lower anal lobe. Anterior to each palidium, on each side, is a patch of hamate, tegillar setae. Ventral surface of prothoracic femora each with a prominent, longitudinal row of 5 spine-like, fossorial setae. Claws long on prothoracic and mesothoracic legs, very short on metathoracic legs.

Diplotaxis adults are small, elongate, black or reddish, shiny beetles, not exceeding an inch in length. They are nocturnal, feeding on the foliage of oak, walnut, rose, apple, apricot, peach, and pear (Essig, 1958). The larvae are more common in sandy soils and are rarely found in cultivated areas. Heit and Henry (1940), in New York, reported severe injury to Norway spruce seedlings by larvae of *Diplotaxis sordida* Say. In central Oregon, larvae of several species are common under harvester ant nests.

TRIBE Dichelonycini

Species belonging to this tribe are most numerous in the states west of the Rocky Mountains. The genus *Coenonycha* is restricted to that region, but the genus *Dichelonyx* also occurs in eastern United States.

Key to Genera

Raster on each side, with dense, triangular patch of very short, stout conical setae ....*Dichelonyx*
Raster with setae rather uniformly distributed............*Coenonycha*

Genus *Dichelonyx* Harris, Third-Stage Larva
(Figs. 196 and 216)

This description is based on the following material and reprinted from my 1949a bulletin:

Seven third-stage larvae of *Dichelonyx backi* Kby. collected at Saskatoon, Saskatchewan, Canada, by Kenneth King. Loaned by the United States National Museum.

Eleven third-stage larvae of *Dichelonyx robusta* Fall and cast skins of 6 third-stage larvae reared to the adult stage. These larvae were part of a lot of 54 third-stage larvae and 17 second-stage larvae dug from soil in a grassy flat near the top of Mt. Hamilton, California, February 22, 1946, by L. W. Saylor and P. O. Ritcher. The larvae were from 1 to 7

inches deep in the soil with most of them being found at depths of between 3 and 5 inches. No. 46-8A.

Three prepupae, 4 third-stage larvae, and cast skins of 2 third-stage larvae of *Dichelonyx muscula* Fall[16] reared to the adult stage. Larvae and prepupae dug from soil beneath *Adenostoma* shrubs in Banning Canyon, 4 miles north of Banning, California, April 5, 1946, by P. O. Ritcher. No. 46-17A.

Two third-stage larvae and cast skins of 3 third-stage larvae of *Dichelonyx vicina* Fall reared to the adult stage. Larvae dug from soil in MacDonald Forest, Benton County, Oregon, July 22, 1954, by P. O. Ritcher. No. 54-10.

Larvae of the genus *Dichelonyx* may be characterized as follows: Head without eye spots. Labrum symmetrical. Epipharynx (Fig. 196) with plegmatia. Plegmata rather short. Proplegmatia consisting of a single proplegma on each side. Clithra present. Epizygum absent. Haptomerum with a transverse row of 2 to 4 heli. Laeophoba absent. Tormae united. Haptolachus without microsensilla. Maxillary stridulatory area with a sparsely set row of 9 to 12 conical, anteriorly directed teeth. Lacinia with 3 unci in a longitudinal row, the distal uncus slightly the largest.

Thoracic spiracles much larger than abdominal spiracles; arms of respiratory plates not constricted. Raster (Fig. 216) lyriform, posteriorly with a broad teges of long hamate setae having curved tips; raster anteriorly, on each side, with a dense, triangular patch of very short, stout conical setae. Anal opening Y-shaped with stem of Y only slightly shorter than the arms of the Y. Ventral anal lobes triangular in shape. Claws long and falcate on prothoracic and mesothoracic legs. Claws of metathoracic legs much shorter.

Key to Larvae of Some Western *Dichelonyx*

1. Anal lobes setose. Raster with stout, sparsely set, rather long conical setae laterad of each dense patch of very short conical setae ....*D. vicina* Fall
   Anal lobes bare. Raster without stout, sparsely set, rather long conical setae laterad of each dense patch of very short conical setae 2
2. Raster with stout setae of two distinct sizes. Raster, on each side, with a posterior, triangular, closely set patch, of very short conical setae and anteriorly with a sparsely set, irregular double or partly triple row of larger subconical setae .... *D. backi* Kirby
   Raster with stout setae of one size only 3
3. Raster, on each side, with a broad triangular patch of short, conical setae which are rather uniformly spaced .... *D. robusta* Fall
   Raster, on each side, with a dense, narrow, triangular patch of short conical setae which are slightly farther apart anteriorly .... *D. muscula* Fall

[16] Through an error in identification of the adult, this species was called *D. truncata* Lec. in my 1949 paper.

Thirty-one species of *Dichelonyx* occur in America north of Mexico (Saylor, 1945). Of these, 8 are eastern species and 23 are western. Adults are elongate, often brightly colored beetles, many with shining metallic green elytra. They are both diurnal and nocturnal and are frequently taken at black light. Some of our common western species feed on needles of ponderosa pine, Douglas-fir, and lodge pole pine. The larvae of some species are grass-root feeders. Sim (1934) described the larva of this genus and figured the raster and epipharynx of a New Jersey species thought to be *D. elongata* (Fab.).

Genus *Coenonycha* Horn, Third-Stage Larva
(Figs. 195 and 215)

This description is based on the following material and reprinted from my 1949a bulletin:

Twelve third-stage larvae of *Coenonycha pascuensis* Potts. These larvae are part of a lot of 26 third-stage, 6 second-stage, and 1 first-stage larvae and 34 adults dug from the soil of a grassy slope 2 miles west of Byron, California, February 20, 1946, by William Barr, Ray Smith, J. W. McSwain, and P. O. Ritcher. No. 46-7B.

Seventeen third-stage larvae of *Coenonycha fusca* McClay. These larvae together with 5 adults were found from 3 to 7 inches deep in the soil beneath sod in a woodland clearing, south of Yosemite National Park, near Bass Lake, California, March 22, 1946, by P. O. Ritcher. No. 46-15B.

Larvae of the genus *Coenonycha* may be distinguished by the following characters: Head without eye spots. Labrum symmetrical. Epipharynx with plegmatia. Plegmata short. Proplegmatia absent. Epizygum present. Haptomerum usually absent. Tormae united. Haptolachus without microsensilla. Maxillary stridulatory area with a sparsely set row of 9 to 13 sharp-pointed, conical teeth, projecting anteriorly. Lacinia with a longitudinal row of 3 unci.

Thoracic spiracles much larger than abdominal spiracles. Abdominal spiracles similar in size. Arms of respiratory plates not constricted. Raster lyriform. Teges consisting posteriorly of long hamate setae with curved tips. Anteriorly, on each side, the setae are less curved and are progressively shorter. Anal opening Y-shaped with stem of Y nearly equal in length to each arm of the Y. Ventral anal lobes triangular in shape. Anal lobes setose. Claws of prothoracic legs long and falcate, those of mesothoracic legs slightly shorter. Claws of metathoracic legs very short.

The genus *Coenonycha* in America north of Mexico contains about 35 species of small, elongate, yellowish-brown to reddish-brown beetles. This genus was revised in 1943 by Cazier and McClay who collected what was known on distribution, biology, and taxonomy.

TRIBE Macrodactylini
Genus *Macrodactylus* Latreille, Third-Stage Larva
(Figs. 193 and 218)

This description is based on the following material and reprinted from my 1949a bulletin:

Two third-stage larvae of *Macrodactylus subspinosus* (Fab.) collected behind the plow in Taylor County, Kentucky, just south of the Marion County line, April 25, 1939, by P. O. Ritcher. No. 39-15AA.

Two third-stage larvae of *M. subspinosus* collected behind the plow in Knox County, Kentucky, halfway between Barbourville and Corbin, April 10, 1939, by H. G. Tilson. No. 39-15-O.

Two third-stage larvae and cast skins of 10 third-stage larvae of *M. subspinosus* (Fab.) reared to the adult stage. Larvae collected from sod land, behind the plow, 7 miles southeast of Mt. Vernon, Kentucky, April 4, 1940, by P. O. Ritcher. No. 40-12FC.

Eighteen third-stage larvae and 5 cast skins of third-stage larvae of *Macrodactylus uniformis* Horn reared to the adult stage. Larvae dug from soil beneath walnut trees growing in a creek bed in the Davis Mountains, north of Ft. Davis, Texas, January 29, 1946, by P. O. Ritcher.

Larvae of the genus *Macrodactylus* may be characterized as follows: Head with eye spots. Labrum symmetrical. Epipharynx (Fig. 193) with plegmatia. Plegmata rather short. Epizygum well developed. Zygum absent. Haptomerum with a transverse, slightly curved row of 4 heli. Laeophoba well developed, fringing the proximal half of the left side of the pedium. Haptolachus with a number of microsensilla. Maxillary stridulatory area with a row of 8 or 9 low teeth. Lacinia with 3 unci in a longitudinal row; middle uncus the smallest. Thoracic spiracles and spiracles on first 5 abdominal segments similar in size; spiracles surrounding the bullae. Raster (Fig. 218) with 2 short, longitudinal, parallel palidia each sparsely set with 5 to 12 compressed, blade-like pali. Tegilla on each side of palidia sparsely set with coarse, hamate setae. Anal opening Y-shaped with stem of Y half as long as each arm of Y. Anal lobes covered with slender, cylindrical setae. Claws of metathoracic legs not conspicuously shorter than claws of prothoracic and mesothoracic legs.

This small genus contains several species of slender, long-legged beetles commonly called rose chafers. Adults damage the foliage, flowers, and fruits of grapes by their feeding and often injure young fruits of apples and peaches. Many other plants are also attacked including pear, cherry, cane berries, truck and field crops, small grains, grasses, and the flowers of roses (Metcalf, Flint, and Metcalf, 1962).

There are two common species, *M. uniformis* Horn, the western rose chafer, occurring in Texas, New Mexico, and Arizona, and *M. subspinosus* (Fat.), the rose chafer, occurring in southeastern Canada and the eastern half of the United States. Both seem to prefer sandy areas. The larvae feed on the roots of grasses, grain, weeds, and other plants (Metcalf, Flint, and Metcalf, 1962) but are of no economic importance. Chittenden and Quaintance (1916) found that *M. subspinosus* has a one year life cycle.

Key to *Macrodactylus* Larvae

Raster with few (1 or 2) or no preseptular, hamate setae..
.... *M. uniformis* Horn
Raster with 4 to 10 preseptular, hamate setae ....
.... *M. subspinosus* (Fab.)

TRIBE Pachydemini
Genus *Phobetus* LeConte, Third-Stage Larva
(Figs. 178, 192, and 221)

This description is based on the following material and reprinted from my 1949a bulletin:

One third-stage larva associated with an adult of *Phobetus comatus sloopi* Barrett. Adult identified by Cazier. Larvae and adult taken behind a plow in old pasture sod, 4 miles west of Fallbrook, California, February 9, 1946, by P. O. Ritcher.

Larvae of this genus may be characterized as follows: Head with a rather dense covering of setae (Fig. 178). Eye spots absent. Epipharynx (Fig. 192) with a group of about 14 heli. Zygum present; epizygum absent. Plegmatia present; proplegmatia absent. Epipharynx with a long phoba bordering each side of the pedium. Haptolochus with numerous microsensilla. Maxillary stridulatory area with a rather closely set row of 14 to 18 truncate teeth. Lacinia with a longitudinal row of 3 stout heli. Scrobis of mandible with a prominent patch of setae. Thoracic spiracles larger than abdominal spiracles which are rather small and nearly alike in size.

Dorsa of abdominal segments 1 to 7 with a dense covering of very short, triangular-shaped asperities. Raster (Fig. 221) lyre-shaped with an oval septula surrounded on each side by 2 curved palidia set with fairly long, stout, depressed pali. Pali of inner palidia pointing mesally, pali of exterior palidia pointing laterally. On each side, anterior to and laterad of each anterior palidium, are a number of short, stout, spine-like setae. Anal opening Y-shaped with ventral stem of Y slightly shorter than the arms of the Y. Claws of prothoracic and mesothoracic legs long; claws of metathoracic very short. Ventral surfaces of femora of prothoracic legs each with a row of prominent fossorial setae.

The writer has examined many larvae of *Phobetus comatus comatus* Lec. from The Dalles, Oregon, and has found their characters in agreement with the above description first published in 1949. The reader is referred to Cazier (1937) for a review of the taxonomy and biology of adults of the known species. Tilden (1944) published biological notes on four California species.

TRIBE Plectrini
Key to Genera

1. Abdominal spiracles similar in size. Raster with a pair of palidia which are parallel anteriorly and widely divergent posteriorly .... genus *Plectris*
   Spiracles on abdominal segments 6 to 8 much smaller in size than those on abdominal segments 1 to 5. Raster with a pair of longitudinal palidia which are not divergent posteriorly (Extra limital, occurs in Cuba) .... genus *Anoplosiagum*

Genus *Plectris* Serv., Third-Stage Larva
(Figs. 182, 190, and 227)

This description is based on the following material and reprinted from my 1949a bulletin:

Five third-stage larvae of *Plectris aliena* Chapin and cast skin of a third-stage larva associated with reared adults. Two larvae collected near Brunswick, Georgia, 1933-1934 by J. F. G. Clark. Loaned by the United States National Museum.

Cast skin of third-stage larva of *Plectris* sp. associated with two adults. Found in soil at Tacnarembo, Uruguay, by H. L. Parker. No. 619-I. Dept. No. 428967. Loaned by the United States National Museum.

Third-stage larva of *Plectris* sp. associated with fragments of adult. Collected at Montevideo, Uruguay, in 1945 (?) by H. L. Parker. No. 961-2. Loaned by the United States National Museum.

Larvae of the genus *Plectris* may be characterized as follows: Labrum symmetrical with 2 or 3 prominent, transverse ridges which are frequently united by a median, longitudinal ridge (Fig. 182). Maxillary stridulatory area with a single row of 12 to 16 low teeth. Lacinia with a longitudinal row of 3 strong unci which are similar in size. Epipharynx (Fig. 190) roughly pentagonal with protruding anterior lobe. Plegmatia present, well developed. Haptomerum with 5 or 6 stout heli, anterior to which is a lightly sclerotized zygum. Epizygum absent. Dexiophoba extending anteriorly from sclerotized plate along the proximal half of the pedium. Laeophoba well developed, consisting of about 7 long, distally forked spines.

Respiratory plates of spiracles curved but not constricted. Prothoracic spiracles much larger than abdominal spiracles. Raster (Fig. 227) with a pair of palidia which are parallel anteriorly and widely divergent posteriorly. Palidia monostichous anteriorly and distichous or tristichous posteriorly along the lower anal lobes. Anal opening Y-shaped; lower anal lobes divided by a short, distinct cleft. Claws of prothoracic and mesothoracic legs long and falcate; claws of metathoracic legs very short.

The larva of *Plectris aliena* Chapin was first described by Sim in 1934 and later described in much more detail by Böving (1936). *P. aliena* is an introduced South American species found at Charleston, South Carolina (Chapin, 1934).

Genus *Anoplosiagum* Blanch.

Anoplosiagum pallidulum Blanch., Third-Stage Larva

This description is based on the following material and reprinted from my 1949a bulletin:

Five third-stage larvae collected at the Cuba Sugar Club Experiment Station, Baragua, Cuba, June 3, 1931, by H. H. Plank. Larvae are pests of sugar cane. Associated, reared adults determined by E. A. Chapin. Larvae loaned by the United States National Museum.

Larvae of this species may be characterized as follows: Maximum width of head 2.38 to 2.94 mm. Head yellow-brown in color, without eye spots. Head, on each side, with a single exterior frontal seta, a single seta in each anterior angle of the frons, 2 or 3 posterior frontal setae, and a transverse row of 6 to 9 anterior frontal setae. Labrum nearly symmetrical. Epipharynx with well-developed plegmatia, each with 10 or 11 plegmata. Proplegmatia also present, each with 8 slightly curved proplegmata. Epizygum present; zygum absent. Haptomerum with 6 stout heli. Haptolachus with about 12 crepidal punctures (microsensilla). Dexiophoba extending along proximal half of right side of pedium. Laeophoba with 6 or 7 stout, branched processes. Maxillary stridulatory area with a row of 12 to 14 more or less truncate teeth, with points projecting slightly forward. Lacinia with a longitudinal row of 3 stout heli. Respiratory plates of spiracles not constricted and not surrounding bullae. Thoracic spiracles much larger than abdominal spiracles. Spiracles on abdominal segments 1 to 5 nearly alike in size and about half the size of the thoracic spiracle. Spiracles on abdominal segments 6 to 8, very small.

Raster with a pair of longitudinal, feebly curved palidia, each sparsely set with 10 to 13 rather short, slightly depressed pali. Preseptular, hamate setae 2 to 4 in number. Anal opening Y-shaped with stem of Y about one third the length of each arm of the Y. Anal lobes setose. Claws of prothoracic legs long and sharp; claws on mesothoracic legs slightly shorter; claws on metathoracic legs minute.

TRIBE Melolonthini

Our most destructive species of white grubs belong to this tribe. The genus *Amphimallon* is represented in eastern United States by the introduced European chafer, *Amphimallon majalis* (Razoumowski). The native genus *Phyllophaga*, which includes a great many species of May beetles, reaches its greatest development in eastern United States. A Cuban May beetle, *Phyllophaga* (*Cnemarachis*) *bruneri* Chapin has become established in Miami, Florida (Woodruff, 1961). In western United States and western Canada, larvae of several species of *Polyphylla* are common and often destructive.

Key to Genera

1. Palidia absent ....a few *Phyllophaga*[17]
   Palidia present 2

2. Anal opening Y-shaped with stem of Y half as long as each arm of the Y. Palidia strongly divergent or strongly divergent posteriorly (Fig. 226) 3
   Anal opening vaguely angulate, V-shaped or Y-shaped; if Y-shaped, stem of Y less than half as long as each arm of the Y. Palidia neither strongly divergent nor strongly divergent posteriorly 4

3. Palidia strongly divergent posteriorly, palia progressively longer posteriorly (Fig. 226) .... genus *Amphimallon*
   Palidia strongly divergent, comb-like posteriorly (Fig. 234) .... *? Phyllophaga sociata* (Horn)

4. Anal opening vaguely angulate or V-shaped, ventral anal lip without sagittal cleft (Fig. 214). Epipharynx (Fig. 191) without distinct epizygum; dexiophoba fringing right side of pedium. Maxillary stridulatory teeth sharp-pointed, much longer than wide (Fig. 183) .... *Polyphylla*
   Anal opening V-shaped or Y-shaped; ventral anal lip with a distinct sagittal cleft or groove (Figs. 217, 219, 220, and 225). Epipharynx with distinct epizygum (Fig. 205). Dexiophoba short. Maxillary stridulatory teeth short and truncate (Fig. 184) .... most *Phyllophaga*

Genus *Amphimallon* Berthold, Third-Stage Larva
(Figs. 185, 206, and 226)

This description is based on the following material and reprinted from my 1949a bulletin:

One third-stage larva of *Amphimallon majalis* (Razoum.). Collected at Newark, New York, April 26, 1942, by H. H. Schwardt.

Four third-stage larvae of *Amphimallon majalis* (Razoum.). Collected at Marbletown, New York, by F. L. Gambrell.

---

[17] Larvae of several Arizona species, as yet undescribed, and the introduced Cuban May beetle, *Phyllophaga* (*Cnemarachis*) *bruneri* Chapin will key out here. The larva of *P. bruneri* has 5 heli and numerous proplegmata (about 30) on its epipharynx (Woodruff, 1961).

Larvae of this genus may be characterized as follows: Head (Fig. 185) without eye spots. Epipharynx (Fig. 206) with distinct epizygum. Haptomerum with a group or a curved transverse row of 5 to 7 stout heli. Plegmatia present; plegmata short. Proplegmatia present but weakly developed, consisting of 15 or more proplegmata. Laeophoba consisting of about 5 stout, branched processes and fringing posterior third of the left side of the pedium. Dexiophoba present, confined to a short distance anterior to the sclerotized plate. Haptolachus with a few microsensilla. Maxilla with a row of 12 or more stridulatory teeth having anteriorly directed points. Lacinia with a longitudinal row of 3 stout heli. Thoracic spiracles slightly larger than spiracles on abdominal segments 1 to 5. Spiracles on abdominal segments 6 to 8 much smaller and nearly alike in size.

Anal opening Y-shaped with stem of Y half the length of each arm of the Y. Anal lobes setose. Raster (Fig. 226) with a pair of longitudinal palidia which are strongly divergent posteriorly. Claws of prothoracic and mesothoracic legs long and falcate. Claws of metathoracic legs reduced in size.

Adults of the introduced *Amphimallon majalis* resemble a tan-colored May beetle (*Phyllophaga*). The elytra, however, are strongly striated. The larvae, which cause severe damage to pastures, lawns, grain, and legumes, have been described in detail by Böving (1942b). Infested areas include parts of New York state and isolated areas in Connecticut, New Jersey, and West Virginia (Anonymous, 1961). Details of the biology were given by Gyrisco et al. (1954). According to these writers, the insect has a one year life cycle with pupation occurring in late May. Adults feed sparingly, if at all.

Genus *Polyphylla* Harris, Third-Stage Larva
(Figs. 179, 183, 191, and 214)

Larvae of this genus may be distinguished by the following characters: Head without eye spots. Frons with a transverse row of 3 or 4 posterior frontal setae on each side. Labrum symmetrical. Epipharynx (Fig. 191) without epizygum; zygum indistinct. Haptomerum with a group of 15 or more heli. Plegmatia present; plegmata short. Proplegmatia present or absent. Dexiophoba extending along much or all of the right side of the pedium. Haptolachus with or without microsensilla. Maxilla (Fig. 179) with a row of 14 or more fairly long, conical stridulatory teeth (Fig. 183). Lacinia with a longitudinal row of 3 stout unci.

Anal slit transverse, more or less angular. Ventral anal lobe not cleft. Raster (Fig. 214) with 2 short, longitudinal, parallel palidia. Preseptular hamate setae numerous. Claws of prothoracic and mesothoracic legs long and falcate, those of mesothoracic leg slightly the smaller. Claws of metathoracic legs minute.

The genus *Polyphylla* contains some 23 valid species (Cazier, 1940) most of which occur in the western United States. Three species, *Polyphylla comes* Casey, *P. occidentalis* (L.), and *P. variolosa* (Hentz) occur in eastern United States.

Key to *Polyphylla* Larvae

1. Proplegmatia present, well developed. Raster with tegilla extending in front of palidia for a distance equal to one half the length of the palidia .... *P. occidentalis* (L.)
   Proplegmatia absent or inconspicuous. Raster with tegilla extending in front of palidia for a distance equal to the length of the palidia 2

2. Haptolachus of epipharynx with crepidal punctures (microsensilla) .... *P. variolosa* (Hentz)
   Haptolachus of epipharynx without crepidal punctures (microsensilla) 3

3. Maximum width of head capsule 7.8 mm or greater.... .... *P. decemlineata* (Say)
   Maximum width of head capsule less than 6.8 mm .... *P. crinita* Lec.

Polyphylla occidentalis (L.), Third-Stage Larva

This description is based on the following material and reprinted from my 1949a bulletin:

One third-stage larva collected at Holland, Virginia, March 21, 1945, by J. M. Grayson. Determined by W. H. Anderson, United States National Museum.

Seven third-stage larvae collected about sedge roots at Clayton, North Carolina, July 18, 1935, by H. R. Johnson. Associated with reared adults. Larvae loaned by the United States National Museum.

Larvae of this species may be characterized as follows: Maximum width of head capsule 6.5 mm. Head light reddish-brown in color, finely reticulate. Haptomerum of epipharynx with about 25 heli. Epipharynx with a pair of large, elliptical proplegmatia each of which has more than 35 fine, curved proplegmata. Each plegmatium with 14 to 16 short plegmata. Haptolachus with about 6 crepidal punctures (microsensilla). Maxilla with a rather irregular, sparsely set row of 18 sharp-pointed stridulatory teeth. Abdominal spiracles progressively smaller in size.

Anal opening broadly V-shaped. Raster with 2 short palidia each sparsely set with 9 to 12 long, sharp, cylindrical pali. Tegilla extending in front of palidia for a distance equal to one half the length of the palidia. Preseptular, hamate setae about 25 to 40 in number. Tegilla occupying slightly less than the caudal half of the area between the lower anal lip and the anterior margin of the last abdominal segment.

Polyphylla variolosa (Hentz), Third-Stage Larva

This description is based on the following material loaned by the United States National Museum and reprinted from my 1949a bulletin:

Ten third-stage larvae, found injuring roots of California privet at Lawrence Harbor, New Jersey, October 12, 1941, by Mrs. M. S. Anderson. Larvae identified by W. H. Anderson, United States National Museum.

Larvae of this species may be distinguished by the following characters: Maximum width of head capsule 6.2 to 6.6 mm. Head light reddish-brown in color, faintly reticulate. Haptomerum of epipharynx with 22 to 25 heli. Epipharynx without proplegmatia, the area covered instead with setae. Each plegmatium with 10 to 12 very short plegmata. Haptolachus with 5 to 10 crepidal punctures (microsensilla). Maxilla with a row of 14 to 18 conical, sharp-pointed stridulatory teeth. Thoracic spiracles slightly larger than spiracles on abdominal segments 1 to 5 which are similar in size. Spiracles on abdominal segments 6 to 8 progressively smaller.

Anal slit curved, only feebly angulate. Raster with 2 short, nearly parallel, longitudinal paladia each sparsely set with 9 to 15 long, sharp, cylindrical pali. Septula narrow. Tegilla extending forward past the palidia for a distance equal to or slightly greater than the length of the palidia. Preseptular setae more than 50 (50 to 70). Tegilla occupying the caudal half or slightly more of the area between the lower anal lip and the anterior margin of the last abdominal segment.

In New York, grubs of *Polyphylla variolosa* caused extensive damage in a forest tree nursery (Heit and Henry, 1940).

Polyphylla decemlineata (Say), Third-Stage Larva
(Figs. 179, 183, 191, and 214)

This description is based on the following material and reprinted from my 1949a bulletin:

Seven third-stage larva and cast skin of a third-stage larva of *Polyphylla decemlineata* (Say) reared to the adult stage. Larvae dug from soil at Greenfield, California, March 28, 1946, by B. B. Richards.

Larvae of this species may be distinguished by the following characters: Maximum width of head capsule 7.8 to 8.66 mm. Head reddish-brown in color, reticulate. Haptomerum of epipharynx with 16 or more heli. Epipharynx (Fig. 191) with or without proplegmatia; proplegmatia, if present, poorly developed and consisting of numerous fine proplegmata. Each plegmatium with 11 to 15 very short plegmata. Haptolachus without crepidal punctures (microsensilla). Maxilla (Fig. 179), with a row of 16 to 18 conical, sharp-pointed stridula-

tory teeth (Fig. 183). Thoracic spiracles slightly larger than spiracles on abdominal segments 1 to 4 which are similar in size. Spiracles on abdominal segments 5 to 8 progressively smaller.

Anal opening broadly V-shaped. Raster (Fig. 214) with 2 short, nearly parallel, longitudinal palidia each sparsely set with 7 to 12 long, sharp, cylindrical pali whose tips frequently almost touch those of the opposite palidium. Tegilla extending in front of palidia for a distance equal to the length of the palidia. Preseptular hamate setae more than 50 in number. Tegilla occupying the caudal half or slightly less than the caudal half of the area between the lower anal lip and the anterior margin of the last abdominal segment.

Polyphylla decemlineata Say, the 10-lined June beetle, is a widely distributed western species occurring from the plains close to the Rocky Mountains to the Pacific coast (Fall, 1928). In western Oregon and Washington, larvae of this species are frequently abundant in sandy areas where they cause severe damage to strawberry plants, coniferous seedlings, roots of fruit trees, corn, table beets, hops, mint, potato tubers, and blueberry plants.

Adults feed at night on needles of conifers such as ponderosa pine (Johnson, 1954). There is a 3 or 4 year life cycle. Pupation occurs in early June at shallow depths of from 5 to 8 inches, in western Oregon (Ritcher, 1958). Adults appear in early July and continue to fly until fall. Downs and Andison (1941) give details of the biology as observed in British Columbia under the name of *P. perversa* Casey (considered a synonym).

Polyphylla crinita Lec., Third-Stage Larva

This description is based on the following material:

Three third-stage larvae reared from eggs found on July 8, 1961, in a rearing cage containing three males and three females. Adults dug from beneath nests of the harvester ant, *Pogonomyrmex owyhei* Cole on June 27, 1961, 10 miles southeast of Sisters, Oregon (Deschutes Co.), No. 61-51A.

Larvae of this species may be distinguished by the following characters: Maximum width of head capsule 6.3 to 6.7 mm. Head yellow to reddish-brown, reticulate; with transverse ridges on either side of anterior part of frons. Haptomerum of epipharynx with 14 to 20 heli, mostly arranged in two irregular semicircles. Acroparia well developed: area between base of haptomerum and anterior margin of epipharynx rather uniformly covered with long, slender setae. Epipharynx with or without proplegmatia; proplegmatia, if present, vague and poorly developed, consisting of about 12 fine proplegmata. Each plegmatium with 12 or 13 short, slightly curved plegmata. Haptolachus without crepidal punctures (microsensilla). Maxilla with a row of 14 to 16 conical, sharp-pointed stridulatory teeth. Respiratory plate of thoracic spiracle with arms only slightly constricted; thoracic spiracle similar in size to spiracles on first four abdominal segments. Spiracles on abdominal segments 5 to 8 progressively smaller in size.

Anal opening broadly V-shaped. Raster with two short, rather sparsely set, longitudinal palidia, each sparsely set with 8 to 12 long, slender palia whose tips meet on the midline of the septula. Tegilla extending past palidia for a distance equal to the length of the palidia. Preseptular tegillar setae 20 to 35 in number on each side. Tegilla extending about half the distance between the lower anal lip and the anterior margin of the last abdominal segment.

Polyphylla crinita Lec. is locally abundant in central Oregon, especially in Deschutes County. Oregon specimens have been identified by Cartwright as *P. modulata* Casey, but according to Fall (1928) this species is a synonym of *P. crinita* Lec. In appearance, adults of this species look like small specimens of *P. decemlineata.* The third-stage larvae of the two species are very close morphologically and can only be distinguished by a difference in head capsule size and slight differences of the epipharynx.

Genus *Phyllophaga* Harris, Third-Stage Larva

(Figs. 176, 177, 180, 184, 188, 189, 198-201, 203-205, 207, 208, 211, 213, 217, 219, 220, 222, 224, and 225)

According to Saylor (1942) the genus *Phyllophaga* includes the subgenus *Phyllophaga (sensu stricto)* plus a number of other subgenera including the subgenus *Listrochelus* and the subgenus *Phytalus.* Of the subgenus *Phyllophaga (sensu stricto)* alone, 152 valid species occur in the United States and Canada (Luginbill and Painter, 1953). Saylor (1939) listed 7 species of the subgenus *Phytalus,* occurring in the southern United States and New Jersey, and (1940) listed 39 species of the subgenus *Listrochelus,* occurring only in the western United States.

Böving in 1942 published his monumental work on the classification of larvae and adults of the genus *Phyllophaga (sensu stricto).* In this publication he described the larvae of 61 species and gave keys for their separation. The key presented in the following pages includes 28 species of which 3 species, *Phyllophaga barda, Phyllophaga karlsioei,* and *Phyllophaga sylvatica,* are not mentioned in Böving's publication. Although my key was originally designed for keying out Kentucky material (Ritcher, 1940 and 1949), it includes many of the species common in the eastern half of the United States and most of the injurious kinds.

I have included mention of the subgenus *Listrochelus* in my key (Ritcher, 1949). Study of reared larvae of *Ph. (Listrochelus) mucoreus* LeConte, *Ph. (Listrochelus) pulcher* (Linell) (loaned by the USNM), and other *Listrochelus* larvae shows that they have the last three pairs of abdominal spiracles reduced in size. In this character they resemble *Phyllophaga sociata* (Horn) which has other characters quite different from known species of the subgenus *Listrochelus* and *Phyllophaga sensu stricto,* (Ritcher, 1962).

Larvae of *Phyllophaga* may be distinguished by the following characters: Head (Fig. 176) without eye spots. Frons with a transverse pair of posterior frontal setae or a single posterior frontal seta, on each side. Labrum slightly asymmetrical. Epipharynx (Fig. 205) with distinct epizygum and zygum. Haptomerum with a group of 6 to 21 heli. Plegmatia present; plegmata rather long. Proplegmatia (Figs. 198-201, 203, and 204) usually present; absent in a few species. Dexiophoba short, extending forward from the sense cone for less than one-third the distance between the sense cone and the heli. Haptolachus often with numerous microsensilla. Maxilla with a row of 10 or more rather short, truncate teeth (Fig. 184).

Anal slit V- or Y-shaped. Stem of Y much shorter than arms of Y. Lower anal lobe usually divided by a sagittal cleft, sometimes divided only by a sagittal groove. Claws of metathoracic legs very small.

Larvae of a number of species of *Phyllophaga (sensu stricto)* annually cause much damage to pastures, lawns, corn, tree seedlings, potatoes, and strawberries in many parts of the eastern United States and Canada. For example, in 1933, Fluke and Ritcher (1934) estimated that 600,000 acres of bluegrass pasture had been destroyed in southern Wisconsin. In addition to the damage caused by larvae, the adults feed voraciously on the foliage of various deciduous trees at night, often completely stripping them over wide areas (Chamberlin and Fluke, 1947). In Kentucky, oats and walnut trees are often stripped during early May in the inner bluegrass region around Lexington (Ritcher, 1949).

In the north-central and eastern states, the most common and most destructive species include *Phyllophaga anxia, Ph. fervida, Ph. fusca, Ph. hirticula, Ph. implicita, Ph. inversa,* and *Ph. rugosa.* Species which are locally abundant are *Ph. arkansana, Ph. bipartita, Ph. crenulata, Ph. ephilida, Ph. fraterna, Ph. futilis, Ph. horni, Ph. micans,* and *Ph. tristis.* For detailed information on the taxonomy, distribution, host plants, and time of flight of adult May beetles, the reader should consult Luginbill and Painter (1953). Details of the biology, distribution, and taxonomy of Kentucky *Phyllophaga* were summarized by Ritcher (1940).

Key to *Phyllophaga* Larvae[18]

1. Spiracles on abdominal segments 6 to 8 much reduced in size (Fig. 233) .... subgenus *Listrochelus*[19]
   Spiracles on abdominal segment 8 reduced in size; spiracles on abdominal segments 6 and 7 as large or larger than spiracles on abdominal segments 1 to 5 .... subgenus *Phyllophaga*......... 2

2. Raster with sparsely set palidia; most pali separated at their bases by a distance greater than the width of a palus at its base (Figs. 220 and 225). Pali often more or less blade-like (compressed) and hooked at tips (Fig. 222) 3
   Raster with closely set palidia (Figs. 217 and 219). Pali depressed or cylindrical; tips never hooked (Fig. 224) 15

3. Epipharynx without proplegmatia 4
   Epipharynx with proplegmatia 5

4. Articulating skin of maxilla with 9 to 17 conical sensilla (Fig. 180). Palidia nearly parallel, each set with 10 to 12 short, hooked pali .... *Ph. quercus*
   Articulating skin of maxilla with setae only; sensilla absent (Fig. 177). Palidia strongly curved, each set with 10 to 13 short, rather conical pali; septula oval (Fig. 225) .... *Ph. tristis*

5. Each palidium set with 13 or fewer pali 6
   Each palidium set with 14 or more pali 7

6. Each proplegmatium with 2 to 4 broad proplegmata. Each palidium with 6 to 10 pali. Three or fewer preseptular hamate setae .... *Ph. praetermissa*
   Each proplegmatium with 5 or 6 indistinct, narrow proplegmata. Each palidium with 9 to 13 pali. Three or more preseptular hamate setae ......*Ph. implicita*

7. Proplegmatia usually well developed with 5 to 10 or more proplegmata on each side 8
   Proplegmata not curved, often indistinct, and may be absent on right side 12

8. Pali not distinctly blade-like but tips turn in (Fig. 220). Preseptular hamate setae usually absent. Each proplegmatium with 7 to 9 proplegmata. Head of third-stage larva less than 4.5 mm wide .... *Ph. ephilida*
   Pali distinctly blade-like (Fig. 222). Preseptular hamate setae numerous. Head of third-stage larva more than 5 mm wide 9

9. Each palidium set with 14 to 20 pali. Dorsoexterior region of mandible usually bare 10
   Each palidium set with 20 to more than 30 pali. Dorsoexterior region of mandible with 3 or more punctures 11

10. Proplegmata narrow, slightly curved (Fig. 200)....*Ph. fusca*
    Proplegmata rather broad, many of them strongly curved .... *Ph. karlsioei*

11. Raster with 4 or more preseptular hamate setae. Right chaetoparia with numerous punctures among the setae .... *Ph. anxia*
    Raster with 1 or 2 preseptular hamate setae. Right chaetoparia with a very few or no punctures among the setae .... *Ph. barda*

12. Proplegmatia narrow, each set with 5 to 8 short proplegmata (Fig. 203) 13
    Proplegmatia indistinct, may be absent on one side 14

13. Dorsoexterior region of mandible with 6 to 15 punctures .... *Ph. fervida*
    Dorsoexterior region of mandible without punctures .... *Ph. vehemens*

14. Each palidium with from 14 to 18 pali .... *Ph. arkansana*
    Each palidium with more than 20 pali .... *Ph. inversa*

15. Epipharynx with fewer than 11 proplegmata on each side 16
    Epipharynx with 11 or more proplegmata on each side (Fig. 198) 25

16. Each proplegmatium with 3 or 4 small, short proplegmata (Fig. 201) .... *Ph. micans*
    Each proplegmatium with 4 to 10 curved proplegmata 17

17. Pali very long and sharp (Fig. 219); at least 4 times as long as the widths of their bases. Each proplegmatium with 8 or 10 narrow proplegmata 18
    Pali shorter, not over 3 times as long as the widths of their bases 19

18. Alar lobe of mesothorax with 17 or fewer setae......*Ph. balia*
    Alar lobe of mesothorax with 30 or more setae......*Ph. futilis*

19. Dorsoexterior region of mandible with fewer than 4 setae (Fig. 207) 20
    Dorsoexterior region of mandible with 5 or more setae (Figs. 208, 211, and 213) 22

20. Proplegmatia with 6 (rarely 7) proplegmata (Fig. 199). Pali fairly long .... *Ph. bipartita*
    Proplegmatia with 7 or more proplegmata. Pali rather short (Fig. 217) 21

21. Basolateral region of mandible with 10 or fewer setae. Head capsule of third-stage larva less than 4.7 mm wide. .... *Ph. fraterna*
    Basolateral region of mandible with 12 to 15 setae. Head capsule of third-stage larva at least 4.8 mm wide .... *Ph. hirticula*

[18] Larvae of many species of *Phyllophaga* are separable by good taxonomic characters. However, study of a large amount of reared material shows considerable overlap and variation in characters commonly used to separate a number of other species, in a few cases even of species whose adults are quite different. Therefore, keys to the *Phyllophaga* fauna of a limited area are apt to prove more accurate for a given area that a key to a great number of species.

[19] Species belonging to this subgenus are found only in the states from Kansas westward.

22. Proplegmata fairly short, usually fewer than 9 on each side; proplegmata in center of row not over 4 times as long as wide. Head capsule of third-stage larva less than 4.9 mm wide .... *Ph. kentuckiana* and *Ph. delata*
Proplegmata long, narrow, and strongly curved; 8 to 11 on each side. Proplegmata in center of row at least 6 times as long as wide. Head capsule of third-stage larva usually 5 mm or more in width 13

23. Dorsoexterior region of mandible with fewer than 10 setae and not over 30 setae and punctures all together (Figs. 188 and 213) .... *Ph. horni*
Dorsoexterior region of mandible with 15 or more setae 24

24. Dorsoexterior region of mandible with 35 or fewer setae and punctures, of which at least two-thirds are setae (Fig. 208). Head capsule of third-stage larva 6 mm or more wide .... *Ph. profunda*
Dorsoexterior region of mandible with 40 or more setae and punctures less than half of which are setae (Fig. 211). Head capsule of third-stage larva less than 5.5 mm wide .... *Ph. rugosa*

25. Maxillary articulating skin with short conical sensilla (Fig. 180). Palidia set with 22 or fewer pali .... *Ph. crenulata*
Maxillary articulating skin without sensilla. Palidia set with 23 to 27 pali 26

26. Proplegmata 11 to 15 on each side. Pali fairly long .... *Ph. prunina*
Proplegmata 20 or more on each side. Pali short...... .... *Ph. sylvatica*

Species of Uncertain Position

?Phyllophaga sociata (Horn), Third-Stage Larva (Figs. 228-235)

This description reprinted from Ritcher (1962), is based on the following material:

Ten third-stage larvae and cast skins of 3 third-stage larvae reared to the adult stage. The larvae were dug from soil beneath nests of *Pogonomyrmex owyheei* Cole, 13 miles southeast of Sisters, Oregon (Deschutes Co.), July 6, 1961, by P. O. Ritcher and David Smith.

The larvae of this species may be distinguished by the following characters: Maximum width of head capsule 3.3 to 3.5 mm. Head (Fig. 231) yellow-brown in color, faintly reticulated. Anterior half of frons with numerous setae; with about 17 setae in a transverse patch near the frontal margin, with 15 to 20 posterior frontal setae on each side, and with 1 long seta at each anterior angle. Dorsoepicranial setae inconspicuous,

2 or 3 on each side. Labrum symmetrical. Epipharynx (Fig. 235) with well-developed epizygum and zygum. Proplegmatia well developed, each elliptical with 17 to 18 proplegma. Proplegma long, narrow, and curved. Haptomerum set with 8 to 10 stout heli, the anterior 4 or 5 in a curved row. No sensilla among the chaetoparia. Haptolachus without crepidal punctures. Pedium with a short laeophoba, anterior to the laeotorma, consisting of 3 to 6 flattened, branched filaments. Dexiophoba brush-like, with about 15 filaments, inserted at base of pedium just anterior to nesium externum (= sclerotized plate).

Dorsoexterior region of mandibles (Figs. 228 and 229) without either setae or pits. Setae in dorsomolar region of left mandible with a cluster of 4 or 5 setae inserted at the base of the molar structure. Dorsomolar region of left mandible with a cluster of 4 or 5 setae near the base of the molar structure. Maxilla (Fig. 232), with a regular row of 13 to 17 truncate stridulatory teeth bordering stipes. Lacinia with a longitudinal row of 3 unci on inner face and an anterior, oblique row of 3 stout, spine-like setae. Galea and lacinia, on inner surface, separated by a nonsclerotized membranous area. Last segment of antenna with a large, ovate sensory spot. Eye spots absent.

Femora of prothoracic legs (Fig. 230) each with a ventral row of 4 stout, fossorial setae (worn down in older specimens). Claws of prothoracic legs unusually long and stout, claws of mesothoracic legs rather long and slender, claws of metathoracic leg much reduced in size. Last 3 pairs of abdominal spiracles much smaller than spiracles on abdominal segments 1 to 5 (Fig. 233). Respiratory plates not surrounding bullae.

Raster (Fig. 234) with 2 widely separated, prominent palidia, diverging posteriorly. Posteriorly, each palidium consists of a comb-like row of 5 to 8 closely set, long, stout, flattened setae; anteriorly each palidium is continued as a sparsely set row of 3 or 4 short, subconical setae. Septula triangular. Laterad of each comb-like portion of each palidia is a patch of 10 to 14 hamate setae. Preseptular hamate setae usually absent. Anal slit Y-shaped with arms of Y about twice as long as stem. Dorsal and ventral anal lobes densely covered with fine, long and short setae. Lobes of lower anal lips bordered anteriorly with a row of 6 to 11 slender hamate setae.

This interesting species was described by Horn in 1878 under the name *Listrochelus sociatus*, in his revision of the species of the genus *Listrochelus* of the United States. Saylor, however, in 1938, removed the species from *Listrochelus* to *Phyllophaga, sensu stricto*, based on studies which he and E. A. Chapin had made preparatory to a revision of the subgenus *Listrochelus* (see also Saylor, 1940). Luginbill and Painter (1953) also included the species in *Phyllophaga, sensu stricto*.

According to M. W. Sanderson (personal communication), *Ph. sociata* belongs in a new group of species, separate from both the subgenus *Listrochelus* and *Phyllophaga sensu stricto*, including *Ph. xerophila* Saylor, *Ph. stohleri* Saylor, *Ph. reevesi* Saylor, *Ph. galeanae* Saylor, and several undescribed species.

Larvae of *Ph. sociata* are quite unique from these other two subgenera in possessing a totally different raster, a different pattern of setation on the head, and a row of strong fossorial setae on each prothoracic leg. In fact, the larval characters of this species are so distinct that it could be placed in a separate genus.

Luginbill and Painter (1953) erroneously listed *Ph. sociata* as a southwestern species. The distribution is given by Leng (1920) as Nevada, Idaho, and Oregon. Based on material in the Hatch collection at the University of Washington and in the collection at Oregon State University, this species has been collected from the following localities: Oregon—Baker Co.: Durkee. Deschutes Co.: 7 miles N. of Tumalo, Redmond, 10 miles SE of Sisters, 13 miles SE Sisters, 15 miles east of Sisters, and 1 mile N. of Cline Falls. Harney Co.: "P" Ranch. Jefferson Co.: Cove State Park (near Culver). Washington—Bent Co.: Prosser. Grant Co.: Moses Lake, Soap Lake, People's Oil Well.

Adults of *Ph. sociata* were collected at black light in central Oregon during June of 1961. All the Hatch specimens (25) were taken during May. My studies show that the species overwinters both in the larval and adult stages, indicating a life cycle of two or three years. Pupation occurs in July and August with transformation to the adult stage occurring 25 to 27 days later (at 25° to 26° C). Adults remain in the soil until the following spring before emerging.

TRIBE Hopliini
Genus *Hoplia* Illiger. Third-Stage Larva
(Figs. 194 and 223)

This description is based on the following material and reprinted from my 1949a bulletin:

Nine third-stage larvae and cast skins of 3 third-stage larvae of *Hoplia cazieri* Boyer reared to the adult stage. Adults identified by Cazier. Larvae dug from soil around grass roots in bottom land near Tessla, California, February 20, 1946, by William Barr, Ray Smith, J. W. McSwain, and P. O. Ritcher. No. 46-7C.

Four third-stage larvae and cast skins of 2 third-stage larvae of *Hoplia equina* Lec. reared to the adult stage. Adults determined by E. A. Chapin. Larvae found at roots of gum hedge at Pittsville, Maryland, April 5, 1941, by W. H. Anderson.

Twenty-three third-stage larvae and cast skins of 7 third-stage larvae of *Hoplia oregona* Lec. reared to the adult stage. Adults identified by Cazier. Larvae dug from low, moist ground with a cover of grass and wild strawberries in the dune area of San Francisco, California, February 26, 1946, by Peter Ting, R. W. L. Potts, and P. O. Ritcher. No. 46-11B.

Larvae of the genus *Hoplia* may be characterized as follows: Head without eye spots. Labrum symmetrical. Maxillary stridulatory area with a row of 9 to 11 short teeth with anteriorly directed points. Lacinia with a longitudinal row of 3 unci. Distal 2 unci fused at their bases; subapical uncus much smaller. Epipharynx (Fig. 194) with plegmatia; plegmata quite short. Proplegmatia present or absent. Epizygum absent. Haptomerum with a single helus. Dexiophoba and laeophoba absent. Haptolachus without microsensilla.

Thoracic spiracles and those on abdominal segments 1 to 3 similar in size; spiracles of abdominal segments 4 to 8 much smaller. Anal opening Y-shaped; stem of Y about half as long as either arm of the Y. Anal lobes setose. Raster (Fig. 223) consisting of a subtriangular teges of approximately 30 to 60 fairly long hamate setae with curved tips. Claws of prothoracic legs large and falcate. Claws of mesothoracic legs much reduced in size. Claws of metathoracic legs minute or absent.

The genus *Hoplia* includes some 16 species as revised by Boyer (1940). The adults are small, robust beetles with elytra gray, brown, or brightly colored yellow or green and black (Essig, 1926). They occur in colonies in sandy areas and are diurnal, feeding on the foliage and flowers of many different plants including orange, apricot, almond, apple, peach, and olive and on the flowers of rose, ceanothus, California poppy, and lupine (Essig, 1958). The larvae have been found feeding on grass roots and the roots of strawberries (Allen, 1959).

Plate XVI

FIGURE 176. *Phyllophaga horni* (Sm.). Head.

FIGURE 177. *Phyllophaga fusca* (Froel.). Left maxilla, ventral view.

FIGURE 178. *Phobetus comatus sloopi* Barrett. Head.

FIGURE 179. *Polyphylla decemlineata* (Say). Left maxilla, dorsal view.

FIGURE 180. *Phyllophaga crenulata* (Froel.). Right maxilla, ventral view.

FIGURE 181. *Maladera castanea* (Arrow). Distal part of lacinia of left maxilla showing unci.

FIGURE 182. *Plectris aliena* Chapin. Head.

FIGURE 183. *Polyphylla decemlineata* (Say). Maxillary stridulatory teeth of right maxilla.

FIGURE 184. *Phyllophaga hirticula* (Knoch). Maxillary stridulatory teeth of right maxilla.

FIGURE 185. *Amphimallon majalis* (Razoum.). Head.

FIGURE 186. *Maladera castanea* (Arrow). Left maxilla, dorsal view.

FIGURE 187. *Serica ligulata* Dawson. Left maxilla, dorsal view.

FIGURE 188. *Phyllophaga horni* (Sm.). Left mandible, dorsal view.

FIGURE 189. *Phyllophaga horni* (Sm.). Left mandible, ventral view.

Symbols Used

A—Antenna
AA—Seta of anterior frontal angle
AC—Acia
AFS—Anterior frontal setae
BLR—Baso-lateral region
CAR—Cardo
DER—Dorsoexterior region
DES—Dorsoepicranial setae
DMS—Dorsomolar setae
DSS—Dorsal sensory spot
E—Epicranium

EFS—Exterior frontal seta
ES—Epicranial suture
F—Frons
FS—Frontal suture
G—Galea
GD—Uncus of galea
L—Labrum
LA—Lacinia
LU—Unci of lacinia
M—Mandible
MP—Maxillary palpus

MAS—Maxillary articulating skin
MO—Molar region
PC—Preclypeus
PSC—Postclypeus
PA—Preartis
PTA—Postartis
SA—Scissorial area
SCR—Scrobis
SD—Maxillary stridulatory area
ST—Stipes
VP—Ventral process

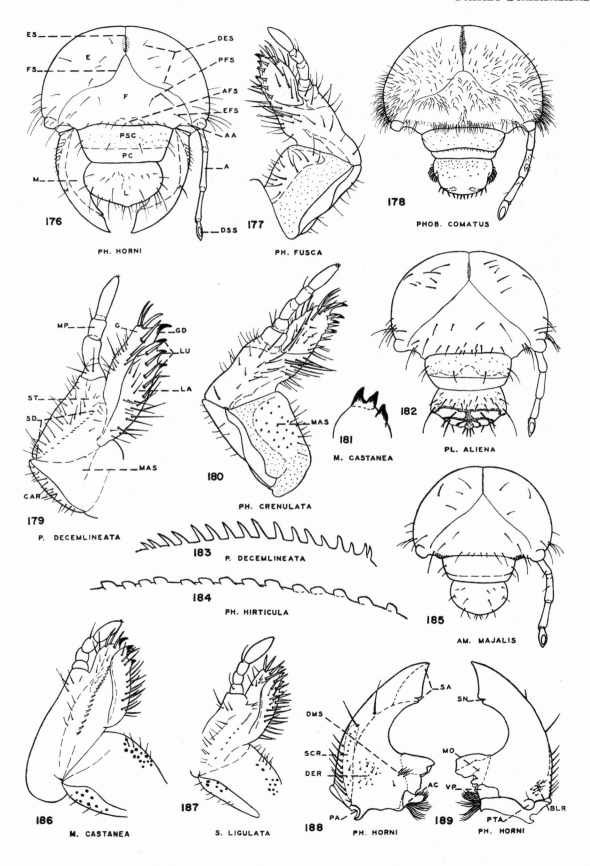

176 PH. HORNI

177 PH. FUSCA

178 PHOB. COMATUS

179 P. DECEMLINEATA

180 PH. CRENULATA

181 M. CASTANEA

182 PL. ALIENA

183 P. DECEMLINEATA

184 PH. HIRTICULA

185 AM. MAJALIS

186 M. CASTANEA

187 S. LIGULATA

188 PH. HORNI

189 PH. HORNI

Plate XVII

Symbols Used

CPA—Chaetoparia
CR—Crepis
DP—Dexiophoba
DX—Dexiotorma
EZ—Epizygum
H—Helus

HM—Haptomerum
LP—Laeophoba
LT—Laeotorma
MIS—Microsensilla
MSS—Macrosensilla
PE—Pedium

PL—Plegmatium
PRL—Proplegmatium
PTL—Pternotorma
SC—Sense cone
SP—Sclerotized plate

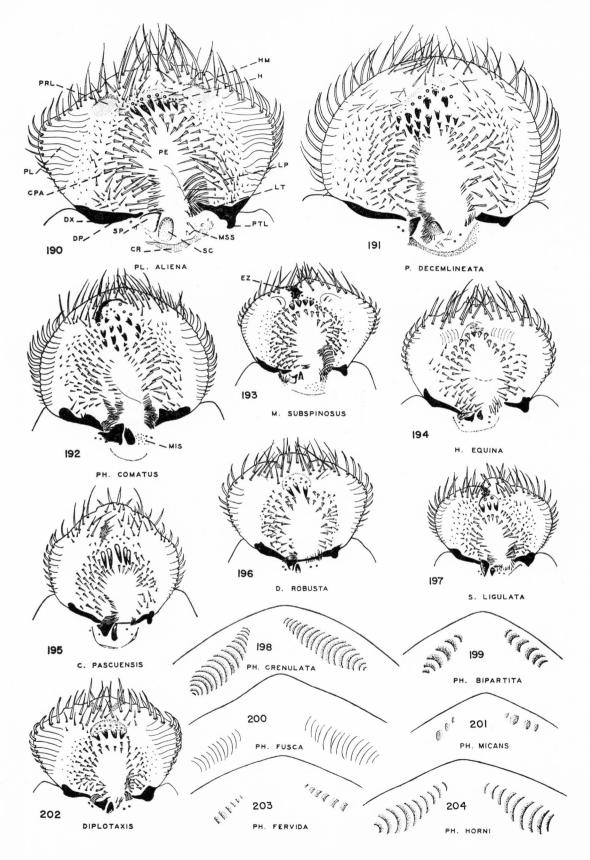

190 PL. ALIENA

191 P. DECEMLINEATA

192 PH. COMATUS

193 M. SUBSPINOSUS

194 H. EQUINA

195 C. PASCUENSIS

196 D. ROBUSTA

197 S. LIGULATA

198 PH. CRENULATA

199 PH. BIPARTITA

200 PH. FUSCA

201 PH. MICANS

202 DIPLOTAXIS

203 PH. FERVIDA

204 PH. HORNI

Plate XVIII

Figure 205. *Phyllophaga hirticula* (Knoch). Epipharynx.

Figure 206. *Amphimallon majalis* (Razoum.). Epipharynx.

Figure 207. *Phyllophaga hirticula* (Knoch). Dorsoexterior region of mandible.

Figure 208. *Phyllophaga profunda* (Blanch.). Dorsoexterior region of mandible.

Figure 209. *Maladera castanea* (Arrow). Venter of tenth abdominal segment.

Figure 210. *Diplotaxis* sp. Venter of tenth abdominal segment.

Figure 211. *Phyllophaga rugosa* (Melsh.). Dorsoexterior region of mandible.

Figure 212. *Serica ligulata* Dawson. Venter of tenth abdominal segment.

Figure 213. *Phyllophaga horni* (Sm.). Dorsoexterior region of mandible.

Figure 214. *Polyphylla decemlineata* (Say). Center portion of raster.

Figure 215. *Coenonycha pascuensis* Potts. Venter of tenth abdominal segment.

Figure 216. *Dichelonyx robusta* Fall. Venter of tenth abdominal segment.

Symbols Used

ASYL—Anal slit	HL—Haptolachus	MSS—Macrosensilla
CPA—Chaetoparia	HM—Haptomerum	PE—Pedium
CR—Crepis	LAL—Lower anal lobe	PL—Plegmatium
DP—Dexiophoba	LPH—Laeophoba	PLA—Palidium
DX—Dexiotorma	LT—Laeotorma	PLM—Plegma
EZ—Epizygum	MIS—Microsensilla	PRL—Proplegmatium

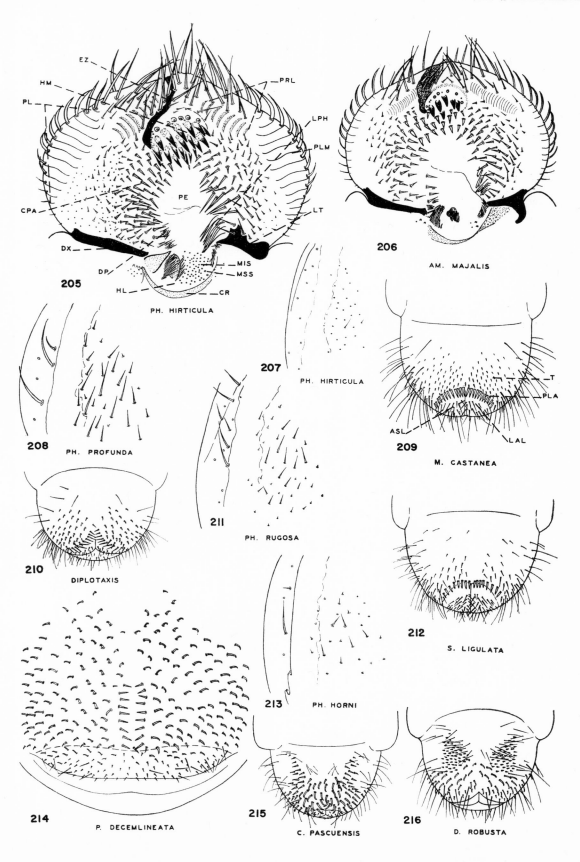

205 PH. HIRTICULA

206 AM. MAJALIS

207 PH. HIRTICULA

208 PH. PROFUNDA

209 M. CASTANEA

210 DIPLOTAXIS

211 PH. RUGOSA

212 S. LIGULATA

213 PH. HORNI

214 P. DECEMLINEATA

215 C. PASCUENSIS

216 D. ROBUSTA

Plate XIX

FIGURE 217.   *Phyllophaga hirticula* (Knoch). Venter of tenth abdominal segment.

FIGURE 218.   *Macrodactylus subspinosus* (F.). Venter of tenth abdominal segment.

FIGURE 219.   *Phyllophaga futilus* (Lec.). Central portion of raster.

FIGURE 220.   *Phyllophaga ephilida* (Say). Central portion of raster.

FIGURE 221.   *Phobetus comatus sloopi* Barrett. Venter of tenth abdominal segment.

FIGURE 222.   *Phyllophaga fusca* (Froel.). Ventral and lateral views of compressed pali.

FIGURE 223.   *Hoplia equina* Lec. Venter of tenth abdominal segment.

FIGURE 224.   *Phyllophaga hirticula* (Knoch). Ventral and lateral views of depressed pali.

FIGURE 225.   *Phyllophaga tristis* (F.). Venter of tenth abdominal segment.

FIGURE 226.   *Amphimallon majalis* (Razoum). Venter of tenth abdominal segment.

FIGURE 227.   *Plectris aliena* Chapin. Venter of tenth abdominal segment.

Symbols Used

B—Barbula	LAL—Lower anal lobe	TE—Tegillum
C—Campus	PLA—Palidium	S—Septula
DAL—Dorsal anal lobe		

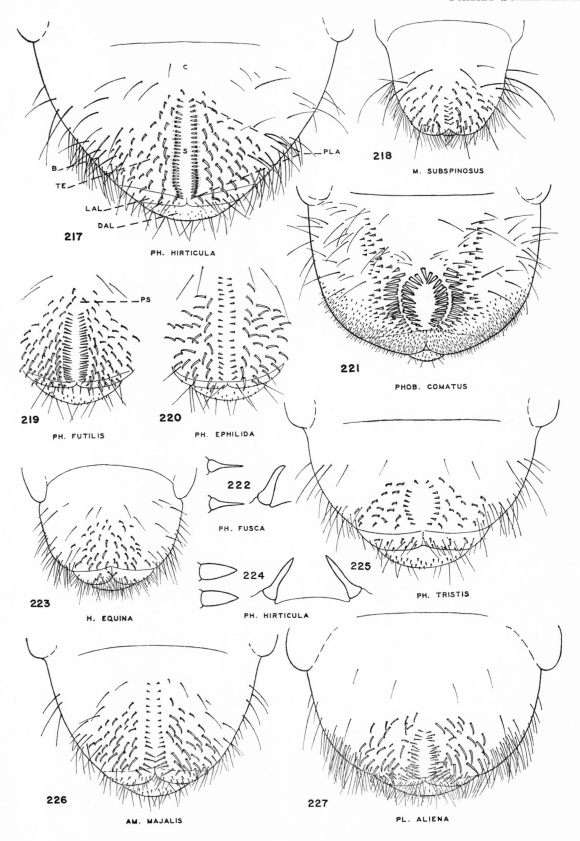

217 PH. HIRTICULA

218 M. SUBSPINOSUS

219 PH. FUTILIS

220 PH. EPHILIDA

221 PHOB. COMATUS

222 PH. FUSCA

223 H. EQUINA

224 PH. HIRTICULA

225 PH. TRISTIS

226 AM. MAJALIS

227 PL. ALIENA

Plate XX

?*Phyllophaga sociata* (Horn)

FIGURE 228.   Left mandible, dorsal view.

FIGURE 229.   Right mandible, dorsal view.

FIGURE 230.   Prothoracic leg, distal portion.

FIGURE 231.   Head.

FIGURE 232.   Left maxilla, dorsal view.

FIGURE 233.   Abdominal segments 5 to 10, left lateral view.

FIGURE 234.   Venter of last abdominal segment.

FIGURE 235.   Epipharynx.

Symbols Used

DX—Dexiotorma
FOS—Fossorial setae
G—Galea
H—Helus
HM—Haptomerum

LA—Lacinia
LP—Laeophoba
LT—Laeotorma
LU—Lacinial unci

PLA—Palidium
PPL—Proplegmatium
S—Septula
TE—Tegillum

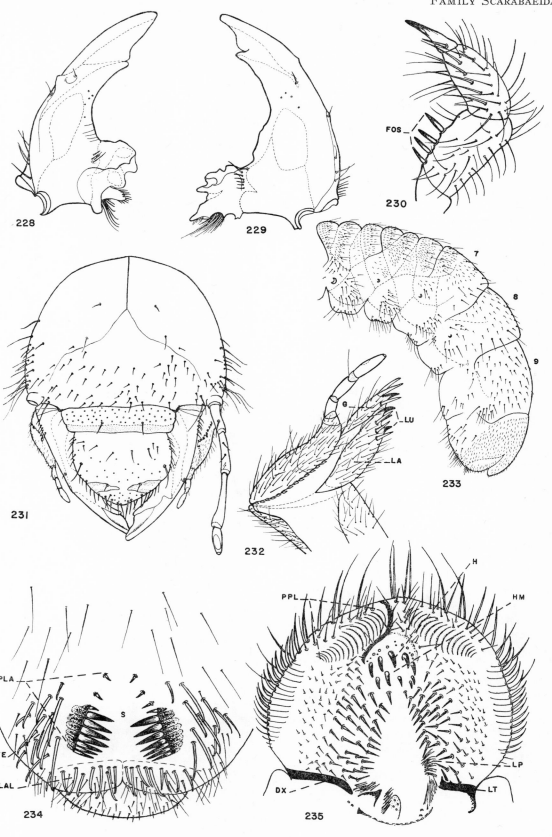

Subfamily Rutelinae

The subfamily Rutelinae includes a great number of small to medium-sized, leaf-feeding beetles, many of which are brilliantly colored with metallic yellow, orange, red, green, blue, or violet. Others are straw-colored, black-spotted or black.

The subfamily reaches its greatest development in the neotropical region where the group is represented by five of the six known tribes (Blackwelder, 1944). In the United States and Canada only the two tribes Anomalini and Rutelini are represented, but there is always the possibility that some species belonging to another tribe may be introduced as were the Japanese beetle, *Popillia japonica* Newman, and the oriental beetle, *Anomala orientalis* (Waterhouse), both belonging to the Anomalini. The injurious Chinese rose beetle, *Adoretus sinicus* Burmeister, belonging to the tribe Adoretini, is well established in Hawaii and from there might be carried to our west coast.

Larvae of many species of Anomalini are subterranean feeders on a great variety of crops, causing much damage to grasses, corn, small grains, strawberries, vegetable crops, and tree seedlings. The adults of several species, such as the Japanese beetle, cause severe injury to fruits and foliage.

Comparatively few larvae of the tribe Rutelini have been studied. Of these, many are feeders on decaying wood (Ohaus, 1934) or decaying vegetation (Viado, 1939). Larvae of *Cotalpa lanigera* (Linn.) feed in the soil and are destructive to raspberry bushes, strawberries, corn, and grass (Davis, 1916). Larvae of *Paracotalpa ursina ursina* (Horn) feed on the roots of *Artemesia* (Ritcher, 1948). The larvae of *Pelidnota punctata* (Linn.), *Parastasia brevipes* (Lec.), and *Plusiotis gloriosa* Lec., feed on decaying wood.

Impetus to the study of larval Anomalini, in particular, has been given by the establishment in eastern North America of two foreign members of the group, the Japanese beetle, *Popillia japonica* Newman, and the oriental beetle, *Anomala orientalis* (Wtrh.), both of which are of considerable economic importance. Böving, in 1921, described the larva of *P. japonica* in detail and compared it with a closely related, native larva which he later found to be *Anomala (Pachystethus) lucicola* (Fab.). Böving (1939) supplemented his earlier studies of the Japanese beetle larva by giving a technical description of the epipharyngeal complex and the raster, using terminology proposed by himself in 1936. Hayes (1927) described the larval anatomy of *Anomala kansana* Hayes and McColloch. Friend, in 1929, published a detailed morphological description of the larva of *Anomala orientalis*. The larva of *Adoretus sinicus* Burm. has been described in detail by Habeck (1962). Several keys to the Anomalini have been published in this country and abroad. Böving (1921) gave Schiodte's (1874) characterization of the tribe and a key of his own, distinguishing between the genera *Cyclocephala, Macrodactylus, Phyllophaga, Anomala, Phyllopertha, Popillia,* and *Strigoderma.* Hayes (1928 and 1929) published a key to genera of the subfamily Rutelinae which included *Strigoderma arboricola, Popillia japonica,* and the genus *Anomala.* Sim (1934) listed characters for distinguishing the larvae of *Popillia japonica* from those of *Anomala orientalis, A. binotata, Strigoderma arboricola, Anomala (Pachystethus) lucicola,* and certain other scarabaeid larvae. Gardner (1935) classified anomaline larvae and included four Indian species of the genus *Anomala.* Descriptions of 14 species of Anomalini have been previously published by the writer (1943 and 1945) and are reprinted in the following pages.

The larvae of only a few North American species have been described which belong to the tribe Rutelini. Davis (1916) described the larvae of *Cotalpa lanigera* (Linn.). Hayes (1925, 1928, and 1929) described the larvae of *Cotalpa lanigera* (Linn.), *Parastasia brevipes* (Lec.), and *Pelidnota punctata* (Linn.). Sim (1934) also described the larva of *Pelidnota punctata* (Linn.). Hayes (1928, 1929) gave a key to the genera of the subfamily Rutelinae which includes the three species mentioned above. The writer redescribed the same three species (1945) and later described the larvae of *Paracotalpa ursina ursina* (Horn) and *Plusiotis woodi* Horn (1948). Descriptions of these five species are reprinted in this publication and the larva of *Rutela formosa* Burm. is described for the first time.

Distinguishing Characters of Subfamily Rutelinae

Larvae of the subfamily Rutelinae may be distinguished from other scarabaeid larvae by the following combination of characters: Mandible with a ventral, oval, stridulatory structure consisting of a number of transverse ridges. Maxilla with a row of anteriorly directed, sharp-pointed, stridulatory teeth. Lacinia of maxilla with 1, 2, or 3 unci; if with 2 unci then the 2 are equal in size. Haptomerum of epipharynx with 2 or more prominent heli in a transverse row or with a raised mound- or beak-like process behind which are grouped 15 or more prominent spine-like setae not arranged in a definite transverse row. Epipharynx without proplegmata; plegmata present or absent; both nesia usually present. Dorsa of abdominal segments 9 and 10 never fused together. Raster with or without palidia. Anal slit transverse, slightly curved. Claws each bearing 2 setae.

Key to Tribes of the Subfamily Rutelinae Based on Characters of the Larvae

1. Last antennal segment with 2 or more dorsal sensory spots (Figs. 259 and 261) .... tribe **Rutelini** page 110
   Last antennal segment with a single, dorsal sensory spot (Figs. 243 and 275) 2

2. Haptomerum of epipharynx with a beak-like process. Palidia absent ....tribe **Anoplognathini**[20]
   Haptomerum of epipharynx with a transverse row of 3 or more heli (Fig. 248) 3

3. Haptomerum of epipharynx with a transverse row of 3 (rarely 2 or 4) prominent heli. Palidia present, monostichous (Fig. 244, 249, and 251). Maxillary stridulatory area with 4 to 7 sharp, recurved teeth (Fig. 313). Lacinia of maxilla with 2 unci................. .... tribe **Anomalini** page 101
   Haptomerum of epipharynx with a dense transverse comb-like row of 6 to 9 heli (Fig. 274). Raster with a subtriangular teges of hamate setae; palidia absent (Fig. 280). Maxillary stridulatory area with 8 or more, sharp recurved teeth. Lacinia of maxilla with 3 unci (Fig. 269)................................tribe **Adoretini**[21]

TRIBE Anomalini

Larvae of the 14 species of Anomalini studied were found to be closely related in structural detail. In fact, based on larval characters, the tribe is more uniform morphologically than the genus *Phyllophaga*. In the latter genus, for example, the epipharynx furnishes a number of characters useful in separating species while in the Anomalini the epipharynges are remarkably similar.

Based on larval characters, it is not possible to separate the genus *Anomala* from the genus *Pachystethus*. Also, the two closely related species, *P. lucicola* and *P. oblivia* are more nearly related to *A. innuba* and *A. nigropicta* than they are to *P. marginata*. For these reasons, species formerly placed in the genus *Pachystethus* are considered here as belonging in the genus *Anomala*.

Larvae of the tribe Anomalini may be distinguished from those of other tribes of the Rutelinae by the following combination of characters: Maxillary stipes with a dorsal row of from 4 to 7 (frequently 6) acute, recurved, anteriorly directed stridulatory teeth and an anterior truncate process, which appose the stridulatory area of the mandible. Lacinia of maxilla with 2 apical unci which are equal in size and arranged longitudinally

___
[20] Species belonging to this tribe are found in Mexico, Central America, and South America. None are known to occur in the United States and Canada.
[21] One species belonging to this oriental tribe has become established in Hawaii.

(a few species have uncus 1 present as a minute tooth anterior to unci 2 and 3). Epipharynx with 3 (rarely 2 or 4) prominent heli and well-developed plegmata; proplegmata absent. Last antennal segment with a single elliptical, dorsal sensory spot. Raster with 2 monostichous palidia. Spiracles on abdominal segments 7 and 8 similar in size and conspicuously larger than spiracles on abdominal segments 1 to 6.

A detailed description of the larva of *Anomala innuba* (Fab.) is reprinted here (Ritcher, 1943) using the terminology devised by Böving (1936). This is followed by a slightly revised key to species of the tribe and short descriptions of the distinguishing characters of the third-stage larva of each species, reprinted from the author's 1943 and 1945 bulletins. After each description is a short account of the more prominent adult characters, distribution, habits, and economic importance.

Third-Stage Larva of *Anomala innuba* (Fab.)

Larva (Fig. 238) cylindrical and typically scarabaeiform. Length of 10 mature, third-instar larvae ranging from 14 to 16 mm, with a mean, mid-dorsal length of 15.24 mm. Range in maximum width of cranium for 10 mature larvae, 2.3 to 2.5 mm, with a mean of 2.43 mm.

Cranium (Fig. 243) narrower than prothorax; surface faintly reticulate, shining, pale yellow-brown. Clypeo-frontal suture (CS) distinct, bounded laterally by the precoilae (PCL). Frontal sutures (FS) fine, whitish, meeting considerably in front of hind margin of head, forming slightly less than a right angle. Epicranial suture (ES) about half the length of one of the frontal sutures, medially with a light brownish enlargement, which, as Hayes (1927) pointed out, marks the site of a lamella for muscle attachment. Anterior part of frons (F) bears 2 prominent, paramedian setae, the anterior frontal setae (AFS)[22] which are equidistant from each other, and the lateral precoilae (PCL). The exterior frontal setae are absent. Each anterior frontal angle (AA) bears 2 setae, the lateral one being the larger. Two other setae, the posterior frontal setae (PFS), are located on the frons, one adjacent to each frontal suture somewhat below the median part of the frons. Epicranium (E) with 3 or 4 setae, the dorso-epicranial setae (DES), borne in a row on each side of the epicranial suture and posterior part of the frons, while laterad of the anterior end of each longitudinal row are 2 fairly long setae. A prominent seta is borne on each side of the epicranium on a level with the posterior frontal setae. A number of long and short

___
[22] The nomenclature of Murayama (1931) is used in describing the frontal and certain other cranial setae.

setae occur on the sides of the epicranium, with 2 especially long setae located near the base of each antenna. Small pores are scattered over the surface of the epicranium and frons.

Clypeus (Fig. 243) trapezoidal; divided transversely into two parts: the proximal postclypeus (PSC) and the distal preclypeus (PC). Two prominent paramedian setae, the anterior clypeal setae, are borne close to the distal end of the postclypeus; and 2 setae, the exterior clypeal setae, of which the proximal is the larger, are borne on each lateral edge. The postclypeus is more heavily sclerotized than the preclypeus and bears a number of scattered, olfactory pores which are absent on the preclypeus.

Labrum (L, Fig. 243) asymmetrical; faintly trilobed; slightly longer and narrower than clypeus; rounded on each lateral margin. Labrum bearing a broad transverse ridge with scattered pores, proximad of which are 2 pairs of setae, and distad of which is a single pair of setae. Apical lobe of labrum with a pair of blunt setae flanked on each side by a smaller, sharppointed seta; lateral lobes bearing 4 or 5 setae near their borders.

Antenna (A, Fig. 243) almost as long as cranium; slender; four-jointed; borne on a short basal piece partially fused to the epicranium. First segment about half as long as second segment; second segment bearing 1 or 2 dorsal setae and a single ventral seta. Subapical (third) segment with 2 small setae, produced apically on inner side into a conical process. Apical joint fusiform, bearing 1 dorsal and 2 ventral sensory spots (compound olfactory organs of McIndoo (1931). One such spot is borne on the inner face of the process of the subapical segment. End of apical segment with approximately 7 olfactory pegs ("tactile points" of Böving (1936)).

Mandibles (Figs. 236, 237, and 239 to 241) asymmetrical; subtriangular in shape; slightly shorter than cranium. Scissorial area (SA) blackish and fairly stout; cutting edge notched and with a small tooth-like projection proximad of the notch. Scrobe (SCR) subtriangular, bounded laterally by 2 carinae converging apically and uniting near the proximal end of the scissorial area adjacent to a single, large seta; scrobe bearing about 5 irregularly placed, prominent setae and approximately 6 minute setae. Longitudinal area laterad of dorsal carina (DC) rugulose and bearing anteriorly a single large seta; also scattered over its surface are about 12 pores, some of which are armed with points.

Mola-bearing part (MO) straw-colored with brown to blackish molar structures. Molar part of left mandible bilobed, with the distal lobe (M_1, Fig. 236) large and concave on its inner surface. The proximal lobe is small (M_2, Fig. 236) and bears a small, rounded tooth-like projection, hidden on the dorsal aspect by the acia (AC), a thin, sclerotized plate fringed distally with numerous coarse setae. Molar part of right mandible consisting of 4 lobes (Fig. 237), the distal one very small and the proximal one (= heel or calx, CA) longer than wide and slightly emarginate posteriorly. At the base of each molar part is a brustia (BR) of long, coarse setae.

On the dorsal surface of each mandible at the base of the anterior molar lobe is a compact row of small, stiff setae, the dorsomolar setae (DMS). The dorsoexterior mandibular region is bare. Caudo-laterad of the scissorial notch is located a group of 3 large pores, the anterior one of which frequently bears a prominent seta.

On the ventral aspect each mandible bears a large, ovate, stridulatory area (STA, Figs. 240 and 241) consisting of many transverse, granular ridges, each ridge consisting of many subquadrate to hexagonal groups of granulae (Fig. 242). At the basolateral angle of the mandible, on the ventral aspect, is located a group of approximately 10 small setae, the basolateral setae (BLS). Near the basomedial angle on the ventral surface of each mandible is found a prominent ventral process (VP)[23] which articulates with the hypopharyngeal sclerome (HSC, Fig. 315).

Maxilla (Figs. 314 and 315) composed of a cardo, stipes, fused galea-lacinia (mala), and maxillary palpus. Cardo (CAR) subquadrate, longer than wide, extending from base of maxilla to the proximal margin of the stipes; divided into 3 sclerites. Stipes (ST) subquadrate in outline on the ventrolateral surface of the maxilla, projecting mesad and uniting with the lacinia of the mala (MA). The stipes bears an occasional seta on its ventral surface, while on its dorsolateral edge, on both maxillae and continuing anteriorly, is found a row of 6 (occasionally 5 or 7) acute, recurved, apically pointing stridulatory teeth (SD, Fig. 313). Distad of the stridulatory teeth is a small, blunt tubercle. Ventral surface of mala composed mainly of the galea which is moderately setose and bears distally a single large, claw-like uncus (UN); dorsal surface of mala lightly sclerotized and moderately setose. Mesal edge of lacinia bearing 3 distinct, longitudinal rows of prominent setae, with 2 unci borne distad of the middle row. Unci of mala surrounded by a circle of about 8 rather short, stout setae. Maxillary palpus (MP) 4-jointed, the obovate fourth joint with a number of sensory pegs at its distal end.

Labium (Figs. 314 and 315) consisting ventrally of a large, subtrapezoidal postmentum (PMP), a terminal

---

[23] Same as "accessory ventral condyle" of Böving and Craighead (1931).

subovate prementum (PRM), and 2 labial palpi (LP). A transverse row of setae is located on the proximal part of the prementum of which 2 setae, separated by a small median seta, are conspicuously larger. A pair of conspicuous setae are located near the meson of the prementum, slightly caudad of the bases of the palpi. On the dorsal aspect is located the fleshy, cushion-like glossa (GL) which is covered posteriorly with many fairly stout setae and distally with many longer, more slender setae.

Hypopharyngeal sclerome (HSC, Fig. 315) located posteriorly to labial glossa (GL); asymmetrical; produced on right side into a strong, blackish, truncate process (TP). Sclerome bearing on each side, a semi-membraneous shoulder or lateral lobe (LL), each bearing a few small setae. Behind each lateral lobe is located a fossa (HFO) which receives the ventral process of the corresponding mandible.

Epipharynx (Fig. 248) broader than long, with rounded lateral margins; apical margin broadly protuberant in region of corypha (CO), indented slightly in region of epizygum (EZ). Corypha and acroparia (ACR) united, bearing a number of long, blunt-tipped setae. Epizygum present, united with right zygum (Z). A plegmatium (PL) present on each lateral margin of epipharynx, each composed of about 18 plegmata. Acanthopariae (ACP) each bearing about 18 flattened setae; each seta set in a socket at the outer end of a plegma, the anterior one or two setae long and straight, the remainder sickle-shaped and decreasing in size posteriorly. Proplegmata absent. Gymnopariae (GP) well developed, bearing a few small, scattered sensilla posteriorly. Chaetopariae (CPA) well developed, covered with many setae which are large and fairly coarse near the pedium (PE), becoming smaller and finer as they approach the gymnopariae; setae strongest on the distal portion of the right chaetoparia, especially in the region of the haptomerum (H). No sensilla interspersed between the setae of the chaetopariae. Haptomerum consisting of a sclerotized zygum, about 10 sensilla, and 3 prominent, backward projecting heli (H). Curved zygum incomplete, with left portion apparently separated from right portion. Sensilla oval, of two sizes, and arranged in a curved, irregular, transverse row anterior to the bases of the heli. The three heli large and stout; arranged in a transverse row. Pedium ovate, longer than wide. Body of laeotorma (LT) pistol-shaped with apotorma absent. Epitorma (EPT) divides pedium into a basal two-thirds and a distal third. Dexiotorma (DX) fairly broad, slightly sinuate, broadest at the base. Dexiophoba (DP) poorly developed, represented by several sclerotized nodules or 1 or 2 closely appressed spines anterior to the sense cone. Haptolachus (HL) complete. Crepis (CR) in the form of a lightly sclerotized, transverse, arcuate bar, expanded laterally.

Two sensilla on the right expanded area of the crepis. Two sclerotized nesia are present. The nesium closer to the dexiotorma, called the sclerotized plate (SP), is blade-like and rather sharply pointed; the other, called the sense cone (SC) by Hayes (1928), is conical and bears 3 sensory pores at its apex. A row of 3 to 7 small, slender setae is mesad of the laeotorma, while between the right end of this row and the sense cone is a group of 4 to 11 small spines with bulbous bases. This row of setae and adjacent spines may be called a mesophoba. About 4 sharp setae are borne at the right end of the crepis, between the base of the sclerotized plate and the adjacent dexiotorma. Two prominent sensilla are located in front of the crepis toward the center of the haptolachus.

Legs (Fig. 238) well developed with the prothoracic pair shorter than the mesothoracic and the mesothoracic shorter than the metathoracic pair. Each leg composed of a long cylindrical coxa, a short trochanter, a clavate femur, and a fused tibio-tarsus. Numerous setae scattered over legs with the largest setae occurring on the ventral surfaces of the femora. Setae sparse on coxae; dorsal surfaces of trochanters and femora practically bare. Bases of tarsal claws subcylindrical and straw-colored, each bearing two setae. Distal parts of tarsal claws brown to blackish in color and slightly curved; those of the prothoracic and mesothoracic legs acuminate and approximately equal in length, those of metathorax only half as long.

Body (Fig. 238) consisting of 3 thoracic and 10 abdominal segments. Prothorax, mesothorax, and metathorax, each with three dorsal areas: the prescutum, scutum, and scutellum. A pair of spiracles located on the prothorax. Prescutum and scutum of prothorax, prescutum and scutellum of mesothorax and metathorax each with a transverse row of slender, fairly long setae; a few short setae interspersed in the row of long setae on the scutellum of the metathorax.

Three dorsal areas, the prescutum (PRSC), scutum (SCU), and scutellum (SCL) clearly defined on each of the first 6 abdominal segments (Fig. 316). Prescutum of first abdominal segment bearing a transverse row of alternately long and short setae; scutum and scutellum of first abdominal segment and the 3 dorsal annulets of the second to sixth abdominal segments each bearing a transverse patch of small setae interspersed posteriorly with a row of long, fine setae which become more numerous laterally on the scuta. Dorsa of seventh to ninth abdominal segments not divided into annulets, each segment with two transverse rows of long, slender setae; a few short setae interspersed among the 6 to 8 long setae of each caudal row. An irregular double row of slender setae borne near cephalic margin of tenth abdominal segment, anterior to an oval impressed line, open posteriorly, which van

Emden (1941) has called the subcircular furrow (SF). Posterior two-thirds of dorsum densely clothed with long and short, slender, caudally pointing setae.

Abdominal, spiracle-bearing sclerites bearing from 2 to 8 long and short setae. Pleural lobes (PLL) of first to ninth abdominal segments with from 4 to 10 long and short setae. Sterna of abdominal segments each with a sparse, transverse row of fairly long setae.

Anal slit transverse, arcuate. Anus (Fig. 244) bordered dorsally by several irregular rows of moderately stout, short setae; anus bordered ventrally, on lower anal lip, by a transverse group of 34 to 54 long and short, straight setae. Lower anal lip (LAL) bearing, in addition, a patch of about 13 scattered, hamate setae similar to those of the tegilla. To each side of the anus is borne a barbula (B) of long, prominent setae contiguous to the setae clothing the caudo-dorsal part of the tenth abdominal segment.

Raster (Fig. 244) made up of a septula, a pair of palidia, a pair of tegilla, and a campus. Septula (S) elongate, dilated posteriorly, extending anteriorly from the base of the lower anal lip (LAL) to about the midpoint of the venter of the tenth abdominal segment. Each palidium (PLA) composed of a fairly straight, monostichous row of 8 to 12 slender, tapering, depressed pali. Pali sparsely set, separated at their bases by a distance greater than the basal width of a palus (P). Palidia diverging posteriorly, with the tips of opposing pali almost meeting at the anterior end of the septula and separated at posterior end of the septula by a distance slightly greater than the length of a palus. Tegilla (T) extending from the barbulae (B) to the palidia and united anterior to the septula into a continuous patch of hamate setae. Preseptular, hamate setae (PHS) 17 to 33 in number. Campus (C) broad, occupying anterior third of venter of tenth abdominal segment; bare except for about 4 long, slender setae.

Spiracles (Fig. 238) consisting of 1 pair of thoracic spiracles and 8 pairs of abdominal spiracles. Thoracic spiracles and seventh and eighth pairs of abdominal spiracles nearly equal in size and approximately twice as large as the spiracles of the first 6 abdominal segments; concavity of thoracic spiracles facing posteriorly, that of abdominal spiracles facing anteriorly. Each spiracle (Fig. 247) with C-shaped respiratory plate (RSP) surrounding a large, oval bulla (BU); spiracular slit (SS) curved. Thinly covered "holes" of respiratory plate elongate-oval in outline and arranged in irregular transverse rows. Trabeculae branched, moderately developed. Thoracic spiracle ranging from 0.22 to 0.25 mm in length, measured dorsoventrally, and from 0.13 to 0.2 mm wide. Bullar opening slightly constricted (ratio of width of bullar opening to dorsoventral diameter of bulla 2:3 or greater). Respiratory plate of thoracic

spiracle with a maximum number of 11 to 14 "holes" in a transverse series.

Key to Larval Anomalini

1. Venter of last abdominal segment bearing 14 or more preseptular, hooked setae (Figs. 244 and 249); pali depressed or conical, never compressed (Figs. 245 and 246) 2
   Venter of last abdominal segment bearing 12 or fewer preseptular, hooked setae (Figs. 251 and 253); pali frequently compressed and bent inwardly at tips (Fig. 254) 8

2. Septula shaped like an equilateral triangle, palidia strongly diverging posteriorly; each palidium with 5 to 7 (rarely 8) long, ensiform pali (Fig. 249) .... *Popillia japonica* Newman
   Septula oblong; each palidium with 6 to 16 pali (Fig. 244) 3

3. Dorsa of seventh to ninth abdominal segments with numerous small and medium-sized setae and each bearing, in addition, a caudal, transverse row of about 6 very long setae; transverse crest of labrum consisting of about 6 rugosities, anterior of which are 3 rugosities arranged in the form of a triangle (Fig. 255) ....*Strigoderma arboricola* (Fab.)
   Dorsa of seventh to ninth abdominal segments practically bare except for 2 transverse rows of setae (Fig. 257); transverse crest of labrum not strongly rugose 4

4. Dorsoexterior region of mandible with 8 to 16 punctures .... *A. minuta* Burm.
   Dorsoexterior region of mandible bare 5

5. Posterior frontal setae consisting of two transverse groups of 2 or 3 setae (Fig. 255) 6
   Posterior frontal setae consisting of 2 single setae (Fig. 243) 7

6. Lower anal lip bearing fewer than 40 (range 20 to 38) long and short, straight setae ....*Anomala oblivia* Horn
   Lower anal lip bearing 42 or more (range 43 to 84) long and short, straight setae........*Anomala lucicola* (Fab.)

7. Palidia incurved at their anterior ends, otherwise parallel, each with 10 to 16 pali; anterior frontal angle of cranium with 3 or 4 setae. Exterior frontal setae present (Fig. 255) ....*A. nigropicta* Casey
   Palidia fairly straight, diverging posteriorly at approximately a 30° angle, each with 8 to 11 (rarely 12) pali (Fig. 244); anterior frontal angle of cranium with 2 setae (Fig. 243). Exterior frontal setae absent (Fig. 243) ....*A. innuba* (Fab.)

8. Dorsa of seventh to ninth abdominal segments rather uniformly covered with many medium-sized setae and each bearing, in addition, a caudal transverse row of 5 or 6 very long setae (Fig. 258)...................... 9
   Dorsa of seventh to ninth abdominal segments with majority or all of setae confined to two transverse rows or bands (Figs. 256 and 257) 10

9. Palidia parallel, each set with 4 to 6 pali..................................
....*Strigodermella pygmaea* (Fab.)
Palidia slightly diverging, each set with 11 to 15 pali....
.... *A. orientalis* Wtrh.

10. Dorsa of seventh to ninth abdominal segments bare
except for two transverse, sparse, single rows of
setae; 1 or 2 setae borne on each anterior frontal
angle of cranium (Fig. 243); preseptular hooked
setae 5 to 12 in number; dexiophoba usually absent
(Fig. 250) ....*Anomala marginata* (Fab.)
Dorsa of seventh to ninth abdominal segments each
with a few scattered setae, especially on ninth
segment, in addition to two transverse, sparse,
single rows of setae (Fig. 256); 3 to 5 setae borne
on each anterior frontal angle of cranium; preseptular hooked setae none to 6 in number (Fig.
253); dexiophoba usually present (Fig. 252) 11

11. Palidia slightly diverging posteriorly, each set with 16
to 25 pali; palidia usually extending anteriorly for
one-fourth their length beyond the adjacent rows
of large, hamate, tegillar setae................*A. binotata* Gyll.
Palidia parallel, each with 7 to 14 pali and not extending beyond the adjacent rows of hamate, tegillar
setae (Fig. 253) 12

12. Opening of respiratory plate of thoracic spiracle not
constricted or only slightly constricted (ratio of
width of opening to dorsoventral diameter of bulla
more than 7.5:10) .... *A. ludoviciana* Schffr.
Opening of respiratory plate of thoracic spiracle much
constricted (ratio of width of opening to dorso-
ventral diameter of bulla less than 6:10) 13

13. Respiratory plate of thoracic spiracle with 19 or
more "holes" in some transverse series ....
....*A. kansana* Hayes & McC.
Respiratory plate of thoracic spiracle with a maximum
of 15 to 17 "holes" in a transverse series.................
.... *A. flavipennis* Burm.

Genus *Anomala* Samduelle
Anomala binotata Gyll.
(Fig. 252)

This description is based on the following material
and reprinted from my 1943 bulletin:

Four reared third-stage larvae, 1 reared second-
stage larva, and exuviae of 2 third-stage larvae
reared to the adult stage (loaned by W. P. Hayes).

Maximum width of head capsule 3.2 to 3.4 mm.
Exterior frontal setae present. Posterior frontal setae
3 to 6 on each side in a transverse group. Dorsoepicranial setae 4 to 7 on each side. Dorsoexterior region
of mandible bare. Epipharynx (Fig. 252) with well-
developed dexiophoba of 9 to 20 spines. Thoracic
spiracle ranging from .43 to .48 mm in length and from
.30 to .34 in width. Respiratory plate with a maximum
of 20 to 22 "holes" in a transverse series. Bullar opening much constricted. Dorsa of seventh to ninth abdominal segments with most of setae confined to 2
transverse rows. Tenth abdominal segment with sub-

circular furrow present. Raster with palidia slightly
diverging posteriorly; septula oblong. Pali in each row
16 to 25 in number and slightly compressed. Preseptular hamate setae none to 2 in number, usually absent.
Lower anal lip bearing 66 to 72 long and short, straight
setae.

This is a rather large, native species common in
Connecticut, Illinois, Indiana, Kansas, Kentucky, New
York, Ohio, and Virginia. The coloration of the adult
is quite distinctive. The legs and underside of the
abdomen are black. Head and thorax black with a
violet-green irridescence. Scutellum black, elytra straw-
colored, margined with black; each elytron bearing a
median, black splotch which sometimes is confluent with
the black-colored area of the humeral angle. Elytra also
bearing a number of longitudinal rows of black punctures. Rarely individuals are found with the elytra entirely black. Adult 10 to 12 mm long and 5 to 7 mm
wide. Flight period in May in Ohio and Kentucky.

Anomala flavipennis Burm.
(Figs. 253, 254, and 256)

This description is based on the following material
and reprinted from my 1943 bulletin:

Ten field-collected third-stage larvae and exuviae
of 8 third-stage larvae reared to the adult stage.

Maximum width of head capsule 2.8 to 3.1 mm.
Exterior frontal setae present. Posterior frontal setae
4 to 5 on each side. Dorsoepicranial setae 9 to 13 on
each side. Dorsoexterior region of mandible bare.
Epipharynx with well-developed dexiophoba of 6 to 12
spines. Thoracic spiracle ranging from .37 to .39 mm
in length and from .25 to .27 mm in width. Respiratory
plate with a maximum of 15 to 17 "holes" in a transverse series. Bullar opening much constricted. Dorsa of
seventh to ninth abdominal segments with most of the
setae confined to 2 transverse rows. Tenth abdominal
segment with subcircular furrow present and pigmented.
Raster with 2 rows of parallel pali; septula oblong.
Pali in each row 10 to 14 in number and much compressed with tips bent inwardly. None to 6 preseptular
hamate setae. Lower anal lip with 42 to 46 long and
short, straight setae.

The larva of this species has many characters in
common with the larvae of *A. kansana* and *A. ludoviciana*. *A. flavipennis* is a native species common in
Kentucky and farther south. Adult distinctive with reddish legs, abdomen, head, thorax, and scutellum. Elytra
light yellow-brown, narrowly margined with red. Adult
9 to 12 mm long and 5 to 7 mm wide. Larvae sometimes fairly abundant in pasture land in southern Kentucky. Flight period in June and July in Kentucky.

Anomala innuba (Fab.)
(Figs. 236 to 248 and 313 to 316)

This description is based on the following material and reprinted from my 1943 bulletin:

Fifteen field collected third-stage larvae and exuviae of 25 third-stage larvae reared to the adult stage.

Maximum width of head capsule 2.3 to 2.5 mm. Exterior frontal setae absent. A single posterior frontal seta on each side. Dorsoepicranial setae 3 to 4 on each side. Dorsoexterior region of mandible bare. Epipharynx with poorly developed dexiophoba of 0 to 2 spines. Thoracic spiracle ranging from .22 to .25 mm in length and from .13 to .2 mm in width. Respiratory plate with a maximum of 11 to 14 "holes" in a transverse series. Bullar opening constricted. Dorsa of seventh to ninth abdominal segments each with 2 transverse rows of setae only. Tenth abdominal segment with subcircular furrow present. Raster with palidia diverging posteriorly at a 30° angle; septula oblong. Pali in each row 8 to 12 in number, slender, and slightly depressed. Preseptular, hamate setae 17 to 33 in number. Lower anal lip bearing 34 to 54 long and short, straight setae. (Larva described in greater detail on pages 84, 86-88.)

A. innuba is a small, common, widely distributed species of very variable color which is frequently misidentified in collections as *A. undulata* Melsh., a synonym, or as *A. nigropicta,* a closely related, valid species. Lighter-colored adults with straw-colored legs and venters, straw colored elytra spotted with black, and straw-colored thoraces each with a central black area. In some specimens the black markings on the elytra form 2 prominent, irregular transverse bands. Other specimens are almost entirely black. Adult 7 to 8.5 mm long and 4 to 4.5 mm wide. Larvae found in bluegrass and other pasture land in Kentucky. Pupation occurs in the spring. Adults feed on heads of timothy and wheat, often in the daytime. Flight period in June in Kentucky; common at lights.

Anomala kansana Hayes and McColloch

This description is based on the following material and reprinted from my 1943 bulletin:

Two, reared third-stage larvae donated by W. P. Hayes.

Maximum width of head capsule 3.1 to 3.2 mm. Exterior frontal setae present. Posterior frontal setae 3 on each side in a transverse row. Dorsoepicranial setae 8 to 10 on each side. Dorsoexterior region of mandible bare. Epipharynx with well-developed dexiophoba. Thoracic spiracle ranging from .39 to .43 mm in length and from .29 to .32 mm in width. Respiratory plate with a maximum of 19 to 22 "holes" in a transverse series. Bullar opening much constricted. Dorsa of seventh to ninth abdominal segments with most of setae confined to 2 transverse rows. Tenth abdominal segment with subcircular furrow present. Raster with parallel palidia; septula oblong. Pali in each row 9 to 12 in number and slightly compressed. Preseptular, hamate setae 1 to 4 in number. Lower anal lip with approximately 50 long and short, straight setae.

A. kansana is a large, native species common in Kansas. Hayes and McColloch (1924) consider the larvae capable of doing some damage to wheat. Adult with reddish-brown venter, legs, thorax, and scutellum. Head dark red-brown, often with a metallic luster. Lighter-colored specimens with yellowish elytra and each elytron margined with brown and marked with 3 rather vague, longitudinal, brown stripes. Darker specimens with elytra brown or blackish with a caudolateral, yellowish-brown band. Adult 10.5 to 12 mm long and 5.5 to 7 mm wide. See Hayes and McColloch (1924) for an account of the biology of this species.

Anomala ludoviciana Schffr.

This description is based on the following material and reprinted from my 1943 bulletin:

Exuviae of 5 third-stage larvae reared to the adult stage (donated by T. R. Chamberlain).

Maximum width of head capsule approximately 2.1 mm. Exterior frontal setae present. Posterior frontal setae 5 to 6 on each side. Dorsoepicranial setae 9 to 12 on each side. Dorsoexterior region of mandible bare. Epipharynx with well-developed dexiophoba of 6 to 11 spines. Thoracic spiracle ranging from .26 to .32 mm in length and from .17 to .20 mm in width. Respiratory plate with a maximum of 13 to 14 "holes" in a transverse series. Bullar opening wide, only slightly or not at all constricted. Dorsa of seventh to ninth abdominal segments with most of setae confined to 2 transverse rows. Tenth abdominal segment with subcircular furrow present. Raster with parallel palidia; septula oblong. Pali in each row 7 to 12 in number and compressed. Preseptular, hamate setae 1 to 5 in number. Lower anal lip bearing 41 to 51 long and short, straight setae.

A. ludoviciana is a small, native species. According to Leng (1920), it is found in Louisiana, Kansas, Indiana, and Florida. Chamberlin, of the Federal White Grub Laboratory, has reared adults from a number of larvae collected in Wisconsin. Adult with red-brown to blackish legs, venter, head, thorax, and scutellum. Head with metallic luster. Elytra straw-colored with a number of faint, longitudinal rows of black punctures. Elytra frequently margined laterally with brown or red-brown. Adult 7 to 9 mm long and 4 to 5 mm wide.

Anomala minuta Burm.

This description is based on the following material and reprinted from my 1943 bulletin:

One field-collected third-stage larva and exuviae of 4 third-stage larvae, reared to the adult stage (loaned by the U. S. National Museum).

Maximum width of head capsule approximately 2.3 mm. Exterior frontal setae present. Posterior frontal setae 3 on each side. Dorsoepicranial setae 5 to 7 on each side. Dorsoexterior region of mandible with 8 to 16 punctures. Epipharynx with well-developed dexiophoba of about 5 spines. Thoracic spiracle ranging from .30 to .32 mm in length and about .19 mm in width. Respiratory plate with a maximum of 12 to 13 "holes" in a transverse series. Bullar opening slightly constricted. Dorsa of seventh to ninth abdominal segments with 2 transverse rows of setae only. Tenth abdominal segment with subcircular furrow present. Raster with palidia slightly diverging posteriorly; septula oblong. Pali in each row 8 to 12 in number, slender, slightly depressed. Preseptular, hamate setae 14 to 18 in number. Lower anal lip bearing 46 to 55 long and short, straight setae.

A. minuta is a small, native species occurring in the southeastern states. Adult with yellow-brown venter. Legs yellow-brown with red-brown tarsi. Thorax with discal portion brown, margined laterally and posteriorly with yellow-brown. Elytra yellow-brown suffused with brown. Tarsal claws not cleft. Adult approximately 7 mm long and 4 mm wide.

Anomala nigropicta Casey

This description is based on the following material and reprinted from my 1943 bulletin:

One reared third-stage larva, 2 reared second-stage larvae, and exuviae of 4 third-stage larvae reared to the adult stage.

Maximum width of head capsule approximately 2.8 mm. Exterior frontal setae present. A single posterior frontal seta on each side. Dorsoepicranial setae 3 to 4 on each side. Dorsoexterior region of mandible bare. Epipharynx with poorly developed dexiophoba of 2 to 3 spines. Thoracic spiracle ranging from .28 to .32 mm in length and from .19 to .24 mm in width. Respiratory plate with a maximum of 14 to 15 "holes" in a transverse series. Bullar opening moderately constricted. Dorsa of seventh to ninth abdominal segments with 2 transverse rows of setae only. Tenth abdominal segment with subcircular furrow present. Raster with palidia convergent anteriorly but parallel posteriorly; septula oblong. Pali in each row 10 to 16 in number, slender, slightly depressed. Preseptular, hamate setae 17 to 29. Lower anal lip with 50 to 52 long and short, straight setae.

A. nigropicta is a medium-sized, rather common, native species of variable color, closely related to and frequently confused with *A. innuba*. It is known by the writer to occur in Illinois, Indiana, Ohio, and Kentucky, and according to Leng (1920) it also occurs in North Carolina and Florida. Adult straw-colored with black markings. Venter straw-colored to blackish. Legs straw-colored with tarsi red to red-brown in color. Disc of thorax dark, reddish-brown; lateral margins of thorax straw-colored. Head reddish, lighter in color than thorax, scutellum straw-colored to black. Elytra of lightest-colored individuals entirely straw-colored with no black markings; elytra of darkest individuals with 2 irregular, transverse, black bands. Specimens of intermediate coloration have a characteristic median black spot formed by 2 adjacent spots, one on the caudal third of each elytron. Adult 8 to 10 mm long and 4 to 5 mm wide. Larvae found in bluegrass and other pasture land in Kentucky. Pupation occurs in the fall. Flight period from late April until early June in Kentucky; common at lights.

Anomala orientalis (Wtrh.)
(Figs. 251 and 258)

This description is based on the following material and reprinted from my 1943 bulletin:

Fifty field-collected third-stage larvae (donated by J. P. Johnson).

Maximum width of head capsule 2.3 to 3.0 mm. Exterior frontal setae present. Posterior frontal setae 3 on each side in a transverse row. Dorsoepicranial setae 6 to 7 on each side. Dorsoexterior region of mandible bare. Epipharynx with well-developed dexiophoba of 4 to 10 spines. Thoracic spiracle ranging from .24 to .27 mm in length and from .15 to .18 mm in width. Respiratory plate with a maximum of 9 to 11 "holes" in a transverse series. Bullar opening only slightly constricted. Dorsa of seventh to ninth abdominal segments covered with medium-sized setae and each bearing in addition a transverse row of 5 to 6 very long setae. Tenth abdominal segment with subcircular furrow present but vague. Raster with palidia slightly diverging posteriorly; septula oblong. Pali in each row 11 to 15 in number and slightly compressed. Preseptular, hamate setae 1 to 4 in number. Campus of venter of tenth abdominal segment frequently almost obliterated by the presence of numerous long and short, straight setae. Lower anal lip bearing 33 to 49 long and short, straight setae.

A. orientalis, known as the oriental beetle, is a native of Japan. It became established in this country prior to 1920 and now occurs in parts of Connecticut, New Jersey, and New York. The larva is primarily a pest of lawn turf. Adult color very variable, ranging from straw to black. According to Friend (1937), specimens with thoraces with 2 separate black marks and elytra with either 1 black bar or entirely straw-colored are most common. Adult 8 to 11 mm long and 4 to 5.5 mm wide.

Anomala lucicola (Fab.)
(Fig. 257)

This description is based on the following material and reprinted from my 1943 bulletin:

Six field-collected third-stage larvae and exuviae of 4 third-stage larvae reared to the adult stage (loaned by the U. S. National Museum).

Forty-three field collected third-stage larvae and exuviae of 6 third-stage larvae reared to the adult stage (donated by N. S. Franklin).

Maximum width of head capsule 2.5 to 2.7 mm. Exterior frontal setae present. Posterior frontal setae 2 to 3 on each side in a transverse row. Dorsoepicranial setae 3 to 6 on each side. Dorsoexterior region of mandible bare. Epipharynx with dexiophoba of 0 to 7 spines. Thoracic spiracles ranging from .28 to .34 mm in length and from .21 to .23 mm in width. Respiratory plate with a maximum of 9 to 12 "holes" in a transverse series. Bullar opening moderately to much constricted. Dorsa of seventh to ninth abdominal segments with most of setae confined to 2 transverse rows. Tenth abdominal segment with subcircular furrow present but vague. Raster with palidia slightly diverging posteriorly; septula oblong. Pali in each row 7 to 12 in number, slender, slightly depressed. Preseptular, hamate setae 16 to 26 in number. Lower anal lip bearing 43 to 84 long and short, straight setae.

A. lucicola is a medium-sized native species, closely related to *A. oblivia* in both the adult and larval stages. The species occurs in Connecticut, Indiana, Kentucky, Massachusetts, Michigan, New Jersey, and New York. In the latter state, Heit and Henry (1940) reported severe injury by the larvae to larch seedlings in a forest tree nursery.

According to H. J. Franklin, the species is occasionally very abundant locally on cranberry bogs in Massachusetts. Adult color ranges from straw to black. Venter and legs red-brown to black. Thorax with discal portion black and with marginal and caudal portions straw-colored. Adult 8 to 8.5 mm long and 4.5 to 5 mm wide.

Anomala marginata (Fab.)
(Fig. 250)

This description is based on the following material and reprinted from my 1943 bulletin:

Twelve field-collected larvae and exuviae of 16 third-stage larvae reared to the adult stage.

Maximum width of head capsule 3.6 to 3.9 mm. Exterior frontal setae present. Posterior frontal setae 1 to 2 on each side. Dorsoepicranial setae 4 to 5 on each side. Dorsoexterior region of mandible bare. Epipharynx with dexiophoba usually absent. Thoracic spiracle ranging from .44 to .52 mm in length and from .30 to .36 mm in width. Respiratory plate with a maximum of 21 to 27 "holes" in a transverse series. Bullar opening much constricted. Dorsa of seventh to ninth abdominal segments with 2 transverse rows of setae only. Tenth abdominal segment with subcircular furrow present. Raster with palidia slightly diverging posteriorly; septula oblong. Pali in each row 13 to 20 in number, slightly compressed. Preseptular, hamate setae 5 to 12 in number. Lower anal lip bearing 32 to 46 long and short, straight setae.

A. marginata is a large, very distinctive, native species found in the southeastern states. The larva is sometimes fairly common in steep, wild-grass pastureland in eastern Kentucky. Adult predominantly yellow to yellow-brown with a dorsal, greenish irridescence. Thorax red-brown dorsally, with straw-colored to yellowish lateral margins. Head red-brown. Adult 11 to 14 mm long and 6.5 to 8.5 mm wide. Flight period in July in Kentucky.

Anomala oblivia Horn

This description is based on the following material and reprinted from my 1943 bulletin:

Two reared second-stage larvae (donated by C. E. Heit) and 4 reared third-stage larvae (donated by the U. S. National Museum).

Maximum width of head capsule 2.4 to 2.5 mm. Exterior frontal setae present. Posterior frontal setae 2 to 3 on each side in a transverse group. Dorsoepicranial setae 3 to 5 on each side. Dorsoexterior region of mandible bare. Epipharynx with dexiophoba of 3 to 7 spines. Thoracic spiracle ranging from .22 to .24 mm in length and from .14 to .17 mm in width. Respiratory plate with a maximum of 8 to 9 "holes" in a transverse series. Bullar opening slightly constricted. Dorsa of seventh to ninth abdominal segments with most of setae confined to 2 transverse rows. Tenth abdominal segment with subcircular furrow. Raster with palidia slightly diverging posteriorly; septula oblong. Pali in each row 9 to 14 in number and slightly depressed.

Preseptular hamate setae 14 to 17 in number. Lower anal lip bearing 20 to 38 long and short, straight setae.

A. oblivia, or the pine chafer, as it is commonly called, is a native species closely related to *A. lucicola*. The writer has seen specimens from Massachusetts, Michigan, New York, and Kentucky. Adult with yellow-brown to reddish-brown elytra. Thorax and head yellow or red-brown to black with faint irridescense. Adult 7 to 9 mm long and 4 to 5 mm wide. The male genitalia appears to be identical with that of *P. lucicola*. The adults of the two species can be separated, however, by color differences and by differences in the tarsal claws. As pointed out by Horn (1884), adults of *A. oblivia* have the anterior claws of the front and middle leg cleft at the tip, very unequally on the front leg and very feebly on the middle. *A. lucicola* has the same claws deeply cleft at their tips with the two portions nearly equal. Larvae of the two species are also separable.

Genus *Popillia* Serv.
Popillia japonica Newman
(Fig. 249)

This description is based on the following material and reprinted from my 1943 bulletin:

Fifty field-collected third-stage larvae (donated by J. P. Johnson).

Maximum width of head capsule 2.9 to 3.1 mm. Exterior frontal setae present. Posterior frontal setae 3 on each side in a transverse row. Dorsoepicranial setae 5 to 7 on each side. Dorsoexterior mandibular region bare. Epipharynx with dexiophoba of 2 to 5 spines. Thoracic spiracle ranging from .31 to .34 mm in length and from .22 to .24 mm in width. Respiratory plate with a maximum of 11 to 14 "holes" in a transverse series. Bullar opening much constricted. Dorsa of seventh to ninth abdominal segments with most of setae confined to 2 transverse bands. Tenth abdominal segment with subcircular furrow present. Raster with short palidia strongly diverging posteriorly; septula shaped like an equilateral triangle. Pali in each row 5 to 8 in number, very long and slightly compressed. Preseptular, hamate setae 35 to 57 in number. Lower anal lip bearing 19 to 29 long and short, straight setae.

P. japonica, known as the Japanese beetle, was first found in this country in 1916. At present it is found in eastern Canada and in most of the states east of the Mississippi River, with local colonies in Missouri and California (Fleming, 1962). The adult feeds on the foliage and fruits of a great variety of plants, while the larvae cause severe damage to turf, vegetables, and nursery stock by feeding on the roots (Fleming and Metzger, 1936). Adult a lustrous, irridescent, metallic

green, with reddish-bronze elytra. Five tufts of white hairs are located on each side of the abdomen and 2 patches of white hairs are borne on the abdomen caudad of the elytra. Adult 10 to 12 mm long and 5.5 to 7 mm wide.

Genus *Strigoderma* Burmeister
Strigoderma arboricola (Fab.)
(Fig. 255)

This description is based on the following material and reprinted from my 1943 bulletin:

One reared third-stage larva and exuvia of one third-stage larva reared to the adult stage (loaned by the U. S. National Museum).

Eight field-collected larvae (donated by C. B. Dominick).

Maximum width of head capsule 2.8 to 3.1 mm. Exterior frontal setae present. Posterior frontal setae 2 to 4 on each side in a transverse group. Dorsoepicranial setae 7 to 9 on each side. Dorsoexterior region of mandible bare. Epipharynx with a dexiophoba of 2 to 8 spines. Labrum with a transverse crest consisting of about 6 rugosities. Thoracic spiracle ranging from .34 to .41 mm in length and from .23 to .29 mm in width. Respiratory plate with a maximum of 13 to 16 "holes" in a transverse series. Bullar opening moderately to much constricted. Dorsa of seventh to ninth abdominal segments with a sparse covering of medium-sized setae and each dorsum bearing in addition a transverse row of about 6 very long setae. Tenth abdominal segment with subcircular furrow absent. Raster with palidia slightly diverging posteriorly; septula oblong. Each palidium set with 5 to 12, fairly long, cylindrical pali. Preseptular, hamate setae 20 to 35 in number. Lower anal lip bearing 33 to 41 long and short, straight setae.

S. arboricola is a widely distributed, native species, known to occur in Kansas, Wisconsin, Illinois, Indiana, New York, New Jersey, Kentucky, North Carolina, and Virginia, and probably present in other adjacent states. In Illinois, the species is abundant in sandy areas near Havana, on the Illinois River.

Hayes (1921) gave an account of the life cycle of the species. A paper by Hoffman (1936) describes the biology and ecology of the species in Minnesota. There, adults are common in wild rose blossoms near sand dunes or in bogs. According to Miller (1943), larvae of this species cause severe damage to peanuts in Virginia by feeding on the pods and other underground parts of the plants. Injury was worst in low spots and in dark soils high in organic matter. Adults are in flight in eastern Virginia from May 15 to June 10, and cause great damage to rose blossoms. The beetle is also common on peanut foliage and blossoms.

The adult is elongate with blackish venter, legs, thorax, and head. Thorax and head hairy and with a greenish-purple irridescence. Elytra straw to yellowish-orange in color. Adult 8.5 to 12.5 mm long and 4.5 to 6.5 mm wide.

Genus *Strigodermella* Casey
Strigodermella pygmaea (Fab.), Third-Stage Larva
(Figs. 307 to 312)

This description is based on the following material and reprinted from my 1945a bulletin:

> Three, third-stage larvae and cast skin of a fourth third-stage larva reared to the adult stage. Larvae collected by R. J. Sim. Reared adult determined as this species by Chapin of the United States National Museum. One larva was presented by the Federal Japanese Beetle Laboratory of Moorestown, New Jersey, and the remainder were loaned by the United States National Museum.

Maximum width of head capsule of third-stage larva 1.7 mm. Surface of cranium smooth, light yellow-brown. Frons (Fig. 307) bearing on each side 3 posterior frontal setae, a single exterior frontal seta, a single anterior frontal seta, and a single anterior angle seta. Dorsoepicranial setae about 6 on each side. Labrum (Fig. 307) symmetrical, wider than long, with a proximal transverse row of about 10 setae. Haptomerum of epipharynx (Fig. 308) with a curved zygum and 3 prominent heli; epizygum absent. Plegmata present. No sensilla found among the setae of the chaetoparia. Dexiophoba composed of about 4 closely appressed spines. Haptolachus complete. Left mandible with a blade-like scissorial portion anterior to the scissorial notch and a small scissorial tooth posterior to the same notch. Lacinia of maxilla with 2 well-developed, terminal unci. Maxillary stridulatory area consisting of a row of 4 sharp-pointed recurved teeth and a small anterior process. Last antennal segment (Fig. 307) with a single oval, dorsal sensory spot.

Thoracic spiracle .1 to .136 mm long and .06 to .07 mm wide. Respiratory plate with no more than 7 oval "holes" along any diameter (Fig. 312). Lobes of respiratory plate not constricted. Spiracles of abdominal segments 1 to 6, inclusive, similar in size, much smaller than those of abdominal segments 7 and 8 which are similar in size. Dorsa of abdominal segments 7 to 9, inclusive, covered with slender setae; each dorsum posteriorly with about 6 long setae (among the other setae). Venter of abdominal segment 10 (Fig. 309) with palidia and tegilla. Each palidium consisting of a single sparsely set, longitudinal row of 4 to 6 long, slightly compressed, medianly directed pali. Palidia parallel, septula rectangular. Tegillar setae hamate with bent tips; 11 preseptular, hamate setae. Lower anal lip with a patch of large, hamate setae and a caudal fringe of about 25 long, slender setae. Claws each bearing 2

setae. Claws of the metathoracic legs (Fig. 311) much shorter than those of the prothoracic and mesothoracic legs (Fig. 310) which are falcate and about equal in size.

Strigodermella pygmaea is a small native species occurring in eastern United States. Specimens have been examined from New Jersey, Virginia, Maryland, North Carolina, and Florida. Adult with elytra ranging from a uniform straw color to brown or black with straw-colored patches. Thorax and head brown or black with a slight greenish-violet iridescence. Legs and venter red-brown to black. Adult 4.3 to 5.3 mm long and 2.3 to 2.9 mm wide. Practically nothing is known about the larvae except that they live in the soil. The species is common in sandy areas. Adults are in flight in June and July; often taken at lights or by sweeping low vegetation.

TRIBE Rutelini

Larvae of the tribe Rutelini may be characterized as follows: Stipes of maxilla with a dorsal row of 5 to 14 conical, sharp-pointed or recurved stridulatory teeth and an anterior process. Lacinia with a single terminal uncus or with 2 unci usually of similar size fused at their bases. Haptomerum of epipharynx moundlike or beak-like and/or with a group of 15 or more spine-like setae. Epipharynx with one or both nesia present (sclerotized plate absent in some genera). Proplegmata absent, plegmata present or absent. Dorsal surface of last antennal segment with 2 or more dorsal sensory spots. Raster with or without palidia; palidia, if present, polystichous.

Key to Some Genera and Species of the Tribe Rutelini, Based on Characters of the Larvae

1. Venter of last abdominal segment without palidia (Figs. 278 and 279).. 2
   Venter of last abdominal segment with polystichous palidia; palidia (and septula) extending longitudinally across the lower anal lip (Figs. 281 and 282) 6

2. Venter of last abdominal segment without teges or teges consisting of 2 or 3 very small, stout setae (Fig. 277) ....*Rutela formosa* Burm.
   Venter of last abdominal segment with well-developed teges of 30 or more setae 3

3. Head rather densely covered with setae (Fig. 259). Haptomerum of epipharynx with a beak-like process (Fig. 273). Maxillary stridulatory area with from 4 to 6 sharp teeth (Fig. 268). Claws at least moderately long (Figs. 265 and 266) 4
   Head sparsely covered with setae (Fig. 261). Haptomerum of epipharynx elevated but not beak-like (Figs. 272 and 276). Haptolachus incomplete; sclerotized plate absent (Figs. 272 and 276). Maxillary stridulatory area with from 8 to 14 sharp teeth (Fig. 267). Claws short (Fig. 270 and 271)................ 5

4. Labrum bearing 2 prominent, transverse, crescent-shaped ridges (Fig. 259). Spiracles of abdominal segments 7 and 8 larger than spiracles of abdominal segments 1 to 6 .... *Paracotalpa ursina* (Horn)
   Labrum without prominent ridges. Spiracles of abdominal segments 7 and 8 smaller than those of abdominal segments 1 to 6 ....*Cotalpa lanigera* (Linn.)

5. Labrum (and epipharynx) much wider than long. Epipharynx without epizygum (Fig. 276) .... *Pelidnota punctata* (Linn.)
   Labrum only slightly wider than long. Epipharynx with epizygum (Fig. 272) .... *Plusiotis*

6. Left mandible with 2 scissorial teeth. Epipharynx with plegmata present but rather inconspicuous. Venter of last abdominal segment with a prominent septula between the polystichous palidia, anterior to the lower anal lip (Fig. 281).............*Parastasia brevipes* (Lec.)
   Left mandible with 3 scissorial teeth (Fig. 2). Epipharynx without plegmata. Venter of last abdominal segment with septula vague between palidia, anterior to the lower anal lip (Fig. 282) ....*Macraspis lucida* (Ol)[24]

Genus *Paracotalpa* Ohaus

Paracotalpa ursina ursina (Horn), Third-Stage Larva
(Figs. 259, 263, 265, 266, 268, 273, and 278)

This description is based on the following material and reprinted from my 1948 paper:

Four, third-stage larvae and 5 cast skins of third-stage larvae collected February 20, 1946, above Tessla, California, in Corral Hollow, by William Barr, Ray Smith, J. W. McSwain, and the writer. The third-stage larvae were found from 6 to 12 inches deep in the soil beneath *Artemesia* bushes and 3 of the larvae were feeding on the larger roots. The cast larval skins were found in pupal cells from 8 to 10 inches deep in the soil. Seventeen adult beetles were collected at varying depths, ranging from close to the pupal cells to within 1 or 2 inches of the surface.

Three, third-stage larvae reared by the writer in August 1946, from eggs laid by the above adults. No. 46-7E.

Maximum width of head capsule of third-stage larva 4.8 to 5.8 mm. Surface of cranium smooth, yellowish-brown, and rather uniformly covered with numerous setae (Fig. 259). Clypeus with a prominent transverse ridge. Labrum wider than long and bearing 2 prominent, transverse, crescent-shaped ridges (Fig. 259). Epipharynx (Fig. 273) slightly wider than long. Haptomerum with a beak-like process behind which is a group of about 30 spine-like setae. Heli absent. Epizygum and zygum absent. Chaetoparia with almost no sensilla among the chaetae. Proplegmata and plegmata absent. Haptolachus complete with a small spine-like

[24] This species does not occur in the United States but is found in Mexico, Central America, and northern South America (Blackwelder, 1944).

sclerotized plate adjacent to the dexiotorma and 2 median sense cones. Each mandible (Fig. 263) with a blade-like portion anterior to the scissorial notch and a single, small, scissorial tooth posterior to the notch. Lacinia of maxilla with 2 unci which are alike in size and shape. Maxillary stridulatory area consisting of 5 or 6 sharp-pointed, anteriorly directed, recurved teeth and a prominent distal, truncate process (Fig. 268). Last antennal segment with 3 to 5 round to oval, dorsal sensory spots (Fig. 259).

Thoracic spiracles ranging from .41 to .52 mm in length and from .267 to .28 mm in width. Respiratory plate with a maximum of about 17 to 22 round to oval "holes" along any diameter. "Holes" not in definite rows. Distance between the two lobes of the respiratory plate equal to or only slightly less than the dorsoventral diameter of the bulla. Thoracic spiracles and spiracles of abdominal segments 2 to 6 inclusive very similar in size; spiracles of first abdominal segment noticeably smaller, those of abdominal segments 7 and 8 noticeably larger.

Dorsum of seventh abdominal segment with 2 vague annulets, a prescutum and a scutum. Scutum with a dense covering of short, stout setae among which is a posterior transverse row of long setae. Prescutum with a sparse transverse band of long and short setae. Dorsa of abdominal segments 8 and 9 each with 2 well-separated, sparsely set, transverse bands of long and short setae. Dorsum of tenth abdominal segment with a rather conspicuous sclerotized, dorsal, impressed line. Dorsal impressed line emarginate anteriorly. Venter of tenth abdominal segment without palidia (Fig. 278). Teges consisting of a transverse patch of from 34 to 38 fairly short hamate setae with curved tips. Lower anal lip covered with similar hamate setae and with a caudal fringe of 17 to 38 long, cylindrical setae. Claws (Figs. 265 and 266) sharp-pointed, each bearing 2 setae. Claws of metathoracic legs slightly shorter than those of mesothoracic legs; claws of mesothoracic legs shorter than those of prothoracic legs.

The adult of *Paracotalpa ursina* (Horn), commonly called the little bear, is a robust, steel blue or greenish black beetle with reddish elytra, steel blue pronotum, and a body covered with long yellow hairs (Essig, 1958). According to the same writer, the species flies on dull days and is often very numerous in southern California. Larvae have been found feeding on roots of sage brush (*Artemesia*) in central California.

Genus *Plusiotis* Burmeister

Plusiotis woodi Horn, Third-Stage Larva
(Figs. 261, 262, 264, 267, 270-272, and 279)

This description is based on the following material and reprinted from my 1948 paper:

Two third-stage larvae reared from 2 second-stage larvae collected January 29, 1946, in the Davis Mountains, north of Ft. Davis, Texas, by the writer in an area suggested by H. A. Scullen of Oregon State University. The larvae were dug from rich soil and debris at the base of a walnut tree in a stream bed. Remains of 10 adults were found in the surface debris.

Maximum width of head capsule of third-stage larva 6.95 to 7.25 mm. Surface of cranium smooth, reddish-brown; labrium, clypeus, and anterior half of frons reddish-black. Frons (Fig. 261) bearing, on each side, a patch of 4 or 5 posterior frontal setae, a single exterior frontal seta, 3 or 4 anterior frontal setae, and a single seta in each anterior frontal angle. Dorsepicranial setae 3 or 4 on each side. Labrum (Fig. 261) slightly wider than long, nearly symmetrical. Epipharynx (Fig. 272) with an epizygum and a raised haptomeral process behind which is a group of about 30 stout, spine-like setae. Chaetoparia with a few sensilla among the chaetae. Acanthoparia poorly developed. Plegmata and proplegmata absent. Haptolachus incomplete. Sclerotized plate absent. Left mandible (Fig. 262) with 2 scissorial teeth anterior to the scissorial notch and a single scissorial tooth posterior to the same notch. Right mandible with only a single scissorial tooth anterior to the scissorial notch. Lacinia of maxilla (Fig. 264) with 2 terminal unci which are fused together at their bases; dorsal uncus with a single seta. Maxillary stridulatory area (Fig. 267) consisting of a row of 8 or 9 sharp-pointed teeth and a wide, anterior, truncate process. Last antennal segment (Fig. 261) with 4 or 5 oral, dorsal sensory spots.

Thoracic spiracles ranging from .8 to .86 mm in length and .56 to .62 mm in width. Respiratory plate with a maximum of about 32 oval to round "holes" along any diameter. Distance between the two lobes of the respiratory plate much less than the dorsoventral diameter of the bulla. Spiracles of abdominal segments 1 to 8 similar in size. Thoracic spiracles distinctly larger than any abdominal spiracles.

Dorsum of seventh abdominal segment with two vaguely defined annulets, a prescutum and a scutum. Prescutum with a transverse lens-shaped patch of short, stout setae interspersed caudally with a few long setae. Scutum with a sparsely set transverse row of long and short setae anterior to which are a few short, stout setae. Dorsa of abdominal segments 8 and 9 each with 2 widely separated, sparsely set transverse rows of long setae among which are a very few short setae. Dorsum of tenth abdominal segment on each side with a sparse covering of long setae and short stout setae; median longitudinal area bare. Dorsal impressed line absent. Venter of tenth abdominal segment (Fig. 279) without palidia. Teges consisting of a transverse patch of 47 to 55 fairly short, stout setae. Lower anal lip clothed anteriorly with similar setae, posteriorly with a fringe of 62 to 71 stout cylindrical setae. Claws (Figs. 270 and 271) short, each with 2 setae.

Adults of *Plusiotis woodi* have smooth, pale green elytra and green and purple legs. They have been taken in numbers in the Davis Mountains, north of Ft. Davis, Texas, by H. A. Scullen, where they were feeding on walnut.

The writer has examined many larvae of *Plusiotis gloriosa* but is unable to distinguish them as yet from larvae of *P. woodi*. *P. gloriosa* is very common in the vicinity of the Southwest Research Station of the American Museum of Natural History, near Portal, Arizona. The larvae are abundant in decaying sycamore logs. Pupation occurs in the soil, in June.

Genus *Rutela* Latreille
Rutela formosa Burm., Third-Stage Larva
(Figs. 277, 302, and 304)

This description is based on the following material:

Two, third-stage larvae associated with 4 pupae and an adult, Miami Beach, Florida (?), Coconut Grove. Hopkins U. S., 14970, Leng 259, adult identified by E. A. Chapin (loaned by the U. S. National Museum).

Two, third-stage larvae collected from old stump at Cayamas, Cuba, February 1904, by E. A. Schwarz.

Maximum width of head capsule 3.8 to 4.7 mm. Surface of cranium smooth, but finely reticulated, light yellow-brown. Frons, on each side, with a single large, anterior frontal setae (plus 2 or 3 smaller setae), a single large seta in each anterior frontal angle, 1 or 2 exterior frontal setae, and 2 or 3 posterior frontal setae arranged in a transverse row. Dorsoepicranial setae 2 to 3 on each side. Eye spots absent. Labrum symmetrical, wider than long; anterior margin broadly rounded. Epipharynx (Fig. 302) with a small, raised, haptomeral process behind which is a group of about 25 stout setae. Epizygum, proplegmatia, and plegmatia all absent. Chaetoparia well developed, but with no sensillae among the chaetae. Haptolachus incomplete; sclerotized plate (nesium externum) absent; sensory cone (nesium internum) poorly developed, with 4 sensory pits. Left mandible with 2 scissorial teeth anterior to the scissorial notch and a single tooth posterior to the same notch. Scissorial area unusually short; not as long as the molar area. Left mandible with an irregular, longitudinal row of 7 setae. Maxilla with vestigial lacinial unci, represented by a small sclerotized area bearing a single small seta. Stipes with a row of 8 to 10 sharp-pointed, conical stridulatory teeth and an anterior truncate process. Last antennal segment with 3 oval, dorsal sensory spots.

Thoracic spiracles ranging from 1.4 to 1.8 mm in length and .7 to 1.2 mm in width. Respiratory plate with a maximum of 17 to 18 "holes" along any diameter. "Holes" irregular oval in outline, often with 1 to 4 inwardly projecting points, not in definite rows. Distance between the two lobes of the respiratory plate less than the dorsoventral diameter of the bulla. Respiratory plates of abdominal spiracles almost surrounding bullae. Spiracles on segments 1 to 6 similar in size, smaller than prothoracic spiracles and those on abdominal segments 7 and 8.

Dorsum of seventh abdominal segment not divided into annulets, anteriorly with a few short, stout setae among which are a few very long setae. Dorsa of abdominal segments 8 and 9, each with 2 widely separated, very sparsely set, transverse rows of long setae. Dorsum of abdominal segment 10 almost bare. Dorsal impressed line absent. Venter of tenth abdominal segment without palidia (Fig. 277). Teges absent or at most consisting of 2 or 3 very small, stout, setae, similar to those covering the lower anal lip. Hamate setae absent. Anal slit curved. Claws poorly developed, terminal portion absent. Claws of prothoracic and mesothoracic legs with 2 setae; claws of metathoracic leg with 2 or 3 setae. One seta near apex of each claw.

The adult of *Rutela formosa* is a brightly colored, shiny metallic green and yellow beetle. This West Indian species has apparently become established in Florida. The larvae are found in decaying wood.

Genus *Cotalpa* Burmeister

Cotalpa lanigera (Linn.)., Third-Stage Larva
(Figs. 283, 286, 290, 292-294, 296, 301, and 305)

This description is based on the following material and reprinted from my 1945a bulletin:

One, third-stage larva and cast skin of one third-stage larva reared to the pupal stage. Both larvae collected at Onekama, Michigan, by William P. Hayes. Loaned by William P. Hayes of the University of Illinois.

One, third-stage larva, collected May 26, 1943, in Breathitt County, Kentucky, by P. O. Ritcher, from sandy soil in bottomland, behind the plow.

Maximum width of head capsule of third-stage larva 6.2 to 6.96 mm. Surface of cranium slightly roughened, especially rugose on the anterior part of the frons. Cranium a light red-brown. Anterior half of frons bearing many setae, making it difficult to distinguish primary setae (Fig. 290). Posterior and exterior frontal setae consisting of a patch of about 20 to 25 setae, on each side, which is confluent with the patch of about 6 to 8 setae in the adjacent anterior angle and transverse patch of about 8 to 15 anterior frontal setae, with about 12 dorsoepicranial setae on each side. Setae numerous laterad of the frontal sutures and behind the antennal support. Epipharynx (Fig. 301) almost as long as wide and with a beak-like haptomeral process behind which are borne a number of spine-like setae similar to those of the chaetoparia. Chaetoparia strongly developed with few or no sensilla among the setae. Proplegmata and plegmata absent. Haptolachus with a very small sclerotized plate and well-developed, bilobed sense cone. Each mandible (Fig. 292) with a blade-like portion anterior to the scissorial notch and a single, small, scissorial tooth posterior to the notch. Lacinia of maxilla with 2 terminal unci which are alike in size and shape (Figs. 293 and 294). Maxillary stridulatory area (Fig. 296) consisting of a row of 4 to 5 large, sharp-pointed, recurved teeth and a wide, truncate anterior process. Last antennal segment with 4 to 8 oval, dorsal sensory spots (Fig. 290).

Thoracic spiracles ranging from .72 to .82 mm in length and from .5 to .54 mm in width. Respiratory plate with a maximum of about 25 to 35 oval "holes" along any diameter. "Holes" not in definite rows. Distance between the two lobes of the respiratory plate somewhat less than the dorsoventral diameter of the bulla. Spiracles of abdominal segments 1 to 4 inclusive, similar in size; those of abdominal segment 5 smaller, those of abdominal segments 6 to 8 inclusive, very much smaller.

Dorsum of abdominal segment 7 divided into 2 annulets. Each annulet with a narrow transverse band of setae, the anterior band consisting of a number of short setae and a caudal row of longer setae, the posterior band consisting of a number of long setae interspersed with short setae. Dorsum of abdominal segment 8 with 2 widely separated, sparsely-set, transverse rows of setae, most of which are long. Dorsum of abdominal segment 9 similarly clothed except that the posterior group of setae is about 3 rows wide. Dorsal impressed line present on abdominal segment 10. Venter of abdominal segment 10 (Fig. 305) without palidia. Teges consisting of a broad patch of moderately short, hamate setae with curved tips, which covers the caudal half of the area between the lower anal lip and the caudal margin of the abdominal segment (Fig. 305). Lower anal lip covered with similar hamate setae and with a caudal fringe of 18 to 35 long, cylindrical setae. Claws (Figs. 283 and 286) falcate, sharp-pointed, each with 2 proximal setae.

Cotalpa lanigera, called the goldsmith beetle, is a large light-colored species occurring in Canada and from Maine to Florida westward through Minnesota, Iowa, and eastern Kansas (Leng, 1920; Saylor, 1940; and Hayes, 1925). Adult oval in shape, straw-colored to yellow with yellow-red legs. Surface of pronotum, head, legs, and venter with an irridescent green luster.

Venter with white, wool-like hairs. Adult 20.5 to 23 mm long and 12.5 to 13 mm wide. According to Hayes (1925) and Saunders (1879), the adults are active at night and frequently attracted to lights. In the daytime the beetles remain inactive on their food plants, such as willow and cottonwood trees, from which they may sometimes be beaten in large numbers. Saunders (1879) found adults feeding on the foliage of pear, oak, poplar, hickory, silver maple, and sweet gum trees. Hayes (1925) reports them in flight from May 24 to July 22 at Manhattan, Kansas. Larvae live in sandy soil and sometimes cause severe injury to raspberry bushes, strawberries, corn, and grasses (Davis, 1916).

Genus *Parastasia* Westwood
Parastasia brevipes (Lec.)., Third-Stage Larva
(Figs. 281, 285, 289, 291, 298, 300, and 303)

This description is based on the following material and reprinted from my 1945a bulletin:

Five cast skins of 5 third-stage larvae. Skins found with 5 pupae, July 6, 1940, near Lexington, Kentucky, by P. O. Ritcher, in cells in the dead heart of a large chinquapin oak which blew down in a storm several days before. Reared to the adult stage.

Twenty-five, third-stage larvae (also 11 second-stage and 23 first-stage larvae) collected February 25, 1940, 8 miles north of Lexington, Kentucky, by W. S. Webb and P. O. Ritcher, in burrows in a large, decaying sycamore log. Associated with reared adults.

Maximum width of head capsule of third-stage larva 4.6 to 4.9 mm. Surface of cranium rather smooth, light yellow-brown, with reddish-brown clypeus and labrum and reddish-black mandibles. Frons (Fig. 291) bearing, on each side, 1 or 2 posterior frontal setae, a single exterior frontal seta, a single anterior frontal seta, and a single anterior angle seta. Dorsoepicranial setae 4 to 7 on each side. Labrum symmetrical, slightly wider than long. Haptomerum of epipharynx (Fig. 300) raised, with a curved zygum behind which are borne 25 to 30 stout, spine-like setae. Epizygum absent. Chaetoparia with no sensilla among the setae. Plegmata present but rather inconspicuous. Proplegmata absent. Haptolachus complete. Both mandibles (Fig. 289) with only 2 scissorial teeth, one anterior to and one posterior to the scissorial notch. Lacinia of maxilla (Fig. 295) with a single terminal uncus. Galea with a ventral row of 4 or 5 very large, stout setae posterior to the terminal uncus. Maxillary stridulatory area (Fig. 298) consisting of a row of 6 to 9 sharp-pointed teeth and a wide, anterior truncate process. Last antennal segment (Fig. 285) with 2 to 4 oval, dorsal sensory spots.

Thoracic spiracles ranging from .58 to .74 mm in length and from .44 to .54 mm in width. Respiratory plate with a maximum of about 20 to 25 oval to round, irregularly margined "holes" along any diameter. Distance between the two lobes of the respiratory plate much less than the dorsoventral diameter of the bulla. Spiracles of abdominal segments 1 to 6 inclusive, similar in size, noticeably smaller than those of abdominal segments 7 and 8 which are similar in size.

Dorsum of abdominal segment 7 (Fig. 303) with 2 vague annulets, the prescutum and the scutum. Prescutum with a rather dense patch of short, stout setae and a caudal, sparsely set row of long setae. Scutum with a sparsely set row of long setae anterior on which, on the median part of the scutum, are a few, short, stout setae. A number of short, stout setae are borne laterally on the scutum. Dorsa of abdominal segments 8 and 9 each with 2 widely separated, sparsely set, transverse patches of long and short, slender setae. Venter of abdominal segment 10 (Fig. 281) with palidia. Each palidium composed of a longitudinal patch of medianly directed, spine-like setae with the setae along the inner edge of each patch arranged in a fairly definite row. Palidia (and septula) extending across the lower anal lip and forward to the mid-point of the segment. Lower anal lip with about 15 very long, medianly directed setae. Claws short, each bearing 2 setae.

Parastasia brevipes is a fairly large, wood-inhabiting species which occurs from New York to Florida and westward to Kansas (Blackwelder, 1939; Leng, 1920; and Hayes, 1929). Adult elongate-oval with reddish-black elytra, body, and legs. Pronotum and head blackish. Anterior margin of clypeus with 2 teeth. As Casey (1915) has pointed out, the species has dynastid affinities with the adult resembling *Ligyrus* in general appearance. Adult 16 to 18 mm long and 8.5 to 10 mm wide. Larvae burrow and feed in dead wood. Near Lexington, Kentucky, 5 pupae were found in tunnels in the dead heart of a giant chinquapin oak and a great many larvae were found in the harder parts of a large, decaying sycamore log. Adults appear in July and early August. The winter is passed in the larval stage and all three instars may be found together in the same log. The species probably has a two-year life cycle.

Genus *Pelidnota* MacLeay
Pelidnota punctata (Linn.)., Third-Stage Larva
(Figs. 276, 284, 287, 288, 297, 299, and 306)

This description is based on the following material and reprinted from my 1945a bulletin:

Three, third-stage larvae and 2 cast skins of 2 third-stage larvae (1 associated with adult and one with pupa). All 5 larvae collected November 30,

1942, at Lexington, Kentucky, by P. O. Ritcher, from burrows in the decayed base of a soft maple tree. No. 42-6C.

One, third-stage larva and 1 cast skin of third-stage larva reared to the adult stage. The 2 larvae collected February 28, 1938, at Morganfield, Kentucky, by Hans Shacklette, from the decaying roots of an apple tree. No. 38-36.

Three, third-stage larvae collected March 26, 1938, at Lexington, Kentucky, by P. O. Ritcher.

One cast skin of third-stage larva reared to the adult stage. Larva collected May 18, 1943, at Lexington, Kentucky, by Claude Jones from a pile of decaying wood shavings. No. 43-2M.

Maximum width of head capsule of third-stage larva 6.1 to 6.9 mm. Surface of cranium smooth, yellowish-brown, with reddish-black clypeus and labrum and reddish-black to black mandibles. Frons bearing, on each side, 4 to 7 posterior frontal setae, 1 to 3 exterior frontal setae, 3 to 4 anterior frontal setae, and 2 anterior angle setae. Dorsoepicranial setae 5 to 7 on each side. Labrum (and epipharynx) much wider than long. Haptomerum of epipharynx (Fig. 276) with a curved zygum behind which are about 20 stout, spine-like setae. Epizygum absent. Chaetoparia with no sensilla among the setae. Proplegmata and plegmata absent. Haptolachus incomplete; sclerotized plate absent. Left mandible (Fig. 288) with 2 scissorial teeth anterior to the scissorial notch and a single scissorial tooth posterior to the notch. Lacinia of maxilla (Fig. 297) with 2 terminal unci which are fused at their bases. Maxillary stridulatory area (Fig. 299) consisting of a row of 8 to 14 sharp-pointed teeth and a wide, anterior, truncate process. Last antennal segment with 4 to 5 oval, dorsal sensory spots.

Thoracic spiracles range from .84 to .94 mm in length and from .61 to .68 mm in width. Respiratory plate with a maximum of about 35 oval to round "holes" along any diameter. Distance between the two lobes of the respiratory plate much less than the dorsoventral diameter of the bulla. Spiracles of abdominal segments 1 to 6 inclusive, similar in size, noticeably smaller than those of abdominal segments 7 and 8 which are similar in size.

Dorsum of abdominal segment 7 anteriorly with a transverse patch of very short, stout setae and a caudal sparsely set row of long setae; posteriorly with a sparsely set, transverse row of long, slender setae. Dorsa of abdominal segments 8 and 9 each with 2 widely separated, sparsely set rows of long and short, slender setae. Dorsal impressed line absent on abdominal segment 10. Venter of abdominal segment 10 (Fig. 306) without palidia. Teges consist of 9 to 21 sparsely set, short, curved, hamate setae, interspersed with a very few, long, cylindrical setae. Lower anal lip anteriorly with a patch of similar hamate setae, posteriorly with a caudal fringe of 44 to 52 long, cylindrical setae. Claws short, each with 2 setae.

Pelidnota punctata, called the spotted pelidnota, is a large, rather common, native species occurring from Iowa to Connecticut and southward through Kansas, Kentucky, and Virginia (Leng, 1920; Hayes, 1925; and Blatchley, 1910). Adult oval with light brown to reddish-brown elytra and pronotum. Pronotum with a black spot on each side and each elytron with 3 lateral black spots. Venter and legs dark reddish-black with violet-green iridescence. Adult 19 to 26 mm long and 10 to 13.5 mm wide. Larvae feed in decaying wood or in soil in the vicinity of decaying wood. Larvae often found in decaying stumps or under rotten logs, sometimes in the decayed bases of standing trees. Taken from soft maple, oak, hackberry, and apple wood in Kentucky. Hayes (1925) found larvae in elm wood. J. Hoffman (1936), in Kansas, in January and March of 1933, collected 80 larvae from the rotten wood of elm, sycamore, and walnut stumps; most of them were found in elm. A few were found feeding in the center of old stumps but most were in tunnels in the large, lateral roots, 6 to 8 inches below the surface of the soil.

Adults are in flight, at Lexington, Kentucky, from June 27 to August 13; often taken at lights. The adults feed on the foliage of cultivated and wild grapes. The winter is passed in the larval stage and the species appears to have a two-year life cycle.

Plate XXI

Anomala innuba

FIGURE 236.   Left mandible, dorsal view.

FIGURE 237.   Right mandible, dorsal view.

FIGURE 238.   Third-stage larva, left, lateral view.

FIGURE 239.   Left mandible, lateral view.

FIGURE 240.   Right mandible, ventral view.

FIGURE 241.   Left mandible, ventral view.

FIGURE 242.   Portion of mandibular stridulatory area.

FIGURE 243.   Head, dorsal view.

FIGURE 244.   Venter of tenth abdominal segment.

FIGURE 245.   Palus, ventral view.

FIGURE 246.   Palus, lateral view.

FIGURE 247.   Abdominal spiracle.

FIGURE 248.   Epipharynx.

Symbols Used

A—Antenna
AA—Anterior frontal angle
AC—Acia
ACP—Acanthoparia
ACR—Acroparia
AFS—Anterior frontal seta
ASP—Asperities
B—Barbula
BLS—Basolateral setae
BR—Brustia
BU—Bulla
C—Campus
CA—Calx
CO—Corypha
CPA—Chaetoparia
CR—Crepis
CS—Clypeo-frontal suture
DC—Dorsal carina
DEMR—Dorsoexterior mandibular region
DES—Dorsoepicranial setae
DMS—Dorsomolar setae

DX—Dexiotorma
E—Epicranium
EPT—Epitorma
ES—Epicranial suture
EZ—Epizygum
F—Frons
FS—Frontal suture
GP—Gymnoparia
H—Helus
HL—Haptolachus
HM—Haptomerum
HS—Hamate setae
L—Labrum
LAL—Lower anal lip
LT—Laeotorma
M—Mandible
M_{1-3}—Molar lobes
MO—Mola bearing part
P—Palus
PA—Preartis
PC—Preclypeus

PCL—Precoila
PE—Pedium
PFS—Posterior frontal setae
PHS—Preseptular hamate setae
PL—Plegmatium
PLA—Palidia
PSC—Postclypeus
PTA—Postartis
PTL—Pternotorma
RSP—Respiratory plate
S—Septula
SA—Scissorial area
SC—Sense cone
SCR—Scrobe
SF—Subcircular furrow
SP—Sclerotized plate
SS—Spiracular slit
STA—Stridulatory area
T—Tegilla
VP—Ventral process
Z—Zygum

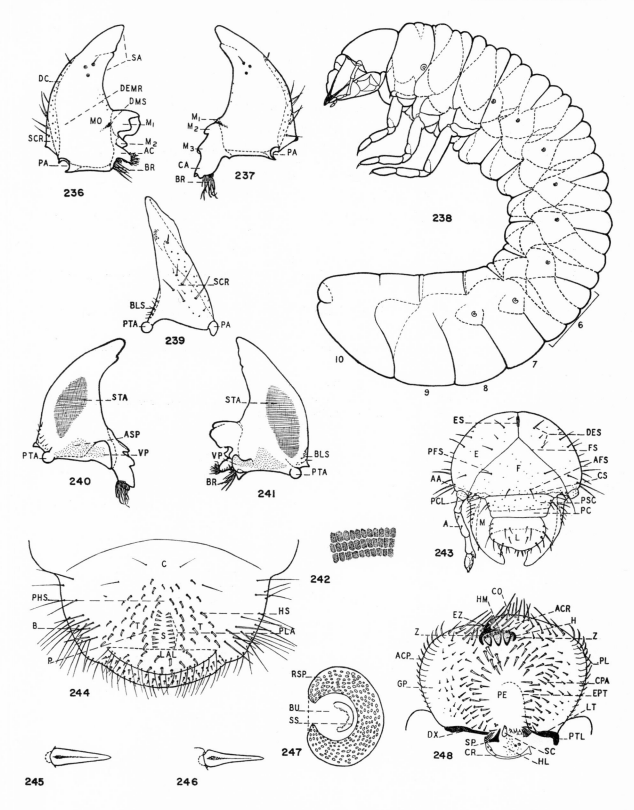

Plate XXII

FIGURE 249. *Popillia japonica* Newm. Venter of tenth abdominal segment.

FIGURE 250. *Anomala marginata* (Fab.). Epipharynx.

FIGURE 251. *Anomala orientalis* Wtrh. Venter of tenth abdominal segment.

FIGURE 252. *Anomala binotata* Gyll. Epipharynx.

FIGURE 253. *Anomala flavipennis* Burm. Venter of tenth abdominal segment.

FIGURE 254. *Anomala flavipennis* Burm. Venter view (upper figure) and lateral view (lower figure) of palus.

FIGURE 255. *Strigoderma arboricola* (Fab.). Head, dorsal view.

FIGURE 256. *Anomala flavipennis* Burm. Dorsal view of eighth, ninth, and tenth abdominal segments.

FIGURE 257. *Anomala lucicola* (Fab.). Lateral view of seventh to tenth abdominal segments.

FIGURE 258. *Anomala orientalis* Wtrh. Lateral view of seventh to tenth abdominal segments.

Symbols Used

AFS—Anterior frontal seta   DP—Dexiophoba   PFS—Posterior frontal setae
DES—Dorsoepicranial setae   EFS—Exterior frontal seta

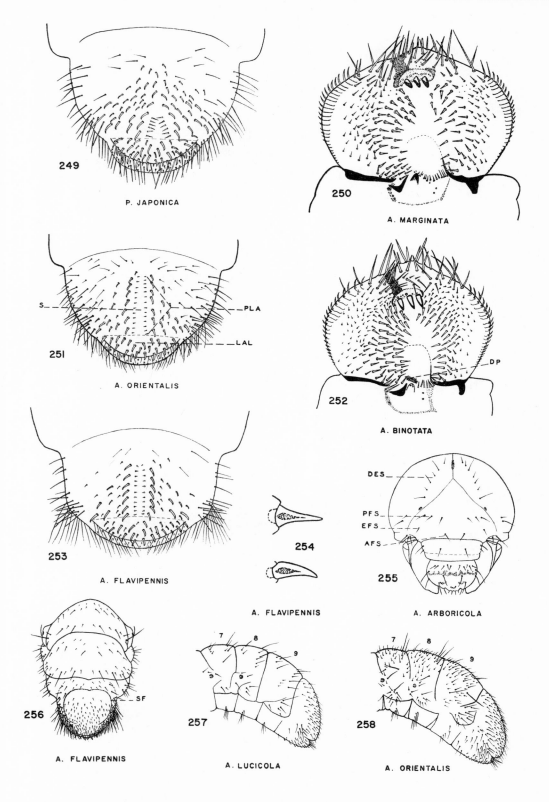

249 P. JAPONICA

250 A. MARGINATA

251 A. ORIENTALIS

S — — — PLA

— LAL

252 A. BINOTATA

DP

253 A. FLAVIPENNIS

254 A. FLAVIPENNIS

255 A. ARBORICOLA

DES — —
PFS — —
EFS — —
AFS — —

256 A. FLAVIPENNIS

SF

257 A. LUCICOLA

7   8   9

258 A. ORIENTALIS

7   8   9

Plate XXIII

FIGURE 259.   *Paracotalpa ursina ursina* (Horn). Head, dorsal view.

FIGURE 260.   *Macraspis lucida* (Ol.). Scissorial area of left mandible, dorsal view.

FIGURE 261.   *Plusiotis woodi* Horn. Head, dorsal view.

FIGURE 262.   *Plusiotis woodi* Horn. Scissorial area of left mandible, dorsal view.

FIGURE 263.   *Paracotalpa ursina ursina* (Horn). Scissorial area of left mandible, dorsal view.

FIGURE 264.   *Plusiotis woodi* Horn. Distal part of left maxilla.

FIGURE 265.   *Paracotalpa ursina ursina* (Horn). Claw of left mesothoracic leg.

FIGURE 266.   *Paracotalpa ursina ursina* (Horn). Claw of left metathoracic leg.

FIGURE 267.   *Plusiotis woodi* Horn. Maxillary stridulatory area.

FIGURE 268.   *Paracotalpa ursina ursina* (Horn). Maxillary stridulatory area.

FIGURE 269.   *Adoretus sinicus* Burm. Distal part of left maxilla.

FIGURE 270.   *Plusiotis woodi* Horn. Claw of left metathoracic leg.

FIGURE 271.   *Plusiotis woodi* Horn. Claw of left mesothoracic leg.

FIGURE 272.   *Plusiotis woodi* Horn. Epipharynx.

FIGURE 273.   *Paracotalpa ursina ursina* (Horn). Epipharynx.

Symbols Used

A—Antenna	EFS—Exerior frontal setae	LT—Laeotorma
ACP—Acanthoparia	ES—Epicranial suture	LU—Unci of lacinia
AFS—Anterior frontal setae	EZ—Epizygum	PE—Pedium
CPA—Chaetoparia	F—Frons	PFS—Posterior frontal setae
CR—Crepis	FS—Frontal suture	PTT—Pternotorma
DES—Dorsoepicranial setae	GU—Uncus of galea	SC—Sensory cone
DX—Dexiotorma	H—Haptomerum	SP—Sclerotized plate
E—Epicranium	L—Labrum	

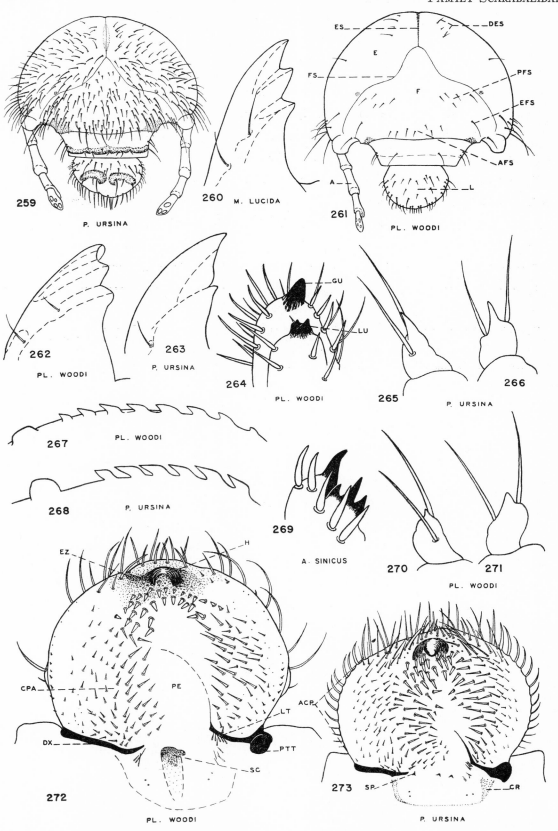

259 P. URSINA

260 M. LUCIDA

261 PL. WOODI

262 PL. WOODI

263 P. URSINA

264 PL. WOODI

265 P. URSINA

266

267 PL. WOODI

268 P. URSINA

269 A. SINICUS

270   271 PL. WOODI

272 PL. WOODI

273 P. URSINA

Plate XXIV

FIGURE 274.   *Adoretus sinicus* Burm. Epipharynx.

FIGURE 275.   *Adoretus sinicus* Burm. Last antennal segment, dorsal view.

FIGURE 276.   *Pelidnota punctata* (Linn.). Epipharynx.

FIGURE 277.   *Rutela formosa* Burm. Venter of last abdominal segment.

FIGURE 278.   *Paracotalpa ursina ursina* (Horn). Venter of last abdominal segment.

FIGURE 279.   *Plusiotis woodi* Horn. Venter of last abdominal segment.

FIGURE 280.   *Adoretus sinicus* Burm. Venter of last abdominal segment.

FIGURE 281.   *Parastasia brevipes* (Lec.). Venter of last abdominal segment.

FIGURE 282.   *Macraspis lucida* (Ol.). Part of venter of last abdominal segment.

Symbols Used

DSS—Dorsal sensory spot
DX—Dexiotorma
HE—Helus
LAL—Lower anal lip

LT—Laeotorma
PE—Pedium
PL—Plegmatium
PLA—Palidium

S—Septula
SP—Sclerotized plate
T—Teges

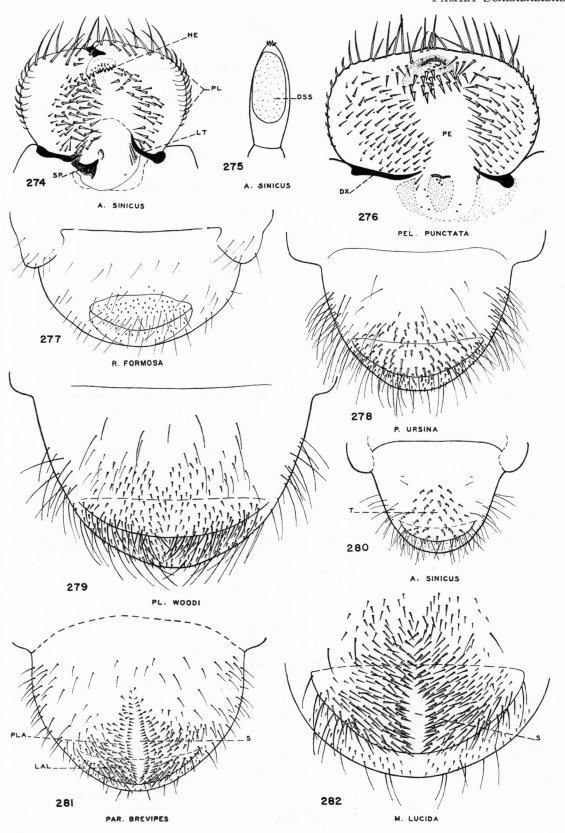

274 A. SINICUS

275 A. SINICUS

276 PEL. PUNCTATA

277 R. FORMOSA

278 P. URSINA

279 PL. WOODI

280 A. SINICUS

281 PAR. BREVIPES

282 M. LUCIDA

Plate XXV

FIGURE 283. *Cotalpa lanigera* (Linn.). Claw of left mesothoracic leg, side view.

FIGURE 284. *Pelidnota punctata* (Linn.). Claw of left mesothoracic leg, side view.

FIGURE 285. *Parastasia brevipes* (Lec.). Last antennal segment, dorsal view.

FIGURE 286. *Cotalpa lanigera* (Linn.). Claw of left mesothoracic leg, dorsal view.

FIGURE 287. *Pelidnota punctata* (Linn.). Claw of left mesothoracic leg, dorsal view.

FIGURE 288. *Pelidnota punctata* (Linn.). Scissorial area of left mandible.

FIGURE 289. *Parastasia brevipes* (Lec.). Scissorial area of left mandible.

FIGURE 290. *Cotalpa lanigera* (Linn.). Head, dorsal view.

FIGURE 291. *Parastasia brevipes* (Lec.). Head, dorsal view.

FIGURE 292. *Cotalpa lanigera* (Linn.). Scissorial area of left mandible.

FIGURE 293. *Cotalpa lanigera* (Linn.). Left maxilla, ental view.

FIGURE 294. *Cotalpa lanigera* (Linn.). Distal portion of left maxilla showing unci.

FIGURE 295. *Parastasia brevipes* (Lec.). Distal portion of right maxilla showing unci.

FIGURE 296. *Cotalpa lanigera* (Linn.). Maxillary stridulatory area, much enlarged.

FIGURE 297. *Pelidnota punctata* (Linn.). Distal portion of right maxilla showing unci.

FIGURE 298. *Parastasia brevipes* (Lec.). Maxillary stridulatory area, much enlarged.

FIGURE 299. *Pelidnota punctata* (Linn.). Maxillary stridulatory area, much enlarged.

FIGURE 300. *Parastasia brevipes* (Lec.). Epipharynx.

FIGURE 301. *Cotalpa lanigera* (Linn.). Epipharynx.

FIGURE 302. *Rutela formosa* Burm. Epipharynx.

Symbols Used

AA—Setae of anterior frontal angle
ACP—Acanthoparia
ACS—Anterior clypeal seta
AFS—Anterior frontal seta
CPA—Chaetoparia
CR—Crepis
DES—Dorsoepicranial setae
DSS—Dorsal sensory spots
DX—Dexiotorma
E—Epicranium
ECS—Exterior clypeal setae

EFS—Exterior frontal seta
ES—Epicranial stem
EZ—Epizygum
F—Frons
FS—Frontal suture
SP—Sclerotized plate
GP—Gymnoparia
GU—Uncus of galea
H—Haptomerum
HL—Haptolachus
L—Labrum
LT—Laeotorma

MU—Unci of lacinia
PC—Preclypeus
PE—Pedium
PFS—Posterior frontal seta
PL—Plegmatium
PSC—Postclypeus
PTT—Pternotorma
S_{1-n}—Scissorial teeth
SC—Sense cone
SN—Scissorial notch
Z—Zygum

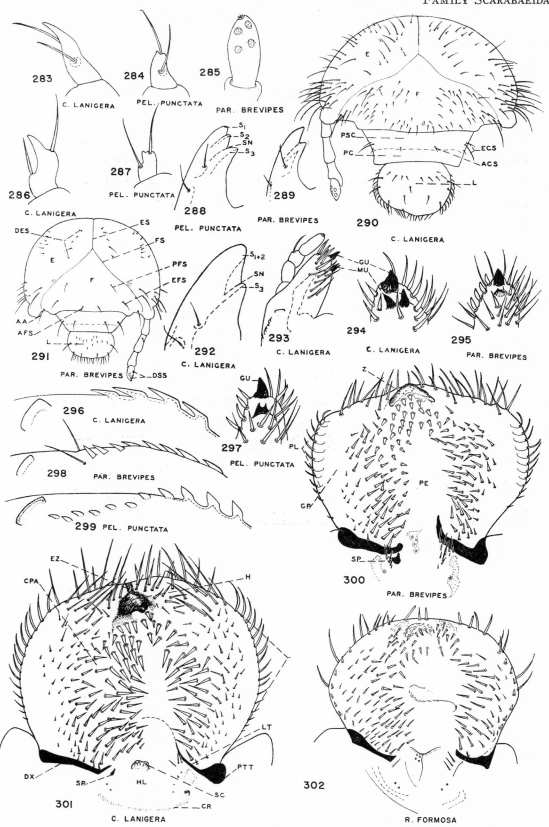

283 C. LANIGERA

284 PEL. PUNCTATA

285 PAR. BREVIPES

286 C. LANIGERA

287 PEL. PUNCTATA

288 PEL. PUNCTATA

289 PAR. BREVIPES

290 C. LANIGERA

291 PAR. BREVIPES

292 C. LANIGERA

293 C. LANIGERA

294 C. LANIGERA

295 PAR. BREVIPES

296 C. LANIGERA

297 PEL. PUNCTATA

298 PAR. BREVIPES

299 PEL. PUNCTATA

300 PAR. BREVIPES

301 C. LANIGERA

302 R. FORMOSA

Plate XXVI

FIGURE 303. *Parastasia brevipes* (Lec.). Abdominal segments 7 to 10, lateral view.

FIGURE 304. *Rutela formosa* Burm. Head, dorsal view.

FIGURE 305. *Cotalpa lanigera* (Linn.). Venter of last abdominal segment.

FIGURE 306. *Pelidnota punctata* (Linn.). Venter of last abdominal segment.

FIGURE 307. *Strigodermella pygmaea* (Fab.). Head, dorsal view.

FIGURE 308. *Strigodermella pygmaea* (Fab.). Epipharynx.

FIGURE 309. *Strigodermella pygmaea* (Fab.). Venter of last abdominal segment.

FIGURE 310. *Strigodermella pygmaea* (Fab.). Claw of mesothoracic leg, side view.

FIGURE 311. *Strigodermella pygmaea* (Fab.). Claw of metathoracic leg, side view.

FIGURE 312. *Strigodermella pygmaea* (Fab.). Thoracic spiracle (under oil immersion).

FIGURE 313. *Anomala innuba* (Fab.). Maxillary stridulatory area, much enlarged.

FIGURE 314. *Anomala innuba* (Fab.). Right maxilla and labium, ventral view.

FIGURE 315. *Anomala innuba* (Fab.). Right maxilla and labium, dorsal view.

FIGURE 316. *Anomala innuba* (Fab.). Fifth abdominal segment, left lateral view.

Symbols Used

CAR—Cardo	MA—Mala	SCU—Scutum
GL—Glossa	MP—Maxillary palpus	SD—Stridulatory teeth
HE—Helus	PLL—Pleural lobe	SPR—Spiracle
HFO—Hypopharyngeal fossa	PMP—Postmentum	ST—Stipes
HSC—Hypopharyngeal sclerome	PRA—Parartis	STN—Sternum
LAC—Labacoria	PRM—Prementum	T—Teges
LL—Lateral lobe	PRSC—Prescutum	UN—Uncus
LP—Labial palpus	SCL—Scutellum	

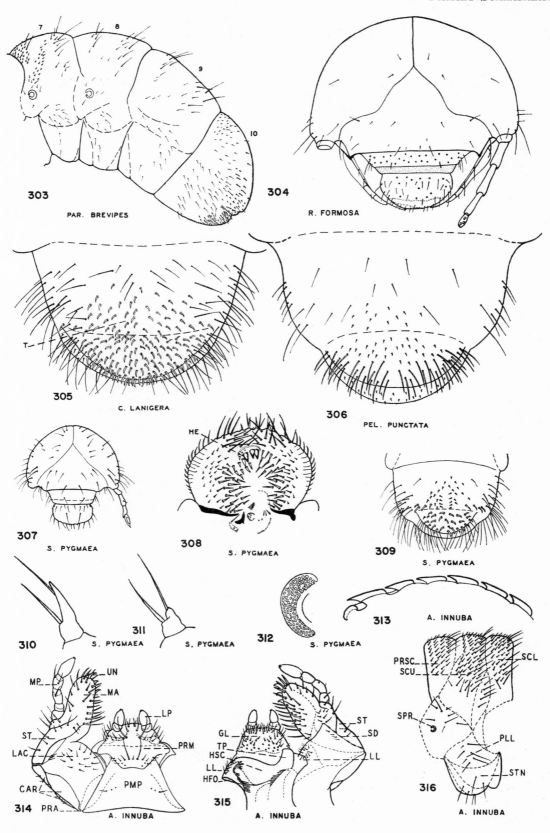

303 PAR. BREVIPES

304 R. FORMOSA

305 C. LANIGERA

306 PEL. PUNCTATA

307 S. PYGMAEA

308 S. PYGMAEA

309 S. PYGMAEA

310 S. PYGMAEA

311 S. PYGMAEA

312 S. PYGMAEA

313 A. INNUBA

314 A. INNUBA

315 A. INNUBA

316 A. INNUBA

Subfamily Dynastinae

The scarabaeid subfamily Dynastinae contains a great many genera and species of small to very large beetles ranging in color from yellow-brown, brown, and dark red to black. A few species are spotted but most are of a fairly uniform color. The group is poorly represented in Canada and the northern part of the United States but abundantly represented by numerous genera and species in the southern part of the United States, in Mexico, and farther south.

Males of genera such as *Strategus, Xyloryctes,* and *Dynastes* frequently bear prominent "horns" or tubercles on the head or thorax, or both, which together with the large size of the beetles has given them such popular names as "rhinoceros," "elephant," or "unicorn" beetles. In general, the bizarre species are not injurious since their larvae live in decaying wood or in the humus of woodlands. The injurious species found in the United States, usually smaller and not ornamented with conspicuous tubercles, inhabit the soil where their larvae are often serious pests of sod, corn, small grains, sugar cane, rice, and vegetable crops. In the genera *Euetheola* and *Ligyrus* the adults themselves are injurious to crops. A few genera, *Ligyrodes* for example, breed in rich organic matter such as is found in old straw stacks or at the edges of manure piles and cattle droppings.

In the American literature one or more species of dynastid larvae have been briefly described by Baerg (1942); Davis (1916); Hayes (1925, 1928, and 1929); Johnson (1941); Phillips and Fox (1924); Sim (1934); Titus (1908); and Weiss (1921). Hayes in 1929 also published a key to six genera of dynastid larvae. In 1944, the writer published a study of 20 species of dynastid larvae belonging to 12 genera. Much of this material is reprinted here together with descriptions of *Ancognatha manca* L. C., *Dynastes granti* Horn, *Phileurus didymus* (Linn.) and a slightly revised key to the genera.

For purposes of comparison, the larva of the common injurious species *Cyclocephala immaculata* Oliv. is described in detail, using the terminology of Böving (1936 and 1942). Next, a key is presented to larvae of the genera of the subfamily, followed by keys to the larvae of the species included in each genus. A technical description is given of the distinguishing characters of the third-stage larva of each species followed by a short account of the more prominent adult characters, distribution, habits, and importance. Adults of some species are pictured on page 157.

In arrangement of genera and species, the Leng catalogue (1920) has been followed. The genus *Anastrategus* has been suppressed and the species *splendens* replaced in the genus *Strategus* as suggested by Arrow

(1937). Studies of the larvae suggest that the genera *Cheiroplatys* and *Aphonus* belong in a distinct tribe and perhaps in the same genus. Larvae of both genera show ruteline affinities.

Third-Stage Larva of *Cyclocephala immaculata* Oliv.
(Figs. 317 to 331 and 356)

Larva (Fig. 322) cylindrical and typically scarabaeiform. Length of mature third-stage larva ranging from 22 to 24 mm with a mid-dorsal length of 22.7 mm. Range in maximum width of cranium of mature larva 3.7 to 4.2 mm with a mean of 3.92 mm.

Cranium (Fig. 325) narrower than prothorax, surface reticulate, shining, light reddish-brown. Clypeofrontal suture distinct, bounded laterally by the precoilae (PCL). Frontal sutures (FS) whitish, meeting considerably in front of hind margin of cranium and forming a right angle. Epicranial suture (ES) about half as long as one of the frontal sutures, medially with a brownish enlargement. A pair of setae, the anterior frontal setae (AFS), are borne on the anterior, median part of the frons. The exterior frontal setae (EFS), when present, consist of a single small seta caudad of each precoila. Two large posterior frontal setae (PFS), with the lateral seta of each pair near a frontal suture, are borne on each side of the frons caudolaterad of the anterior frontal setae. Each anterior angle of the frons (AA) bears a single, large seta and frequently an additional small seta. Epicranium with a row of 3 dorsoepicranial setae (DES) on each side of the epicranial suture. Laterad of each of these rows is a row of 5 or 6 setae consisting posteriorly of 3 or 4 very small setae and anteriorly of 2 large setae. A prominent seta is borne on the epicranium posterior to and slightly mesad of each ocellus on a level with the inner seta of each pair of posterior frontal setae. Anterior half of frons slightly roughened and bearing several minute pores.

Clypeus (Fig. 325) trapezoidal with slightly concave lateral margins. Clypeus divided transversely into a large postclypeus (PSC) and a small preclypeus (PC). Two prominent setae, the anterior clypeal setae (ACS), are borne slightly caudad of the distal margin of the postclypeus. A single, large, exterior clypeal seta (ECS) is found near each lateral margin of the postclypeus. The postclypeus has a rugose surface and bears several pores, while the preclypeus is lightly sclerotized and smooth.

Labrum (Fig. 325) oval in outline, slightly asymmetrical, longer and narrower than the clypeus. Two prominent, paramedian setae are on the distal part of the labrum, while 2 large setae are borne on each lateral, labral margin. Apex of labrum with a pair of blunt setae flanked on each side by a smaller, sharp-pointed seta. Laterad of these are 2 large marginal setae. Surface of labrum rugose and pitted.

Antenna (Fig. 325) almost as long as cranium, slender, 4-jointed, borne on a short basal piece partially fused to the epicranium. First segment slightly more than half as long as second segment; third and fourth segments each about equal in length to first segment. Third segment with an apical process on inner margin; apical process with an oval sensory spot on its inner face. Apical segment subfusiform bearing 2 ventral and 2 dorsal, oval sensory spots (DSS). Proximal and distal, dorsal sensory spots about equal in length. Proximal, dorsal sensory spot of 10 individuals ranging from .144 to .179 mm in length with a mean of .155 mm and ranging from .092 to .108 mm in width with a mean of .10 mm. Distal, dorsal sensory spot of 10 individuals ranging from .140 to .188 mm in length with a mean of .168 mm, and ranging from .104 to .129 mm in width with a mean of .118 mm. End of apical segment with about 9 olfactory pegs. Antenna bare of setae.

Mandibles (Figs. 317, 318, and 319 to 324) asymmetrical, subtriangular in outline, slightly shorter than cranium. Scissorial area (SA) blackish and rather stout with cutting edge divided by a notch, the scissorial notch (SN, Fig. 317), into a blade-like anterior portion and a posterior tooth. Caudolaterad of the scissorial notch (SN) is a single, large seta. Inner margin of mandibles, distad of the molar areas, smooth, not toothed. Mola-bearing part (MO) light reddish-brown with blackish molar structures. Molar part of left mandible bilobed with the distal lobe (M_1, Figs. 317 and 319) overhanging the proximal lobe (M_2, Figs. 317 and 319). Distal molar lobe of left mandible concave on inner surface and subtriangular in outline. Proximal molar lobe of left mandible oval in outline and flattened, bearing on its dorsal margin a small tooth-like projection and a sclerotized plate, the acia (AC). At the base of the left molar structure is a brustia of about 9 coarse setae (BR). Molar part of right mandible also bilobed. Distal molar lobe of right mandible (M_1, Figs. 318 and 321) oval to subtriangular in outline with slightly concave inner surface. Proximal molar lobe of right mandible (M_2, Figs. 318 and 321) deeply concave posteriorly with a single tooth-like projection bounding the concavity ventrally and a bilobed process or calx (CA) bounding the concavity dorsally. On the dorsal surface of each mandible, along the base of the molar structure, is a row of 6 to 9 stiff setae, the dorsomolar setae (DMS). Some of these setae are frequently double. Dorsoexterior mandibular region bare. The lateral face of each mandible (LAF) is bounded by 2 carinae: the dorsal carina (DC) and the ventral carina (VC). A longi-

tudinal row of 4 long setae and a few scattered pores are borne on the median portion, while laterad of the dorsal carina is a rugulose, longitudinal area, the scrobis (SCR), which bears a single prominent apical seta and 2 sparse, irregular rows of small pores. On the ventral aspect each mandible bears a large, ovate stridulatory area (STA) consisting of about 40 transverse, granulated ridges of which the proximal 4 ridges are much broader. Distad of each stridulatory area is a group of 5 or 6 small pores, while mesad of each stridulatory area is a compact cluster of about 5 to 7 setae so closely appressed that they frequently appear to be a single large seta. Near the basomedial angle on the ventral surface of each mandible is a prominent process (VP) which slides against the hypopharyngeal sclerome (HSC, Fig. 331).

Maxilla (Figs. 329 and 331) consisting of a cardo, stipes, fused galea-lacinia (mala), and a maxillary palpus. Cardo (CAR) subquadrate, longer than wide, extending from the base of the maxilla to the proximal edge of the stipes; divided into 3 sclerites. Stipes (ST) subquadrate in outline, located on the lateral surface of the maxilla and bounded posteriorly by the cardo and anteriorly by the maxillary palpus (MP). On its ventral surface the stipes bears about 10 large setae, while the lateral surface is practically bare. Dorsally, the most prominent feature of the stipes is a longitudinal row of 7 to 9 blunt, truncate stridulatory teeth (DS) and a larger, distal, blunt tubercle (SDT, Fig. 330). The dorsal portion of the stipes also bears a number of scattered setae anterior to and laterad of the stridulatory teeth. Mesad of the stridulatory teeth and parallel to them is a row of 4 or 5 small setae. Ventral surface of the mala (MA) composed of the galea; distal portion bearing a large terminal uncus (UN) surrounded by a semicircular ring of about 5 stout setae. Dorsolateral surface of the mala lightly sclerotized and bearing a number of small setae. Lacinia with 3 distal unci, the median one much the smallest. Lacinia bearing about 5 irregular, longitudinal rows of coarse setae. Maxillary palpus 4-jointed with the basal joint much the shortest. The obovate fourth segment bears a number of distal, sensory pegs.

Labium (Figs. 329 and 331) composed ventrally of a large subtrapezoidal postmentum (PMP), the subdivided prementum (PRM_1 and PRM_2), and a pair of labial palpi (LP). A single large seta is near the center of each lateral margin of the postmentum and another is in each anterior, lateral angle. A pair of prominent setae is near the anterior margin of the proximal sclerite of the prementum (PRM_2). On the dorsal surface of the glossa are a number of straight, slender setae caudolaterad of the bases of the 2-jointed labial palpi. The bare median portion of the glossa is surrounded by scattered, short, stout setae.

Hypopharyngeal sclerome (Fig. 331), located posterior to the dorsal surface of the labial glossa, asymmetrical, produced on right side into a strong truncate process (TP). Sclerome with a slightly sclerotized shoulder (LL) on each side behind which is the point of contact with the ventral process (VP) of the adjacent mandible.

Epipharynx (Fig. 327) broader than long, with rounded lateral margins; apical margin rounded and slightly crenulate, indented slightly in region of epizygum (EZ). Corypha (CO) and left acroparia (ACR) united. Corypha with 2 pairs of short, blunt setae. Right acroparia set with 8 to 11 long, sharp setae; left acroparia with 6 to 8 similar setae. Haptomerum (H) consisting of a broad, fused epizygum-zygum (EZ-Z) and a deeply notched haptomeral process. Haptomeral process distally with a tooth-like projection on the left side of the notch and a larger, broadly rounded projection on the right side of the notch. Behind the right projection is a transverse row of 4 sensilla, while behind the apex of the left projection is a transverse row of about 3 sensilla (these latter are not shown in Fig. 327). A single sensilla is located at or near the base of the haptomeral notch and another at the base of the left projection. Each acanthoparia (ACP) bears 11 to 13 flattened, sickle-shaped setae which decrease in size posteriorly. Plegmata and proplegmata absent. Gymnoparia (GP) present. Chaetopariae (CPA) well developed, covered with many coarse setae; setae stoutest in region of haptomerum. Numerous sensilla interspersed among the setae of both chaetoparia and a few borne on the pedium (PE). Pedium longer than wide. Laeotorma (LT) pipe-shaped with rounded pternotorma (PTT). Dexiotorma (DX) fairly straight. Haptolachus (HL) complete. Crepis (CR) divided into two parts of which the left part is the larger and bears scattered sensilla. In front of each part of the crepis is a pair of macrosensilla. Two nesia are present. The right nesium of sclerotized plate (SP), adjacent to the dexitorma, is subtriangular in outline and terminates in a blade-like point. The left nesium or sense cone (SC) is anterior to the sclerotized plate near the median line and bears about 4 sensory pores near its apex. Mesad of the laeotorma (LT) and the left part of the crepis is a longitudinal group of 7 to 9 slender setae. Between the base of the sclerotized plate and the right part of the crepis is a row of 3 to 5 slender setae.

Legs (Fig. 322) well developed, with the prothoracic pair shorter than the mesothoracic and the mesothoracic shorter than the metathoracic pair. Each leg consists of a long, cylindrical coxa, a short trochanter, a clavate femur, and a fused tibiotarsus. Procoxa with a sparse, irregular, longitudinal row of setae on each dorsolateral margin, the two uniting ventrally into a sparse patch of setae posterior to the trochanter. Mesothorax and meta-coxae each with 3 sparse, longitudinal rows of setae. Trochanters and femora clothed ventrally with numerous setae; femora with a distal, sparse circular row of setae, in addition. Dorsal surface of trochanters bare and those of femora bare except for the distal band of setae. Tibiotarsi with numerous setae, the stronger being on the ventral surface. Claws simple, each with a cylindrical yellow-brown base and a terminal, slightly curved, blackish, sharp-pointed apex. Base of each claw bears 2 setae. Prothoracic and mesothoracic claws about equal in length; metathoracic claws shorter.

Body (Fig. 322) composed of 3 thoracic and 10 abdominal segments. Prothorax, mesothorax, and metathorax each with 3 dorsal areas, the prescutum, scutum, and scutellum. Prothorax with a pair of spiracles. Scutum of prothorax with a dorsal transverse row of 8 setae; prescuta of mesothorax and metathorax each with a dorsal, transverse row of 6 setae; scutella of mesothorax and metathorax bare.

Three dorsal areas, the prescutum (PRSC, Fig. 328), scutum (SCU), and scutellum (SCL) are clearly defined on each of the first 6 abdominal segments. Prescutum and scutum of first abdominal segment each with a dorsal, sparse, transverse row of long and short setae of which the row on the prescutum is much the shorter; scutellum with a similar transverse row of setae and bearing cephalad of this row a transverse row of short, stout setae. The 3 dorsal annulets of the second to fifth segments inclusive are each clothed with a transverse band of short, stout setae and a caudal row of long, slender setae (Fig. 328). The dorsal annulets of the sixth abdominal segment are similarly clothed but the band of short, stout setae on the scutellum is reduced to a single or double transverse row. Subscuta of abdominal segments 2 to 6 inclusive, each with 1 seta or a dorsoventral pair of setae. Dorsa of seventh to ninth segments entire, not divided into annulets; each dorsum with 2 widely separated transverse rows of long and short, slender setae. Dorsum of tenth abdominal segment with a lightly sclerotized, crescent-shaped, dorsal impressed line (DIP, Fig. 322), whose arms curve toward the ends of the anal slit (ASL, Fig. 326). Between the anterior margin of the tenth abdominal segment and the dorsal impressed line is a sparse, transverse row of slender setae, which is confluent laterally with the barbulae. The area enclosed by the dorsal impressed line of the tenth abdominal segment bears a sparse covering of slender setae.

Straw-colored, subquadrate sclerite, cephalodorsad of the thoracic spiracle with 1 long and 2 short setae in a dorsoventral row. Spiracle-bearing area of first abdominal segment with 1 small seta above spiracle. Spiracle-bearing areas of second to eighth abdominal segments with 3 to 6 setae (usually 4). Pleural lobes of first to eighth abdominal segments inclusive with 3 to 5 setae; sterna of thoracic segments each with a transverse patch of about 20 setae. Eusterna of abdominal segments 1 to 7 each bearing a transverse row of 2 to 8 setae; pedal areas with a single seta, on each side.

Anal slit (ASL) transverse, curved. Anus bordered on upper anal lip by scattered small, short conical setae; anus bordered ventrally by a sparse row of about 12 long and short, straight setae and the hamate setae of the lower anal lip. Lower anal lip (LAL) covered with about 35 hamate setae similar to those of the teges. To each side of the anus is a barbula (B) of several long setae which is contiguous to the long setae clothing the caudodorsal part of the tenth abdominal segment.

Raster (Fig. 326) consisting of a teges (T) and a campus (C). Teges made up of a broad patch of about 30 hamate setae and occupying about half the area between the base of the lower anal lip (LAL) and the caudal margin of the ninth abdominal segment. Distad of the teges is located an irregular transverse row of about 8 straight setae.

Spiracles (Fig. 322) consisting of 1 pair of thoracic spiracles and 8 pairs of abdominal spiracles. Thoracic spiracles and seventh and eighth pairs of abdominal spiracles considerably larger than the spiracles of abdominal segments 1 to 6, inclusive. Concavities of thoracic spiracles facing posteriorly, those of abdominal spiracles facing anteriorly. Each spiracle with a C-shaped respiratory plate which surrounds a large oval bulla; each bulla bears a curved spiracular slit. Thinly covered "holes" of respiratory plate very small, usually separated from each other by distances greater than the diameters of the holes (Fig. 356). "Holes" usually oval or round, not arranged in very definite transverse rows. Trabeculae moderately developed, branched. Thoracic spiracle ranging from .37 to .42 mm in length and .22 to .26 mm in width. Distance between the two lobes of the respiratory plate somewhat to much less than the dorsoventral diameter of the bulla. Respiratory plate of thoracic spiracle with a maximum of 14 to 22 "holes" along any diameter.

Distinguishing Characters of the Subfamily Dynastinae

Dynastid larvae may be distinguished from other scarabaeid larvae by the following combination of characters. Mandible with a ventral, oval, stridulatory structure consisting of numerous, transverse, granular ridges; ridges usually much wider posteriorly. Maxilla with well-developed stridulatory teeth, which, in most genera, are truncate. If the maxillary stridulatory teeth have anteriorly projecting points, then the distal segment of the antenna has a single, dorsal sensory spot. Lacinia of maxilla with an oblique or transverse row of 3 terminal unci which are more or less fused together at their bases. Haptomerum of epipharynx with a raised bilobed or entire ridge—or a tooth-like process; distinct zygum and prominent heli absent. Epipharynx with both nesia

well developed. Both plegmata and proplegmata absent. Ocelli usually present. Raster usually with no palidia (present in the genera *Euetheola* and *Ligyrodes*). Anal slit transverse, curved.

Key to Genera of the Subfamily Dynastinae Based on the Characters of Third-Stage Larvae

1. Dorsal surface of last antennal segment with a single, oval, sensory spot (Fig. 335). Most of the maxillary stridulatory teeth with anteriorly projecting points (Figs. 346 and 347)..........*Aphonus* and *Cheiroplatys*
Dorsal surface of last antennal segment with 2 or more sensory spots (Figs. 336 and 337). Maxillary stridulatory teeth truncate, without anteriorly projecting points (Figs. 349 and 350) 2

2. Dorsal surface of last antennal segment with 2 or 3 (rarely 4 or 5) sensory spots (Fig. 336) 3
Dorsal surface of last antennal segment with 6 or more spots (Fig. 337) 11

3. Inner, concave surface of left mandible, distad of the molar area, smooth (Figs. 317 and 341, S₄ absent)........... 4
Inner, concave surface of left mandible, distad of the molar area, toothed (Figs. 342, 344 and 345, S₄ present) 5

4. Head reddish-black, coarsely punctate. Haptomeral process of epipharynx unnotched (Fig. 359). Prescuta of mesothorax and metathorax bare. Dorsa of abdominal segments 7 to 9, each bare except for a single posterior, transverse row of about 6 long setae (Fig. 372) ....*Xyloryctes jamaicensis* (Drury)
Head yellow-brown to light reddish-brown, surface faintly reticulate and rather smooth (Fig. 325). Haptomeral process of epipharynx notched forming 2 teeth (Fig. 327). Prescuta of mesothorax and metathorax with at least a few setae. Dorsa of abdominal segments 7 to 9 each with at least 2 widely separated, transverse rows of long setae (Fig. 371) .... *Cyclocephala*

5. Raster with palidia (Figs. 369 and 370) 6
Raster without palidia (Figs. 367 and 368) 7

6. Raster with 2 monostichous palidia. Each palidium consisting of a sparse, irregular longitudinal row of 7 to 10 strongly compressed, sickle-shaped pali (Fig. 370). Palidia (and septula) not extending across the lower anal lip .... *Euetheola rugiceps* Lec.
Raster with 2 polystichous paladia. Each palidium consisting of 5 to 7 irregular, longitudinal rows of sharp, cylindrical, medianly directed setae (Fig. 369). Palidia (and septula) extending across the lower anal lip ....*Ligyrodes relictus* (Say)[25]

7. Chaetopariae of epipharynx with many sensilla among the setae (Fig. 364). Dorsum of abdominal segment 7 with 2 widely separated, transverse rows of long, slender setae; short, stout setae absent................ *Dyscinetus morator* (Fab.)
Chaetopariae of epipharynx with few or no sensilla among the setae (Figs. 361 to 363). Dorsum of abdominal segment 7 with short, stout setae in addition to 2 widely separated, transverse rows of long, slender setae (Fig. 374) 8

[25] Cartwright (1959) synonomyzed both *Ligyrodes* and *Ligyrus* under *Bothynus*. Based on larval characters, I believe *Ligyrodes* should stand as a separate genus.

8. Dorsa of abdominal segments 8 and 9 each with nu-
    merous short, stout setae and 2 widely separated,
    transverse rows of long setae .... Tribe **Phileurini**
    Dorsa of abdominal segments 8 and 9 each with 2
    widely separated, transverse rows of long, slender
    setae; short, stout setae absent (Fig. 372) 9
9. Respiratory plate of thoracic spiracle with 16 or fewer
    "holes" along any diameter ..*Oxygrylius ruginasus* (Lec.)
    Respiratory plate of thoracic spiracle with from 20 to
    35 "holes" along any diameter 10
10. Head without prominent pits........*Bothynus gibbosus* (DeG)
    Head coarsely pitted ....*Ancognatha manca* Lec.
11. Claws bearing 2 setae (Fig. 332) ....*Dynastes*
    Claws bearing 3 or 4 setae (Figs. 333 and 334) ....*Strategus*

TRIBE Cyclocephalini
Genus *Cyclocephala* Latr.

Surface of cranium reticulate, fairly smooth, yellow-
brown to light reddish-brown. Ocelli present (Fig. 325).
Lateral margins of labrum (and epipharynx) rounded,
not angulate posteriorly (Figs. 325 and 327). Haptomeral process of epipharynx notched forming 2 unequal teeth (Fig. 327). Inner margin of mandible, distad of the molar area, smooth, not toothed (Figs. 317
and 318). Maxilla with a row of blunt truncate, stridulatory teeth (Fig. 330). Last segment of antenna bearing 2 dorsal sensory spots (Fig. 325). Raster with a
teges consisting of a broad patch of prominent hamate
setae (Fig. 326). Lower anal lip with similar hamate
setae. Palidia absent. Two setae on each claw.

Key to Larvae of the Species of Cyclocephala

1. Spiracles of first abdominal segment much smaller than
    those on abdominal segments 2 to 8. Mesothoracic
    and metathoracic scutella each with a transverse
    band of more than 20 setae. Pleural lobes of abdom-
    inal segments 1 to 8 with 12 to 16 slender setae........
    .... *C. abrupta* Casey
    Spiracles of first 6 abdominal segments similar in size.
    Mesothoracic and metathoracic scutella each with
    none to 2 setae. Pleural lobes of abdomnial seg-
    ments 1 to 8 with fewer than 8 setae (Figs. 328
    and 371) 2
2. Mesothoracic and metathoracic scutella each with 2
    setae. Median region, anterior to dorsal impressed
    line of tenth abdominal segment, bare..*C. pasadenae* Casey
    Mesothoracic and metathoracic scutella with 1 or no
    setae. Median region, anterior to dorsal impressed
    line of tenth abdominal segment, with a single ir-
    regular, transverse row of setae ....
    ....*C. borealis* Arrow and *C. immaculata* Oliv.

Cyclocephala abrupta Casey, Third-Stage Larva

This description is based on the following material
and reprinted from my 1944 bulletin:
    One third-stage larva (and one second-stage
larva) collected June 7, 1935, at Hermiston, Oregon,
by M. C. Lane. Larvae seriously damaging alfalfa

and grain. Associated adult determined by E. A.
Chapin. Larvae were loaned by the United States
National Museum.

Maximum width of head capsule of third-stage
larva 3.8 mm. Exterior frontal setae consisting of a
single seta on each side. Posterior frontal setae, 3 on
each side. A pair of anterior frontal setae on each side.
Each anterior angle of frons with 2 setae. Dorsoepicranial setae 7 to 8 on each side laterad of the epicranial
suture. Chaetoparia of epipharynx with a few or no
sensilla among the setae. Maxilla with about 10 blunt,
truncate stridulatory teeth and a wider, blunt distal process. Last segment of antenna with 2 large, dorsal sensory spots.

Thoracic spiracle .38 mm long and .25 mm wide.
Respiratory plate with a maximum of 15 to 19 "holes"
along any diameter. "Holes" round to oval with fairly
smooth margins and separated by less than the diameter
of a "hole." Distance between the two lobes of the respiratory plate slightly less than the dorsoventral diameter of the bulla. Spiracles of abdominal segment 1 very
small; those of abdominal segments 1 to 4 progressively larger. Spiracles of abdominal segments 4 to 6
equal in size but smaller than those of abdominal segments 7 and 8. Spiracles of abdominal segment 8 smaller
than those of segment 7.

Scutella of mesothorax and metathorax each with a
transverse band of more than 20 long, slender setae.
Dorsum of abdominal segment 7 anteriorly with a transverse row of long, slender setae and 2 or 3 irregular
transverse rows of shorter setae; posteriorly with a
transverse row of long, slender setae and an irregular
row of shorter setae. Dorsa of abdominal segments 8
and 9 each with an anterior and a posterior, transverse,
irregular, double row of setae. Raster with a teges of
about 30 hamate setae.

Cyclocephala abrupta is known to occur in California, Arizona, Oregon, and Illinois. According to Saylor
(1937), it is very common in southern California. In
Illinois, adults have been found in some numbers the latter part of July (Riegel, 1942). Adult 9.5 to 12 mm
long and 5 to 5.5 mm wide; light yellow-brown and
blackish head. Larvae live in the soil. Damage to alfalfa
and grain crops has been reported in Oregon. Adults
are attracted to lights.

Cyclocephala borealis Arrow, Third-Stage Larva
(Fig. 371)

This description is based on the following material
and reprinted from my 1944 bulletin:
    Six cast larval skins associated with adults
reared from larvae collected at Lexington, Kentucky,
by the writer.

Fourteen third-stage larvae collected near New Haven, Connecticut, by J. P. Johnson of the Connecticut Agricultural Experiment Station.

Maximum width of head capsule of third-stage larva 3.64 to 4 mm with a mean of 3.9 mm. Exterior frontal setae consisting of a single small seta on each side. Posterior frontal setae, 2 on each side. One or a pair of anterior frontal setae on each side. Each anterior angle of frons with 1 or 2 setae. Dorsoepicranial setae consisting of 2 or 3 large setae and 1 or 2 minute setae on each side. Chaetoparia of epipharynx with many sensilla among the setae, sensilla more numerous on right chaetoparia. Maxilla with 7 to 9 blunt, truncate stridulatory teeth and a wider distal process. Last segment of antenna with 2 dorsal sensory spots, the proximal of which is the smaller.

Thoracic spiracles .4 to .44 mm long and .246 to .284 mm wide. Respiratory plate of thoracic spiracles with a maximum of 12 to 22 "holes" along any diameter. "Holes" round to oval with fairly smooth margins. Most of "holes" separated by a distance equal to the diameter of a "hole." Distance between the two lobes of the respiratory plate much less than the dorsoventral diameter of the bulla. Spiracles of abdominal segments 1, 7, and 8 similar in size and larger than those of abdominal segments 2 to 6.

Prescuta of mesothorax and metathorax each with a transverse row of 6 to 8 slender setae. Scutella of mesothorax and metathorax usually bare, rarely with a single seta. Dorsa of abdominal segments 7 to 9 with 2 widely separated, transverse rows of slender setae, many of which are very long (Fig. 371). Raster with a teges of about 25 hamate setae.

C. borealis is a common native species occurring east of the Rocky Mountains, occupying much the same range as *C. immaculata*. In Kentucky, the species is common in Kenton County and a few specimens have been taken in the inner bluegrass region. Adult shining, light yellow-brown with a slightly darker pronotum and a reddish to reddish-black head. Elytra of males with a sparse covering of erect hairs. Adults 11 to 12 mm long and 6 to 7 mm wide. Larvae found in the soil of pastureland and lawns, where they feed on the grass roots. Severe injury to lawns by this species has been reported from Ohio and Connecticut, Neiswander (1938) and Johnson (1941). Life cycle and biology are similar to that of *C. immaculata*.

Cyclocephala immaculata Oliv., Third-Stage Larva
(Figs. 317 to 331 and 356)

This description is based on the following material and reprinted from my 1944 bulletin:

Ten cast larval skins associated with adults, reared from larvae collected at Lexington, Kentucky, by the writer.

Five third-stage larvae reared by the writer from eggs laid by females of this species. The virgin females and the males with which they were mated were reared from larvae collected behind the plow in March 1944, at Lexington, Kentucky.

Maximum width of head capsule of third-stage larva 3.7 to 4.2 mm with a mean of 3.92 mm. Exterior frontal setae consisting of a single small setae located caudad of each precoila (Fig. 325). A widely separated pair of posterior frontal setae on each side. A single anterior frontal seta on each side. Three dorsoepicranial setae on each side laterad of the epicranial suture. Epipharynx with numerous sensilla interspersed among the setae of the chaetopariae (Fig. 327). Maxilla with a row of 7 to 9 blunt, truncate stridulatory teeth and a larger, blunt distal process (Fig. 330). Last segment of antenna with 2 dorsal sensory spots, these approximately equal in length (Fig. 325).

Thoracic spiracles and abdominal spiracles 7 and 8 considerably larger than the spiracles of abdominal segments 1 to 6 (Fig. 322). Thoracic spiracle .37 to .42 mm long and .22 to .26 mm wide. Respiratory plate of thoracic spiracle with a maximum of 14 to 22 "holes" along any diameter. Distance between the two lobes of the respiratory plate much to very much less than the dorsoventral diameter of the bulla.

Scutella of mesothorax and metathorax bare. Dorsa of abdominal segments 7 to 9 each bare except for 2 widely separated, transverse rows of long, slender setae. Raster with a teges consisting of a broad patch of about 30 hamate setae (Fig. 326). Lower anal lip bearing about 35 hamate setae similar to those of the teges.

C. immaculata is a native species commonly found in many of the states east of the Rocky Mountains. The species is found throughout Kentucky and is especially abundant in the inner bluegrass region (Ritcher, 1940). Adults shining, light red-brown with a reddish-black head. Discal areas of elytra bare of hairs. Adult 10.5 to 12 mm long and 6 to 7 mm wide. Adult female pictured in Fig. 382. Larvae, found in the soil of pastureland and lawns, frequently cause severe injury by feeding on grass roots. Also found under cattle droppings and in moist soil close to straw stacks and manure piles. In Kentucky pupation occurs during the latter part of May and continues into June. Adults in flight during June and July; strongly attracted to lights. The adults do not feed. Eggs are laid in the soil during June and July.

Cyclocephala pasadenae Casey, Third-Stage Larva

This description is based on the following material and reprinted from my 1944 bulletin:

Three third-stage larvae collected March 21, 1931, at Sacramento, California, by H. H. Keifer, at privet roots. (Loaned by the U. S. National Museum.)

Maximum width of head capsule of third-stage larva 3.6 to 3.9 mm with a mean of 3.78 mm. Exterior frontal setae consisting of a single seta on each side. Posterior frontal setae 2 to 4 on each side. A pair of anterior frontal setae present on each side. Dorsoepicranial setae 3 to 5 on each side.

Thoracic spiracle .32 to .35 mm long and .19 to .2 mm wide. Respiratory plate of thoracic spiracle with a maximum of 9 to 16 round to oval, rather irregularly shaped "holes" along any diameter. Distance between "holes" as great as or greater than the diameter of a "hole." Distance between the two lobes of the respiratory plate slightly less than the dorsoventral diameter of the bulla. Spiracles of abdominal segments 1 to 6 similar in size; those of 7 and 8 larger.

Scutella of mesothorax and metathorax each bare except for 2 short, widely separated setae. Dorsa of abdominal segments 7 to 9 each with setation limited to 2 transverse, widely separated rows. Raster with a teges of about 30 hamate setae.

Cyclocephala pasadenae is a western species found in California, Baja California, and Arizona. According to Saylor (1937), it is probably the most common California species. Adult light yellow-brown with a slightly reddish cast; head reddish. Adult 10.5 to 13 mm long and 5.5 to 7 mm wide. Larvae live in the soil.

Genus *Ancognatha* Er.

Ancognatha manca LeConte, Third-Stage Larva
(Figs. 375 to 377 and 379 to 381)

This description is based on the following material:

Two third-stage larval skins of 3 associated third-stage larvae reared to the adult stage. Larvae collected March 22, 1961, in the Santa Catalina Mountains, Arizona (Hk. Highway, Mile 10), by George Butler and Floyd Werner.

Maximum width of head capsule 5.3 to 5.6 mm. Surface of head (Fig. 375) light reddish brown, coarsely pitted with pits of similar size. Primary frontal setae, on each side, consisting of a single seta in each anterior angle and a single exterior frontal seta. Two epicranial setae, on each side, located toward the base of the epicranial suture. Epipharynx (Fig. 376) with haptomeral process entire, in the form of a raised, oblique ridge. Epizygum present. Chaetoparia well developed, each with a few sensilla anteriorly. Inner margin of left mandible (Fig. 378), anterior to the molar area, with a distinct tooth (this tooth absent on right mandible). Maxilla with a row of 9 or 10 truncate stridulatory teeth and an anterior truncate process. Dorsal surface of last antennal segment with 2 sensory spots. Ocelli present and pigmented.

Thoracic spiracles ranging from .46 to .73 mm in length and from .38 to .53 mm in width. "Holes" of respiratory plate resembling those of *Dynastes* (Fig. 357); irregular in shape with finger-like projections. "Holes" not in definite rows, with a maximum of 20 to 35 "holes" across any diameter. Distance between "holes" as great or greater than the diameter of a "hole." Distance between the lobes of the respiratory plate varying from somewhat less than to much less than the dorsoventral diameter of the bulla. Spiracle of abdominal segments 1 to 8 similar in size.

Dorsum of abdominal segment 7 anteriorly with a few scattered, short, stout setae and a sparse transverse row of long slender setae; posteriorly with a sparse, transverse row of slender setae of variable lengths. Dorsa of abdominal segments 8 and 9 with similar setation except for the absence of short, stout setae. Raster (Fig. 380) with a triangular teges of 36 to 42 hamate setae. Lower anal lobe with about 40 hamate setae and a sparse caudal fringe of long, slender setae. Claws long, each bearing 2 setae.

Ancognatha manca LeConte is a common scarab found in Arizona and New Mexico and ranging into central Mexico (Saylor, 1945). According to the same writer, adults are 15 to 19 mm long and highly variable in color, ranging from all black to black with reddish-brown elytra and legs, to red-brown with light yellow-brown legs, scutellum, and thoracic margins. Adults are attracted to lights, and larvae are found in the soil.

Genus *Dyscinetus* Harold

Dyscinetus morator (Fab.), Third-Stage Larva
(Figs. 340, 364, and 368)

This description is based on the following material and reprinted from my 1944 bulletin:

Four third-stage larvae collected August 5, 1937, at North Bower's Beach, Delaware, by G. H. Bradley, from newly dead sod. Adults determined by E. A. Chapin. Larvae were loaned by the United States National Museum.

Maximum width of head capsule of third-stage larva 4.5 to 4.96 mm with a mean of 4.8 mm. Surface of cranium slightly roughened, covered with many small pores, orange-brown. Exterior frontal setae consisting of a single seta on each side (Fig. 340). A single posterior frontal seta on each side. Anterior frontal setae absent. Dorsoepicranial setae 4 to 7 on each side. Lateral margins of labrum (and epipharynx) sharply angulate posteriorly (Figs. 340 and 364). Haptomeral process of epipharynx crenulate but not deeply notched. Epipharynx with numerous sensilla among the setae of the chaetopariae. Inner margin of both mandibles, distad of their molar areas, with a distinct tooth. Maxilla with a row of about 10 truncate stridulatory teeth. Last segment of antenna with 2 dorsal sensory spots. Ocelli present.

Thoracic spiracles .51 to .57 mm long and .32 to .34 mm wide. Respiratory plate with a maximum of 23 to 34 "holes" along any diameter. "Holes" oval to elongate-oval. Distance between the two lobes of the respiratory plate somewhat to much less than the dorsoventral diameter of the bulla. Spiracles of abdominal segments similar in size.

Dorsa of abdominal segments 7 to 9 each practically bare except for 2 widely separated, sparsely set transverse rows of about 8 slender setae. Raster with a subtriangular teges of about 35 rather short, flattened, slightly curved setae (Fig. 368). Palidia absent. Lower anal lip with a sparse covering of similar setae and a caudal fringe of about 8 long, slender setae. Claws each bearing 2 setae.

Dyscinetus morator, commonly called the rice beetle, occurs in the eastern and southeastern states and, according to Casey (1915), ranges westward to Iowa, Oklahoma, and Louisiana. Adult dull dark reddish-brown, 15 to 17 mm long and 8 to 10 mm wide (Fig. 383). Larvae feed beneath sod and, according to Phillips and Fox (1924), are also found in compost heaps or near pigpens. On June 14, 1892, 30 adults of this species were collected at lights at Pineville, Kentucky.

TRIBE Oryctini

Genus *Ligyrodes* Casey

Ligyrodes relictus (Say.), Third-Stage Larva
(Fig. 369)

This description is based on the following material and reprinted from my 1944 bulletin:

One third-stage larva collected in June 1939, on Long Island, New York, by A. Davis, in fertilizer. No. 260. Larva determined by A. G. Böving. (Loaned by the U. S. National Museum.)

One third-stage larva collected July 20, 1933, at Hampton, Virginia, in an azalea bed in a nursery. (Loaned by the U. S. National Museum.)

Four cast skins of third-stage larvae collected in 1871 at Monroe, Michigan, by Hubbard. No. 511. (Loaned by the U. S. National Museum.)

Cast skin of one third-stage larva collected in July 1933, at Gays Mills, Wisconsin, by the writer, in rich soil under strawy litter. Associated with a reared adult. (Loaned by E. M. Searles of the University of Wisconsin.)

Two third-stage larvae. (Loaned by the Illinois Natural History Survey.)

Maximum width of head capsule of third-stage larva 6.4 to 6.5 mm. Surface of cranium red-colored, covered with numerous prominent pits. Exterior frontal setae consisting of a single seta located caudad of each precoila. Anterior and posterior frontal setae absent. Each anterior angle of frons with 2 or 3 setae. Dorso-epicranial setae 3 to 5 on each side. Left lateral margin of labrum (and epipharynx) angulate. Haptomeral process entire, not notched. Right chaetoparia with a few sensilla. Inner margin of both mandibles, distad of their molar areas, each with a distinct tooth; that of the left mandible truncate, that of the right, conical. Maxilla with a row of 7 to 10 truncate stridulatory teeth and an anterior process which is almost identical in shape and size. Last segment of antenna with 2 to 4 dorsal sensory spots. Ocelli present.

Thoracic spiracles .76 to .84 mm long and .5 to .54 mm wide. Respiratory plate with a maximum of about 40 "holes" along any diameter. Distance between the two lobes of the respiratory plate very much less than the dorsoventral diameter of the bulla. Spiracles of abdominal segments 1 to 8 similar in size. Bullae of all spiracles with an oval, convex central area.

Dorsum of abdominal segment 7 with an anterior, transverse patch of short setae among which is a transverse row of sparse, long, slender setae; posteriorly with 2 sparsely set transverse rows of setae, the anterior of which contains short setae and the posterior, long, slender setae. Abdominal segments 8 and 9 with similar setation except that the short setae in each anterior, transverse patch are sparse or absent. Pleural lobes of abdominal segments 1 to 8 with 5 to 8 setae. Spiracular areas with about 6 setae. Venter of last abdominal segment with 2 polystichous palidia (Fig. 369). Each palidium consisting of a patch of about 5 to 7 very irregular, longitudinal rows of sharp, cylindrical, medially directed, spine-like setae. Palidia (and septula) extending across the lower anal lip. Palidia surrounded laterally and anteriorly by a small number of slightly flattened, tegillar setae which are slightly curved distally. Palidia separated by a conspicious septula which is widest near the base of the lower anal lip. Lower anal lip fringed caudally with about 13 long setae. Claws sharp, each bearing 2 setae.

Ligyrodes relictus is a common, native species, found in the middle western and northeastern states. Adult dark reddish-black, 18 to 22 mm long and 10 to 12 mm wide (Fig. 396). Larvae are found in moist rich soil such as at the edges of old straw stacks and manure piles; they feed on organic matter. Because of these habits the larvae are sometimes called "manure worms" or "muck worms." Pupation occurs the last half of July in Wisconsin. Adults winter in the soil and oviposit the following spring. Adults are attracted to lights.

Genus *Euetheola* Bates

Euetheola rugiceps Lec., Third-Stage Larva
(Fig. 370)

This description is based on the following material and reprinted from my 1944 bulletin:

Three third-stage larvae reared August 16, 1943, by the writer, from eggs laid by a female of this species collected June 24, 1943, at Leitchfield, Kentucky, by R. T. Faulkner. Beetle damages corn by boring into the stem just above prop roots.

Maximum width of head capsule of third-stage larva 4.2 to 4.37 mm. Surface of cranium chestnut brown, covered with numerous prominent pits. Exterior frontal setae consisting of a single small seta just caudad of each precoila. Posterior frontal setae absent. A single large anterior frontal seta present on each side. Each anterior angle of frons with 2 setae. Dorsoepicranial setae 3 to 5 on each side. Right lateral margin of labrum (and epipharynx) broadly rounded posteriorly; left lateral margin slightly angulate. Haptomeral process of epipharynx crenulate but not deeply notched. Right chaetoparia of epipharynx with a very few sensilla among the setae; left chaetoparia with no sensilla. Inner margin of left mandible, distad of the molar area, with a distinct tooth; tooth vague on right mandible. Maxilla with a row of about 9 truncate stridulatory teeth. Last segment of antenna with 2 dorsal sensory spots. Ocelli present.

Thoracic spiracles .47 to .51 mm long and .27 to .3 mm wide. Respiratory plate with a maximum of 16 to 32 "holes" along any diameter. "Holes" oval to elongate-oval, not in definite rows. Distance between the two lobes of the respiratory plate somewhat less than the dorsoventral diameter of the bulla. Spiracles of abdominal segments 2 to 5 similar in size; those of abdominal segments 1, 6, and 7 slightly smaller; those of abdominal segment 8 much smaller.

Dorsum of abdominal segment 7 with an anterior row of 4 long, slender setae and a few scattered, rather stout setae; posteriorly with a row of sparse long and short, slender setae. Dorsa of abdominal segments 8 and 9 each with an anterior pair of long, slender setae and a posterior, sparsely set, transverse row of long and short, slender setae. Raster with paired tegilla, a pair of palidia and a septula (Fig. 370). Palidia nearly parallel, each palidium consisting of a rather sparsely set row of 7 to 10 strongly compressed pali whose tips are slightly hooked. Septula oblong. Preseptular, hamate setae 1 to 4 in number. Lower anal lip with a caudal fringe of 9 to 14 long, slender setae. Claws each bearing 2 setae.

Euetheola rugiceps is commonly called the sugarcane beetle or roughheaded cornstalk beetle because of frequent and sometimes severe injury by the adults to cane and corn in the southeastern states. Adult oval, dull black, 12 to 15 mm long and 7 to 8.5 mm wide (Fig. 384). In Kentucky, damage by the beetle has been confined to the western half of the state and usually occurs from the middle of May through June. Injury is caused by the beetles gnawing holes in the stalks of young corn just below the surface of the soil. Eggs are laid during the latter part of June and the first half of July. According to Phillips and Fox (1924), the larvae feed chiefly on decayed vegetable matter in low or poorly drained soil or in heavy, dark soils of open grasslands. The species overwinters in the adult stage.

Genus *Bothynus* Hope

Bothynus gibbosus (DeGeer), Third-Stage Larva
(Figs. 336, 338, 345, 349, 352, 362, and 374)

This description is based on the following material and reprinted from my 1944 bulletin:

Cast skin of one third-stage larva collected in 1936 at Lexington, Kentucky, by H. G. Tilson, from garden soil. Transformed to an adult July 18, 1936.

Cast skin of one third-stage larva collected in 1940 at Lexington, Kentucky, by the writer. Associated with an adult.

Eight third-stage larvae reared from eggs laid by adults of this species collected at lights during May and June 1943, at Lexington, Kentucky, by the writer.

Maximum width of head capsule of third-stage larva 4.37 to 4.8 mm with a mean of 4.65 mm. Surface of cranium light brown, slightly roughened, reticulate, without prominent pits. Exterior frontal setae usually consisting of a single prominent seta located caudad of each precoila; occasionally with an additional smaller seta (Fig. 338). A single posterior frontal seta usually present on each side. Anterior frontal setae absent. Each anterior angle of the frons with 3 to 5 setae. Dorsoepicranial setae 2 on each side laterad of the epicranial suture. Lateral margins of labrum (and epipharynx) broadly rounded posteriorly (Figs. 338 and 362). Haptomeral process of epipharynx not notched. Chaetoparia of epipharynx with no sensilla among the setae. Inner margin of left mandible, between scissorial and molar areas, with a small but distinct tooth (Fig. 345); tooth vague on right mandible. Maxilla with a row of about 8 truncate stridulatory teeth (Fig. 349). Last segment of antenna with 2 dorsal sensory spots (Fig. 336). Ocelli present.

Thoracic spiracles .49 to .56 mm long and from .3 to .36 mm wide. Respiratory plate with a maximum of 20 to 35 "holes" along any diameter. Distance between the two lobes of the respiratory plate somewhat to much less than the dorsoventral diameter of the bulla. Spiracles of abdominal segments 1 and 8 slightly smaller than those of abdominal segments 2 to 7, which are similar in size.

Dorsum of abdominal segment 7, anteriorly with a transverse patch of sparse, short, rather stout setae in addition to several long and short, slender setae; posteriorly with a transverse row of sparse, long and short, slender setae (Fig. 374). Dorsa of abdominal segments 8 to 9 each with 2 widely separated transverse rows of sparse, long and short, slender setae. Raster with a subtriangular teges of about 50 short, flattened, slightly curved setae. Lower anal lip covered with a similar number of similar setae and a caudal fringe of 7 to 11 long, slender setae. Palidia absent. Claws bearing 2 setae.

Bothynus gibbosus is a common native species, ranging from northern Mexico throughout most of the United States (Saylor, 1946), whose larva superficially resembles that of *Cyclocephala* sp. It may be readily separated, however, by the presence of a tooth on the inner margin of the left mandible between the scissorial and molar areas, which is absent on the mandibles of *Cyclocephala* larvae. Adult oval, reddish-brown to reddish-black, 12 to 16 mm long and 7.5 to 9 mm wide. Pronotum with an anterior dimple. Eggs laid from late May through June in Kentucky. Larvae found from June until August in garden soil or other soil rich in organic matter, where they frequently feed on the roots of growing plants. Adults present during most of the year; strongly attracted to lights. According to Hayes (1917), the adult feeds underground, injuring carrots, sunflowers, celery, sugar beets, potatoes, and several other plants.

Genus *Oxygrylius* Casey
Oxygrylius ruginasus (Lec.), Third-Stage Larva
(Fig. 361)

This description is based on the following material and reprinted from my 1944 bulletin:

> One third-stage larva, and cast skin of one third-stage larva collected at Canton, Mississippi. No. 3993. (Loaned by the U. S. National Museum.)

Maximum width of head capsule of third-stage larva 2.8 to 3 mm. Surface of cranium with numerous small pores, yellow-brown. Exterior frontal setae consisting of a single prominent seta on each side. Posterior and anterior frontal setae apparently absent. Each anterior angle of frons with 2 to 3 setae. Dorsoepicranial setae about 3 on each side. Lateral margins of labrum rounded, not angulate posteriorly (Fig. 361). Haptomeral process of epipharynx consisting of an unnotched ridge. Epipharynx with a few sensilla interspersed among the setae of the right chaetoparia. Inner, concave surface of left mandible, distad of the molar area, with small tooth; tooth apparently absent on right mandible.

Maxilla with a row of 8 to 10 truncate stridulatory teeth and an anterior process. Last segment of antenna with 2 dorsal sensory spots. Ocelli vague.

Thoracic spiracle .23 to .26 mm long (measured dorsoventrally) and .14 to .16 mm wide. Respiratory plate of thoracic spiracle with a maximum of 14 to 16 "holes" along any diameter. Distance between the two lobes of the respiratory plate equal to or somewhat less than the dorsoventral diameter of the bulla. All spiracles similar in size with elliptical, C-shaped respiratory plates.

Abdominal segment 7 with an anterior, sparsely set, transverse patch of about 3 irregular rows of short setae and with 2 long setae; posteriorly with a sparsely set row of about 6 long setae. Abdominal segments 8 and 9 each with 2 widely separated, sparsely set, single transverse rows of slender setae. Raster with a triangular teges of about 40, rather short, somewhat curved setae. Lower anal lip with a fringe of about 13 long setae. Palidia absent. Claws bearing 2 setae.

Oxygrylius ruginasus is a southwestern dynastid occurring in Mississippi, Texas, and Arizona. Adult a shining dark red, oval. Anterior, median part of pronotum with a small excavated portion; anterior margin of this pit with a suggestion of a small horn. General appearance very similar to that of *Bothynus gibbosus*. Adult 13 to 17 mm long and 7.5 to 9.5 mm wide.

Key to Larvae of the Species of *Cheiroplatys* Hope and *Aphonus* LeConte

1. Prescutum of abdominal segment 7 with 2 or 3 irregular rows of short, stout setae in addition to a single transverse row of very long, slender setae (Fig. 374) 2
   Prescutum of abdominal segment 7 with only a single transverse row of long, slender setae (Fig. 371) 3

2. Distance between the two lobes of the respiratory plate of the thoracic spiracles much less than the dorsoventral diameter of the bulla (Fig. 353); first segment of antenna with 1 to 3 setae; maximum width of cranium more than 6 mm .... *Cheiroplatys pyriformis* Lec.
   Distance between the two lobes of the respiratory plate of the thoracic spiracles only slightly less than the dorsoventral diameter of the bulla; first segment of antenna bare of setae; maximum width of head capsule less than 4.5 mm ...... *Aphonus castaneus* (Melsh.)

3. Prescuta of mesothorax and metathorax each with a transverse row of 6 or fewer setae .... *Aphonus densicauda* Casey
   Prescuta of mesothorax and metathorax each with a transverse row of 10 or more setae .... *Aphonus tridentatus* (Say)

Genus *Cheiroplatys* Hope

Cheiroplatys pyriformis Lec., Third-Stage Larva

This description is based on the following material and reprinted from my 1944 bulletin:

Two third-stage larva collected April 8, 1917, at Maxwell, New Mexico, by J. D. Caffrey. (Loaned by the U. S. National Museum.)

Maximum width of head capsule of third-stage larva about 6.5 mm. Surface of cranium reddish-brown with numerous pits most of which bear setae. Frons with a fairly uniform covering of setae; primary setae difficult to distinguish. Lateral margins of labrum and epipharynx broadly rounded, not angulate. Haptomeral process beak-like, not notched. Right chaetoparia with a very few sensilla. Inner margin of mandibles smooth between scissorial and molar areas, not toothed. Maxilla with a row of 8 to 10 stridulatory teeth and a large anterior truncate process which is as long as 2 stridulatory teeth. Most of the stridulatory teeth have anteriorly projecting points. Last segment of antenna with a single, large, oval, dorsal sensory spot.

Thoracic spiracle .45 to .56 mm long and .3 to .33 mm wide. Respiratory plate with a maximum of 13 to 25 oval "holes" across any diameter. Distance between the two lobes of the respiratory plate much less than the dorsoventral diameter of the bulla. Abdominal spiracles of segments 1 to 4 similar in size; those of segments 5 to 8 progressively smaller.

Prescuta of mesothorax and metathorax each with a transverse patch of about 16 setae. Median parts of the scuta of mesothorax and metathorax bare. Scutella of mesothorax and metathorax each with a transverse row of about 18 long and short slender setae. Dorsa of abdominal segments 7 to 9 each with 2 widely separated rows of long and short, slender setae. Cephalad of the anterior row on the dorsum of segment 7 are 2 or 3 rows of short, stout setae. Pleural lobes of abdominal segments 1 to 8 with 6 to 12 setae; spiracular areas of abdominal segments 1 to 6 with 10 to 14 setae. Raster with a broad teges of about 30 flattened, slightly curved setae, which extends slightly less than half the distance from the anterior margin of the lower anal lip to the posterior margin of abdominal segment 9. Lower anal lip covered with setae similar in shape to those of the teges and with a caudal fringe of about 24 long setae. Claws each bearing 2 setae.

Cheiroplatys pyriformis is a southwestern dynastid occurring in New Mexico and Colorado. Adult reddish-black, oval (Fig. 385). Pronotum not excavated. Adult 15 to 18 mm long and 9.5 to 11 mm wide. According to Casey (1915), this genus inhabits the arid mountainous country of the southwest.

Genus *Aphonus* LeConte

Aphonus castaneus (Melsh.), Third-Stage Larva
(Fig. 346)

This description is based on the following material and reprinted from my 1944 bulletin:

Five third-stage larvae collected September 30, 1942, at the Racebrook Country Club at Orange, Connecticut, by J. P. Johnson, from golf course turf. (Loaned by the U. S. National Museum.)

Eleven third-stage larvae collected September 30, 1942, at Orange, Connecticut, by J. P. Johnson. In the Kentucky Agricultural Experiment Station collection.

One third-stage larva collected October 1942, at Springfield, Massachusetts. Determined by A. G. Böving. From the United States Department of Agriculture Japanese Beetle Laboratory.

Maximum width of head capsule of third-stage larva 3.6 to 4.1 mm with a mean of 3.8. Surface of cranium light reddish-brown with numerous pits most of which bear setae. Frons with a uniform covering of setae; primary setae of frons difficult to distinguish. Each anterior angle of frons with 3 to 6 setae. Dorsoepicranial setae 9 to 11 on each side. Lateral margins of labrum and epipharynx broadly rounded, not angulate. Haptomeral process beak-like, unnotched. Chaetopariae with no sensilla. Inner margin of mandibles smooth between scissorial and molar areas, not toothed. Maxilla with a row of sharp-pointed stridulatory teeth and a large anterior truncate process which is as long as 3 stridulatory teeth (Fig. 346). Last segment of antenna with a single, large, oval, dorsal sensory spot. Ocelli present.

Thoracic spiracle .31 to .35 mm long and .18 to .22 mm wide. Respiratory plate with a maximum of 12 to 17 elongate, irregularly margined "holes" along any diameter. Distance between the two lobes of the respiratory plate equal to or slightly less than the dorsoventral diameter of the bulla. Abdominal spiracles of segments 1 to 4 similar in size, those of segments 5 to 8 progressively smaller.

Prescuta of mesothorax and metathorax each with a transverse patch of 16 or more short and long, slender setae. Median parts of scuta of mesothorax and metathorax bare. Scutella of mesothorax and metathorax each with a transverse row of about 17 long and short, slender setae. Dorsa of abdominal segments 7 to 9 each with 2 widely separated rows of long and short, slender setae. Cephalad of the anterior row on the dorsum of segment 7 is a transverse patch of 2 to 4 rows of short, stout setae. A few long and short setae are scattered between the two rows on the dorsum of segment 9. Pleural lobes of abdominal segments 1 to 7 with 7 to 15 setae. Raster with a broad teges of 34 to 50 rather short, flattened, slightly curved setae, which extends slightly less than half the distance from the anterior margin of the lower anal lip to the posterior margin of abdominal segment 9. Lower anal lip covered with setae similar in shape to those of the teges and with a caudal fringe of 20 to 40 long setae. Claws each bearing 2 setae.

Aphonus castaneus is a small species which, according to Casey (1915), ranges from Massachusetts and Rhode Island southward to North Carolina. Adult reddish-brown, 9.5 to 10 mm long and 6 mm wide (Fig. 386). According to Sim (1934), the larvae are found in sandy soil and are rather abundant in the turf of coastal plains golf courses. Johnson (1942) records the curious fact that during a gentle rain on July 8, 1941, R. T. White and P. J. McCabe of the United States Bureau of Entomology and Plant Quarantine, observed a great number of larvae of *Aphonus castaneus* lying on the turf of a Groton, Connecticut, golf course. When the sun came out, the larvae quickly burrowed into the soil.

Aphonus densicauda Casey, Third-Stage Larva
(Figs. 335, 339, 347, 353, 358, and 366)

This description is based on the following material and reprinted from my 1944 bulletin:

Cast skins of 3 third-stage larvae collected under bluegrass sod at Lexington, Kentucky, by the writer, and reared to the adult stage. Adults identified by E. A. Chapin.

One third-stage larva collected by the writer behind the plow February 17, 1938, at Lexington, Kentucky, in bluegrass sod.

Two third-stage larvae collected by the writer November 9, 1942, at Lexington, Kentucky, in bluegrass sod.

One third-stage larva collected by the writer May 3, 1943, at Lexington, Kentucky, in bluegrass sod.

Maximum width of head capsule of third-stage larva 4.8 to 5.2 mm. Surface of head light red-brown, covered with numerous pits many of which bear setae (Fig. 339). Surface of frons very flat. Exterior frontal setae consisting of 1 or 2 large setae and 1 to 5 smaller setae on each side. Primary frontal setae difficult to distinguish because of the large number of setae present. One to 3 small anterior frontal setae on each side. Posterior frontal setae consisting of 1 or 2 large setae on each side and 2 to 6 smaller ones. Each anterior angle of frons with 2, or rarely 3, setae. Dorsoepicranial setae 7 to 9 on each side. Lateral margins of labrum (and epipharynx) rounded, not angulate (Fig. 358). Haptomeral process entire, beak-like, not notched. Chaetopariae with no sensilla among the setae. Inner margin of mandible, between scissorial and molar areas, smooth, not toothed. Maxilla with a row of 9 to 11 stridulatory teeth

and a large anterior truncate process. Most or all of the stridulatory teeth with anteriorly projecting points (Fig. 347). Dorsal surface of last antennal segment with a single large, oval sensory spot (Fig. 335). Ocelli present and pigmented.

Thoracic spiracles .346 to .46 mm long and .24 to .304 mm wide. Respiratory plate with a maximum of 12 to 20 "holes" along any diameter. "Holes" oval to oblong, of irregular shapes (Fig. 355); those near outer edge of respiratory plate with finger-like processes. Distance between the two lobes of the respiratory plate slightly to much less than the dorsoventral diameter of the bulla (Fig. 353). Spiracles of abdominal segments 1 to 4, similar in size; those of abdominal segments 5 to 8, progressively smaller posteriorly. Bullae of spiracles slightly convex.

Prescuta of mesothorax and metathorax each with 4 long setae. Middorsal portions of scuta of mesothorax and metathorax each with a pair of setae. Dorsa of abdominal segment 7 to 9 each with 2 sparse, widely separated, transverse rows of slender setae. Pleural lobes of abdominal segments 1 to 8 with 2 to 5 setae, spiracular areas with 1 to 6 setae. Raster with a teges of 35 to 50 rather short, flattened setae having slightly curved tips (Fig. 366). Teges not quite extending to the midtransverse line of abdominal segment 10; campus anterior to teges totally devoid of setae. Lower anal lip covered with setae similar in shape to those of the teges and fringed caudally with 22 to 25 long setae. Claws each bearing 2 setae.

Aphonus densicauda is a rather uncommon native species described by Casey (1915) from Pennsylvania. It is fairly common at Lexington, Kentucky. Adult oval, reddish-black to black, 14 to 16 mm long and 8 to 9.5 mm wide. Full grown larvae found in pasture land, in or just beneath the sod, from November to May, in Kentucky. Pupation occurs in late May or early June. Adults are found in soil throughout the year.

Aphonus tridentatus (Say), Third-Stage Larva

This description is based on the following material and reprinted from my 1944 bulletin:

One third-stage larva (and 4 second-stage larvae) collected September 1936, in the Huron National Forest, Michigan, by L. E. Yeager, from forest duff. (Loaned by the U. S. National Museum.)

Maximum width of head capsule of third-stage larva about 4.9 mm. Surface of head light red-brown, covered with numerous pits many of which bear setae. Surface of frons flat. Primary frontal setae difficult to distinguish because of the large number of frontal setae. Lateral margins of labrum (and epipharynx) rounded, not angulate. Haptomeral process entire, somewhat beak-like, not notched. Epipharynx with no sensilla among the setae of the chaetopariae. Inner margin of both mandibles, between scissorial and molar areas, smooth, not toothed. Maxilla with a row of 10 to 12 stridulatory teeth and an anterior truncate process. Stridulatory teeth with anteriorly projecting points. Dorsal surface of last antennal segment with a single, large, oval sensory spot. Ocelli present and pigmented.

Thoracic spiracles about .39 mm long and .22 mm wide. Respiratory plate with a maximum of about 20 "holes" along any diameter. "Holes" of irregular shapes; distance between "holes" less than the diameter of the "holes." Distance between the two lobes of the respiratory plate slightly less than the dorsoventral diameter of the bulla. Spiracles of abdominal segments 1 to 5 similar in size; those of abdominal segments 6 to 8, progressively smaller posteriorly.

Prescuta of mesothorax and metathorax each with 10 or more long and short, slender setae. Middorsal portions of scuta of mesothorax and metathorax bare. Dorsa of abdominal segments 7 to 9 each with 2 sparsely set, widely separated, transverse rows of slender setae. Pleural lobes of abdominal segments 1 to 8 with 3 to 6 setae, spiracular areas with 2 to 6 setae. Raster with a teges of about 45 flattened setae having slightly curved tips. Lower anal lip covered with setae similar in shape to those of the teges and fringed caudally with about 25 long setae. Claws each bearing 2 setae.

Aphonus tridentatus is a native species said by Casey (1915) to occur in Wisconsin.[26] The writer has examined specimens from Indiana and Michigan. Adult oval, dark reddish-black to black, 13 to 15 mm long and 7.5 to 9 mm wide. Larvae found in woodland loam.

Genus *Strategus* Hope
Key to Larvae of the Species of *Strategus*

1. Bullae of spiracles each with a raised, knob-like process (Figs. 348 and 351). Dorsa of abdominal segments 8 and 9 each posteriorly with many slender setae (Fig. 373) ....*Strategus julianus* Burm.
  Bullae of spiracles convex but each without a knob-like process. Dorsa of abdominal segments 8 and 9 each posteriorly with a few slender setae (Fig. 374)......... 2

2. Last segment of antenna with about 6 dorsal sensory spots; claws short (Fig. 333)..*Strategus splendens* Beauv.
  Last segment of antenna with 8 to 14 dorsal sensory spots; claws long, curved (Fig. 334) ....
  .... *Strategus antaeus* (Fab.)

---

[26] Casey also lists this species from Arkansas giving Say (1823) as his authority. Say, in his original description (pages 209 and 210) says the species "inhabits Arkansas" and comments "A single specimen was brought from the Arkansa by Thomas Nuttall." The Arkansa Territory of the 1820's included a great deal of territory besides the state of Arkansas and even extended across Kansas into Colorado. Therefore, the locality of Say's specimen is problematical.

Strategus antaeus (Fab.), Third-Stage Larva
(Fig. 334)

This description is based on the following material and reprinted from my 1944 bulletin:

One third-stage larva collected in the winter or spring of 1940-41 at Raleigh, North Carolina, by C. F. Smith.

One third-stage larva collected at Newton Grove, North Carolina, in the fall of 1951, by P. O. Ritcher and H. H. Howden.

Four third-stage larvae, reared from eggs collected in burrows of this species at Newton Grove, North Carolina, in late August and early September 1951, by P. O. Ritcher and H. H. Howden.

Maximum width of head capsule 9 to 10.7 mm. Surface of cranium dark reddish-black, bearing numerous coarse pits. Frons on each side with 1 exterior frontal seta, 3 or 4 setae in each anterior angle; 1 or 2 anterior frontal setae, and a transverse pair of posterior frontal setae. Dorsoepicranial setae, 2 on each side. Left lateral margin of labrum (and epipharynx) slightly angulate. Haptomeral process beak-like, not notched. Chaetoparia with a very few sensilla among the setae. Inner margin of left mandible, distad of the molar area, with a large tooth. Maxilla with a row of 10 to 11 truncate stridulatory teeth and a broad, anterior truncate process. Last segment of antenna with 8 to 14 dorsal sensory spots. Ocelli present but vague.

Thoracic spiracle 1.0 to 1.26 mm long and .85 to 1.15 mm wide. Distance between the two lobes of the respiratory plate very much less than the dorsoventral diameter of the bulla. Spiracles of abdominal segments 1 to 7 similar in size, those of abdominal segment 8 smaller. Spiracular bullae convex but without a knob-like process.

Dorsum of abdominal segment 7 anteriorly with a transverse band of 4 or 5 irregular rows of short, stout setae among which is a transverse row of about 4 to 8 very long setae; posteriorly with a similar but much more sparsely set band of short, stout setae among which is a transverse row of about 16 to 25 very long setae. Dorsa of abdominal segments 8 and 9 each with an anterior and a posterior, sparsely set, transverse row of long setae, and adjacent to each row are a few scattered, very stout setae. Pleural lobes and spiracular areas of abdominal segments 1 to 8 each covered with about 35 to 50 setae. Raster with a teges of about 85 to 100 sharp-pointed, flattened setae. Teges bounded laterally and anteriorly by slender setae. Lower anal lip covered with setae similar in shape to those of the teges and with a caudal fringe of about 25 long, cylindrical setae. Claws long and sharp-pointed, each bearing 4 setae (Fig. 334).

Strategus antaeus is a large dynastid occurring from New Jersey southward to Florida and westward along the gulf coast. Adult a rich, dark red color. Sexes dimorphic. Pronotum of male with 3 prominent horns (Figs. 394 and 395). Adult 24 to 34 mm long and 15 to 17 mm wide. Manee (1908) gives an account of the biology of this species in the sand hill country of North Carolina. According to this writer, the larvae feed in the soil on leaf debris and later on decaying oak roots. A pair of adults was collected July 25, 1917, at Pine Knot, Kentucky.

In 1951, between August 28, 1951, and January 2, 1952, Henry Howden and the writer dug 11 burrows of *Strategus antaeus* in a burned-over pine woodland, 6 miles north of Newton Grove, North Carolina. Each was beneath a conspicuous sandy pushup. Burrows usually extended obliquely, were from 8 to 19 inches in depth, and were provisioned with surface litter. One burrow was straight, one was L-shaped, and the rest had a single branch or were forked. Usually a single egg was found toward the end of each burrow, or each branch of burrows, excavated in August and September. First instars were found in the two burrows excavated on October 12, 1951, and three second instars were found in burrows excavated on January 2, 1952.

Strategus julianus Burm., Third-Stage Larva
(Figs. 348, 351, and 373)

This description is based on the following material and reprinted from my 1944 bulletin:

One third-stage larva collected July 27, 1930, at Harlinges, Texas, by W. D. Wood, in date palm. (Loaned by the U. S. National Museum.)

Two third-stage larvae collected June 1, 1919, at Victoria, Texas, by J. D. Mitchell, in a rotten log. No. L-261. (Loaned by the U. S. National Museum.)

Two third-stage larvae collected January 6, 1909, at Brownsville, Texas, by F. H. Chittenden in rotten ash. (Loaned by the U. S. National Museum.)

One third-stage larva. Locality given as West Indies (probably wrong). (Loaned by the U. S. National Museum.)

Maximum width of head capsule of third-stage larva 10.35 to 12.1 mm with a mean of 11.5 mm. Surface of cranium dark red, bearing numerous pits which are more abundant and deeper on the anterior half of the frons. Exterior frontal setae consisting of 1 or 2 setae caudad of each precoila. Anterior frontal setae consisting of 1 to 3 small setae on each side. Posterior frontal setae, 1 or 2 on each side. Each anterior angle of frons with 3 to 5 setae. Dorsoepicranial setae 4 to 8 on each side. Left lateral margin of labrum (and epipharynx) angulate. Haptomeral process entire, not notched. Chaetopariae with a few distal sensilla among the setae. Inner

margin of left mandible, distad of the molar area, with a distinct tooth. Maxilla with a row of 5 to 8 truncate stridulatory teeth and a prominent anterior process. Last segment of antenna with 7 to 13 dorsal sensory spots. Ocelli present but vague.

Thoracic spiracle 1.82 to 2.4 mm long and 1.33 to 1.54 mm wide. Respiratory plate with a maximum of over 50 small oval, irregularly shaped "holes" along any diameter. Distance between the two lobes of the respiratory plate very much less than the dorsoventral diameter of the bulla. Spiracles of abdominal segments 1 to 8 similar in size. Spiracular bullae each with a knob-like process, each of which bears a more or less distinct nipple (Figs. 348 and 351).

Dorsa of abdominal segments 7 to 9 each with 2 transverse bands of setae which are confluent laterally (Fig. 373). Median portion of each anterior, transverse band consists of scattered small, stout setae and a single, central transverse row of very long setae. Median portion of each posterior transverse band consists of a few small, stout setae, a single transverse row of very long setae and many posterior slender setae of moderate length. Pleural lobes and spiracular areas of abdominal segments 1 to 8 each with a dense covering of slender setae. Raster with a teges of 38 to 55 rather short, sharp, flattened setae. Teges bounded laterally and anteriorly by slender setae. Lower anal lip covered with setae similar to those of the teges and with a caudal fringe of 30 to 105 long, slender setae. Claws each have 3 or 4 setae.

Strategus julianus is a very large, southern dynastid, which, according to Casey (1915), occurs from Vicksburg, Mississippi, to El Paso, Texas, and ranges southward to Central America. Adult a rich mahogany color with slightly darker pronotum. Sexes dimorphic. Female with the median, anterior part of the pronotum excavated and a slight suggestion of an anterior process. Pronotum of male similarly excavated but divided by a longitudinal ridge terminating in an anterior up-curved horn, posteriorly, on the raised part of the pronotum, with a process on each side. Adult 37 to 47 mm long and 19 to 23 mm wide (Fig. 392). Larvae feed in decaying wood. Casey (1915) found many adults of this species at electric lights at Alexandria, Louisiana, on June 1, 1901.

Strategus splendens (Beauv.), Third-Stage Larva
(Figs. 333, 342, 343, and 365)

This description is based on the following material and reprinted from my 1944 bulletin:

Four third-stage larvae collected with adults of this species April 18, 1927, at St. Petersburg, Florida, by J. M. Schultz. (Loaned by the U. S. National Museum.)

Maximum width of head capsule of third-stage larva 9.5 to 10.1 mm. Surface of cranium coarsely pitted, reddish. Exterior frontal setae consisting of a single seta located caudad of each precoila. A single posterior frontal seta present on each side. Anterior frontal setae absent. Dorsoepicranial setae 4 to 7 in number on each side. Lateral margins of labrum (and epipharynx) angulate posteriorly, especially on the left side (Fig. 365). Haptomeral process of epipharynx in the form of an unnotched, ridge-like elevation. Epipharynx with a few sensilla interspersed among the setae of the chaetopariae. Inner margin of each mandible, distad of their molar areas, with a distinct tooth (Figs. 342 and 343). Maxilla with a row of about 10 truncate stridulatory teeth and an anterior process. Last segment of antenna with about 6 dorsal sensory spots. Ocelli present but vague.

Thoracic spiracles 1.2 to 1.4 mm long and .91 to .98 mm wide. Respiratory plate with a maximum of about 45 very small, irregularly shaped, elongate "holes" along any diameter. The two lobes of the respiratory plate are almost contiguous. Spiracles of abdominal segments 1 to 5 similar in size, while those of abdominal segments 6 to 7 are slightly smaller and those of segments 8 are much smaller.

Dorsum of abdominal segment 7 with 3 vague annulets. Prescutum with a transverse patch of 4 to 6 irregular rows of rather short setae and 6 long setae. Scutum with a few scattered short setae and about 18 long setae. Scutellum with a few short setae in a transverse row. Setation of abdominal segments 8 and 9 limited on each segment to 2 sparsely set, widely separated, transverse rows of long and short, slender setae. Pleural lobes of abdominal segments 1 to 8 each with 15 to 30 long setae. Spiracular areas each with 10 to 25 long setae. Raster with a patch of about 65 straight or very slightly curved tegillar setae. Lower anal lip with a fringe of 25 to 30 long, straight setae. Claws short, bearing 3 or 4 prominent setae (Fig. 333).

Strategus splendens is a large, rather rare dynastid occurring from North Carolina to Florida (Casey, 1915). Adult a shining, rich red-brown (mahogany) color. Both sexes with the central, anterior part of the pronotum excavated. Anterior to the pronotal excavation is a short, median horn-like process. Adult 26 to 34 mm long and 14.5 to 17.5 mm wide.

Genus *Xyloryctes* Hope

Xyloryctes jamaicensis (Drury), Third-Stage Larva
(Figs. 341, 359, and 372)

This description is based on the following material and reprinted from my 1944 bulletin:

Two third-stage larvae collected October 5, 1901, at Wilbraham, Massachusetts, by G. Dimmock, from beneath pieces of chestnut bark. No. 2032. (Loaned by the U. S. National Museum.)

Thirteen third-stage larvae collected September 11, 1921, at Urbana, Illinois, by W. P. Flint and C. P. Alexander. From the Illinois Natural History Survey.

Maximum width of head capsule of third-stage larva 7.4 to 8.3 mm. Cranium deep reddish-black. Exterior, posterior, and anterior frontal setae absent. Each anterior angle of the frons with a single large seta. Dorsoepicranial setae absent. About 4 long setae near the antennal support. Epicranium, frons, and clypeus uniformly pitted with coarse punctures. Lateral margins of labrum (and epipharynx) rounded, not angulate posteriorly. Haptomeral process of epipharynx unnotched (Fig. 359). Epipharynx with several sensilla interspersed among the setae of the right chaetoparia. Inner margins of mandibles, between scissorial and molar areas, not toothed (Fig. 341). Each maxilla with a row of 8 to 11 truncate stridulatory teeth and an anterior process which is almost twice as broad as a truncate tooth. Last segment of antenna with 2 dorsal sensory spots. Ocelli present.

Thoracic spiracles .84 to .93 mm long and .63 to .66 mm wide. Respiratory plate with a maximum of 27 to 32 small, oval to oblong, irregularly shaped "holes," along any diameter. Distance between the two lobes of the respiratory plate very much less than the dorsoventral diameter of the bulla. Spiracles of abdominal segment 8 distinctly smaller than those of abdominal segments 1 to 7. Bullae of all spiracles rather strongly convex.

Prescuta of mesothorax and metathorax bare. Dorsum of abdominal segment 7 with a single posterior, transverse row of about 6 long setae, anteriorly sometimes with a few very short setae (Fig. 372). Dorsa of abdominal segments 8 and 9 each bare except for a posterior transverse row of 4 to 6 long setae. Raster with a teges of 5 to 70 slightly flattened and curved setae. Lower anal lip with about the same number of similar, rather short setae and a caudal fringe of 9 to 12 long, cylindrical setae. Claws each bearing 2 setae.

Xyloryctes jamaicensis is a large, northern dynastid occurring in Massachusetts, New York, Wisconsin, Indiana, Illinois, Kentucky, and North Carolina. According to Casey (1915), its range extends to Texas. The species is rather abundant in woodlands near Urbana, Illinois, and at Raleigh, North Carolina. Adult a uniform, dark reddish-black. Sexes dimorphic. Head of male with a prominent, upcurved horn (Figs. 388 and 389). Adult 24 to 30 mm long and 14 to 15 mm wide. Larvae found in deciduous woods under leaf mold. The winter is passed in the full-grown larval stage.

TRIBE *Dynastini*

Genus *Dynastes* Kirby
Key to Larvae of the Species of *Dynastes*

1. Teges with 10 to 14 short, flattened, tapering setae; right chaetoparia of epipharynx with few or no sensilla ....*Dynastes granti* Horn
   Teges with 16 to 35 short, flattened tapering setae (Fig. 367); right chaetoparia of epipharynx with numerous sensilla (Fig. 360) ....*Dynastes tityus* (L.)

Dynastes granti Horn, Third-Stage Larva

This description is based on the following material:

One third-stage larva collected April 1, 1961, northeast of Portal, Arizona, in rotten sycamore, by Mont Cazier.

One third-stage larva collected April 3, 1961, in Jhus Canyon, Chiracahua Mountains, Arizona, from *Platanus racemosa,* by Cazier and Mortenson. No. 61-2B 3.

One third-stage larva collected in 1961, near the American Museum of Natural History, Southwest Research Station, Portal, Arizona, by Mont Cazier.

Maximum width of head capsule 11 to 13 mm. Cranium reddish-black, coarsely pitted, with many of the pits bearing small setae. Primary frontal setae hard to differentiate; anterior frontal setae apparently absent, with 4 or 5 anterior angle setae, 1 or 2 exterior frontal setae, and 2 or 3 posterior frontal setae, on each side. Epicranial setae 6 to 8 in number on each side. Eye spots present, not pigmented. Epipharynx with a raised, beak-like haptomeral process. Chaetoparia with few (2 to 4) or no sensilla. Tormae asymmetrical. Left mandible with prominent tooth in inner margin, between scissorial and molar areas; right mandible with a small tooth in same region. Maxillary stridulatory area with 4 to 7 truncate teeth and an anterior process. Last segment of antenna with 6 to 9 small, round to oval, dorsal sensory spots.

Thoracic spiracles 2.0 to 2.1 mm long and 1.3 to 1.5 mm wide. Respiratory plate with a maximum of 55 to 70 small "holes" along any diameter. The two lobes of the respiratory plate almost contiguous; blackish rim of respiratory plate completely surrounding bulla. Spiracles of abdominal segments similar in size. Dorsa of abdominal segments 7 to 9 each covered with short, stiff setae plus sparsely set anterior and posterior rows of long setae. Raster with a very small, triangular teges of 10 to 14 flattened, tapering setae. Lower anal lip with 52 to 58 setae similar to those of the teges and fringed posteriorly with about 55 long setae. Claws sharp, each bearing 2 setae.

Dynastes granti Horn is a large rhinoceros beetle occurring in Arizona, which closely resembles *D. tityus.* Saylor (1948) gives a key for separating males of the two species, but is not sure of the validity of *granti* which he calls "a weak subspecies." The larvae are found in rotten wood.

Dynastes tityus (L.), Third-Stage Larva
(Figs. 332, 337, 350, 354, 357, 360, and 367)

This description is based on the following material and reprinted from my 1944 bulletin:

Three third-stage larvae collected February 11, 1936, at Louisville, Kentucky, by T. P. Boston, from an old locust tree. Associated with an adult female.

One third-stage larva collected September 14, 1929, in Union County, Kentucky, by E. J. Kilpatrick. Associated with a female pupa.

Exuviae of 2 third-stage larvae collected September 7, 1907, at Prospect, Kentucky, from a wild-cherry stump. In pupal cells with 2 female pupae.

Maximum width of head capsule of third-stage larva 10.1 to 11.2 mm. Surface of cranium dark red, covered with many prominent pits. Exterior frontal setae consisting of a single small seta located caudad of each precoila. Anterior frontal setae absent. Posterior frontal setae, 1 or rarely 2 on each side. Each anterior angle of frons with 3 to 5 setae. Dorsoepicranial setae 2 to 5 in number on each side. Left lateral margin of labrum (and epipharynx) angulate (Fig. 360). Haptomeral process entire. Right chaetoparia of epipharynx bearing a number of sensilla; left chaetoparia with a few sensilla. Inner margin of left mandible, distad of the molar area, with a distinct tooth; rather vague on the right mandible. Maxillary stridulatory area with a row of 5 to 8 truncate stridulatory teeth and a wide, anterior truncate process (Fig. 350). Last segment of antenna with 7 or 8 dorsal sensory spots (Fig. 337). Ocelli present.

Thoracic spiracles 1.6 to 2.1 mm long and .84 to 1.26 mm wide. Respiratory plate with a maximum of over 50 very small, round, irregularly shaped "holes" along any diameter (Fig. 357). The two lobes of the respiratory plate almost contiguous (Fig. 354). Spiracles of abdominal segments 1 to 8 similar in size.

Dorsa of abdominal segments 7 and 8 each with 3 distinguishable annulets. Each prescutum and scutum covered with numerous short, stout setae and among these on each annulet is a sparsely set, transverse, posterior row of long, slender setae. Each scutellum with a sparsely set, irregular, transverse row of shorter, slender setae which is interrupted medianly. Dorsum of abdominal segment 9 covered with numerous short, stout setae among which are 2 widely separated, sparsely set, transverse rows of long, slender setae. Pleural lobes of abdominal segments 1 to 8 each with about 30 to 35 setae. Spiracular sclerites each with 20 to 40 long setae. Raster with a small teges consisting of 16 to 35 short, fairly sharp, straight, flattened setae (Fig. 367). Teges bounded laterally and anteriorly by numerous long, slender setae. Lower anal lip covered with about 60 setae similar in shape to those of the teges and with a caudal fringe of 45 to 70 long, slender setae. Claws sharp, each bearing 2 setae (Fig. 332).

Dynastes tityus is a very large insect of striking appearance commonly called the rhinoceros beetle. It occurs throughout the southeastern states, ranging as far north as New Jersey, Pennsylvania, and Indiana and as far west as Arkansas, Texas, and Missouri (Chittenden, 1899; and Casey, 1915). The beetle is fairly common in Kentucky. Color of adult variable, ranging from yellow-green or gray with mahogany spots to a uniform dark-mahogany color. Sometimes with one elytron spotted, the other mahogany colored.[27] Legs and head blackish. Sexes dimorphic. The male has a large, upcurved, horn-like protuberance on the head and a large, anteriorly projecting, slightly down-curved, median, horn-like protuberance on the pronotum. Slightly below and to each side of the pronotal horn is a much smaller, horn-like protuberance. Adult 40 to 60 mm long (including the "horns" of the male) and 20 to 27 mm wide (Figs. 390 and 391). Larvae feed on decaying wood of wild cherry, black locust, oak, pine, willow, and other trees. Adults feed on the sap of ash trees at scars made in the bark; they are also attracted to decaying peaches, plums, pears, and apples (Manee, 1915). Pupation occurs in late summer. The adults hibernate in their pupal cells in decaying wood (see footnote). Eggs are laid the following summer. The fact that large larvae also overwinter suggests a 2-year cycle.

TRIBE Phileurini

Key to Larvae of the Species of Phileurini

1. Last segment of antenna with 2 dorsal sensory spots 2
   Last segment of antenna with 3 to 5 dorsal sensory spots 3

2. Entire frons coarsely pitted..*Archophileurus cribrosus* (Lec.)[28]
   Anterior two thirds of frons with small, shallow pits..
   .... *Phileurus castaneus* Hald.

---

[27] On December 27, 1938, near Stamping Ground, Kentucky, 15 pupal cases of *Dynastes tityus* were dug from an old wild-cherry stump by a dog which had treed a rabbit. Fourteen pupal cells were opened and found to contain 7 males and 7 females. Of these, 8 individuals were spotted, 5 had one elytron spotted and the other a solid dark-mahogany color, and 1 beetle was of a uniform dark-mahogany color.

[28] Based on material sent me recently by Howden. The cast skin is too fragmentary for a description of the larva.

3. Entire frons coarsely pitted. Stridulatory area of mandible with broad posterior ridges ....*Phileurus illatus* (Lec.)
   Anterior half of frons with numerous, shallow pits. Stridulatory area of mandible with ridges of similar width .... *Phileurus didymus* (Linn)

Genus *Phileurus* Latreille

Phileurus castaneus Hald., Third-Stage Larva

This description is based on the following material and reprinted from my 1944 bulletin:

Cast skin of one third-stage larva collected by P. O. Ritcher in August 1943, at Lexington, Kentucky, from a cavity in a dead basswood tree. Reared adult determined by E. A. Chapin of the United States National Museum.

Maximum width of head capsule of third-stage larva about 4.35 mm. Frons with small, shallow pits scattered sparsely and irregularly over the anterior two thirds. Exterior frontal setae probably consisting of a single seta on each side. A single posterior frontal seta on each side. Anterior frontal setae absent. Left lateral margin of labrum (and epipharynx) somewhat angulate; right lateral margin rounded. None or perhaps a very few sensilla interspersed among the setae of the chaetopariae. Inner margin of left mandible, distad of the molar area, with a small tooth. Maxilla with a row of about 7 truncate stridulatory teeth and an anterior truncate process which is twice as broad as a stridulatory tooth. Last segment of antenna with 2 dorsal sensory spots. Dorsa of abdominal segments 7 to 9 each bearing a number of short setae and in addition 2 sparsely set, transverse, widely separated rows of long, slender setae. Palidia absent. Claws each with 2 setae.

According to Cazier (1939) *Phileurus castaneus* Hald. is the correct name for the species listed in the Leng catalogue (1920) as *Phileurus floridanus* Casey. The species is fairly common in collections, and the writer has seen specimens from Kansas, Texas, Mississippi, Indiana, and Kentucky. Adult oblong, flattened above, and resembling *Passalus cornutus* superficially. Adult black, 18.5 to 22 mm long and 9 to 11 mm wide (Fig. 387). Larvae live in wood. Adults are frequently attracted to light.

Phileurus illatus Lec., Third-Stage Larva
(Figs. 344 and 363)

This description is based on the following material and reprinted from my 1944 bulletin:

Two third-stage, larvae collected in Galiuro Mountain, Arizona, by H. G. Hubbard, in *Dasylirion*. No. 841. Tentative identification. (Loaned by the U. S. National Museum.)

One third-stage larva collected June 26, 1897, in Santa Rita Mountains, Arizona, by H. G. Hubbard, in the trunk of *Dasylirion*. (Loaned by the U. S. National Museum.) U. S. D. A. No. 7330.

One third-stage larva, associated with an adult, collected August 21, 1958, seven miles southeast of Rodeo, Hildalgo County, New Mexico, in association with *Bumelia languinosa,* by E. G. Linsley and P. D. Hurd.

Maximum width of head capsule of third-stage larva 4.8 to 6 mm. Surface of cranium and frons prominently pitted, reddish-brown. Exterior frontal setae consisting of a group of 1 large and 1 to 3 small setae on each side. A single posterior frontal setae is borne on each side. Anterior frontal setae absent. Dorsoepicranial setae 9 to 14 in number on each side, laterad of the epicranial suture. Lateral margins of labrum (and epipharynx) sharply angulate, posteriorly (Fig. 363). Haptomeral process of epipharynx in the form of an unnotched ridge-like elevation. Few or no sensilla interspersed among the setae of the chaetopariae. Inner margin of mandible, distad of the molar area, toothed (Fig. 344). Maxilla with a row of 7 to 9 blunt stridulatory teeth and an anterior process. Last segment of antenna with 3 or 4 dorsal sensory spots. Ocelli present.

Thoracic spiracles .42 to .63 mm long and .28 to .44 mm wide. Respiratory plate with a maximum of 18 to 24 small, oval, irregularly margined "holes" along any diameter. Distance between the two lobes of the respiratory plate slightly to much less than the dorsoventral diameter of the bulla. Spiracles similar in size; bullae slightly convex.

Dorsa of abdominal segments 7 to 9 each bearing a number of short setae and 2 transverse, widely separated, rows of long, slender setae. Raster with a teges of 15 to 22 setae. Tegillar setae frequently slightly curved at tips but not hamate. Lower anal lip with a fringe of about 20 long setae. Claws each with 2 setae.

According to Casey (1915), *Phileurus illatus* occurs in Arizona and southern California. Adult oblong, flattened, black, 20 to 24 mm long and 9.5 to 11 mm wide. In the male, the tubercles in front of the eyes are es-

pecially prominent. Color, size, and shape very similar to that of *Ph. castaneus.* Larvae live in wood.

Phileurus didymus (Linn.), Third-Stage Larva (Fig. 378)

This description is based on the following material:

Two third-stage larvae and one adult collected July 30, 1958, at Simojovel, Chiapas, Mexico, in a decayed stump of *Spondias mombin* L. by J. A. Chemsak. Loaned by the University of California.

Maximum width of head capsule 6.7 to 8 mm. Cranium reddish, finely shagreened. Anterior two thirds of frons with scattered, small, shallow pits. Frons, on each side, with one anterior frontal seta, 1 to 3 setae in each anterior angle, one exterior frontal seta, and a transverse pair of posterior frontal setae. Dorsoepicranial setae 2 or 3, on each side. Lateral margins of labrum (and epipharynx) broadly rounded, not angulate. Haptomeral process of epipharynx ridge-like, not notched. Chaetoparia well developed, without sensillae. Inner margin of left mandible (Fig. 378), distad of molar area, with a prominent, triangular tooth; right mandible with a similar but much smaller tooth. Stridulatory area of each mandible with transverse ridges that are similar in width. Maxilla with a row of 7 to 8 truncate stridulatory teeth and a prominent anterior process. Last segment of antenna with 3 to 5 round or oval, dorsal sensory "spots."

Thoracic spiracles 1.1 to 1.2 mm long and .67 to .74 mm wide. Respiratory plate with a maximum of 24 to 28 subquadrate to circular "holes" along any diameter. Distance between the two lobes of the respiratory plate somewhat less than dorsoventral diameter of bulla.

Dorsa of abdominal segments 8 and 9 covered with numerous short setae; each posteriorly also with a sparsely set, transverse row of longer setae. Raster with 2 adjacent patches of 45 to 60 short, stout setae; setae of each patch directed somewhat laterally. Lower anal lobe entire, covered with short, stout setae similar to those of the raster; fringed posteriorly with longer, slender setae. Claws each with 2 setae.

Phileurus didymus (Linn.) is a large, widely distributed species which occurs in Mexico, Central America, South America, and several of the West Indies (Blackwelder, 1944). Both larvae and adults are found in decayed wood.

Plate XXVII

Cyclocephala immaculata Oliv.

Figure 317.   Left mandible, dorsal view.

Figure 318.   Right mandible, dorsal view.

Figure 319.   Third-stage larva, left lateral view.

Figure 320.   Molar region of left mandible, median view.

Figure 321.   Left mandible, lateral view.

Figure 322.   Molar region of right mandible, median view.

Figure 323.   Right mandible, ventral view.

Figure 324.   Left mandible, ventral view.

Figure 325.   Head, cephalic view.

Figure 326.   Venter of tenth abdominal segment.

Figure 327.   Epipharynx.

Figure 328.   Fifth abdominal segment, left lateral view.

Figure 329.   Right maxilla and labium, ventral view.

Figure 330.   Maxillary stridulatory area.

Figure 331.   Right maxilla, labium and hypopharyngeal sclerome, ental view.

Symbols Used

AA—Anterior frontal angle
AC—Acia
ACP—Acanthoparia
ACR—Acroparia
ACS—Anterior clypeal seta
AFS—Anterior frontal seta
ASL—Anal slit
ASP—Asperities
B—Barbula
BR—Brustia
C—Campus
CA—Calx
CAR—Cardo
CO—Corypha
CPA—Chaetoparia
CR—Crepis
CS—Clypeofrontal suture
DC—Dorsal carina
DES—Dorsoepicranial setae
DIP—Dorsal impressed line
DMS—Dorsomolar setae
DSS—Dorsal sensory spots
DX—Dexiotorma
E—Epicranium
ECS—Exterior clypeal seta
EFS—Exterior frontal setae
ES—Epicranial suture
EUS—Eusternum
EZ—Epizygum
F—Frons

FS—Frontal suture
G—Galea
GL—Glossa
GP—Gymnoparia
H—Haptomeral process
HL—Haptolachus
HSC—Hypopharyngeal sclerome
L—Labrum
LA—Lacinia
LAF—Lateral face of mandible
LAL—Lower anal lip
LL—Lateral lobe
LP—Labial palpus
LT—Laeotorma
M—Mandible
M_{1-2}—Molar lobes
MA—Mala
MO—Mola bearing part
MP—Maxillary palpus
O—Ocellus
PA—Preartis
PC—Preclypeus
PCL—Precoila
PE—Pedium
PEA—Pedal area
PFS—Posterior frontal setae
PLL—Pleural lobe
PMP—Postmentum
PRA—Parartis
PRM_1—Prementum, distal sclerite

PRM_2—Prementum, proximal
    sclerite
PRSC—Prescutum
PSC—Postclypeus
PSCL—Postscutellum
PTA—Postartis
PTT—Pternotorma
S_{1-4}—Scissorial teeth
SA—Scissorial area
SC—Sense cone
SCL—Scutellum
SCR—Scrobis
SCU—Scutum
SD—Stridulatory teeth
SDT—Tubercle anterior to stridu-
    latory teeth
SN—Scissorial notch
SP—Sclerotized plate
SPA—Spiracular area
SPR—Spiracle
SSC—Subscutum
ST—Stipes
STA—Stridulatory area
STL—Sternellum
T—Teges
TP—Truncate process
UN—Uncus
VC—Ventral carina
VP—Ventral process
Z—Zygum

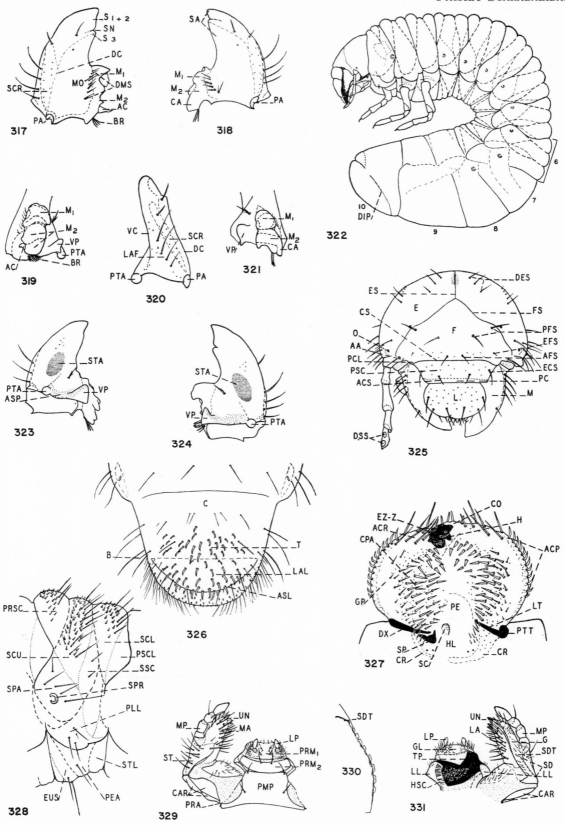

317

318

319

320

321

322

323

324

325

326

327

328

329

330

331

Plate XXVIII

Figure 332.   *Dynastes tityus* (L.). Claw of left mesothoracic leg, side view.

Figure 333.   *Strategus splendens* Beauv. Claw of left mesothoracic leg, side view.

Figure 334.   *Strategus antaeus* (Fab.). Claw of left mesothoracic leg, dorsal view

Figure 335.   *Aphonus densicauda* Casey. Distal segment of antenna, dorsal view.

Figure 336.   *Bothynus gibbosus* (DeG.). Diistal segment of antenna, dorsal view.

Figure 337.   *Dynastes tityus* (L.). Distal segment of antenna, dorsal view.

Figure 338.   *Bothynus gibbosus* (DeG.). Head, dorsal view.

Figure 339.   *Aphonus densicauda* Casey. Head, dorsal view.

Figure 340.   *Dyscinetus morator* (Fab.). Head, dorsal view.

Figure 341.   *Xyloryctes jamaicensis* (Drury). Left mandible, dorsal view.

Figure 342.   *Strategus splendens* Beauv. Left mandible, dorsal view.

Figure 343.   *Strategus splendens* Beauv. Scissorial area of right mandible, dorsal view.

Figure 344.   *Phileurus illatus* Lec. Left mandible, dorsal view.

Figure 345.   *Bothynus gibbosus* (DeG.). Left mandible, dorsal view.

Figure 346.   *Aphonus castaneus* (Melsh). Maxillary stridulatory area.

Figure 347.   *Aphonus densicauda* Casey. Maxillary stridulatory area.

Figure 348.   *Strategus julianus* Burm. Thoracic spiracle.

Figure 349.   *Bothynus gibbosus* (DeG.). Maxillary stridulatory area.

Figure 350.   *Dynastes tityus* (L.). Maxillary stridulatory area.

Figure 351.   *Strategus julianus* Burm. Thoracic spiracle, side view.

Figure 352.   *Bothynus gibbosus* (DeG.). Thoracic spiracle.

Figure 353.   *Aphonus densicauda* Casey. Thoracic spiracle.

Figure 354.   *Dynastes tityus* (L.). Thoracic spiracle.

Figure 355.   *Aphonus densicauda* Casey. Structure of respiratory plate of thoracic spiracle (highly magnified).

Figure 356.   *Cyclocephala immaculata* Oliv. Structure of respiratory plate of thoracic spiracle (highly magnified).

Figure 357.   *Dynastes tityus* (L.). Structure of respiratory plate of thoracic spiracle (highly magnified).

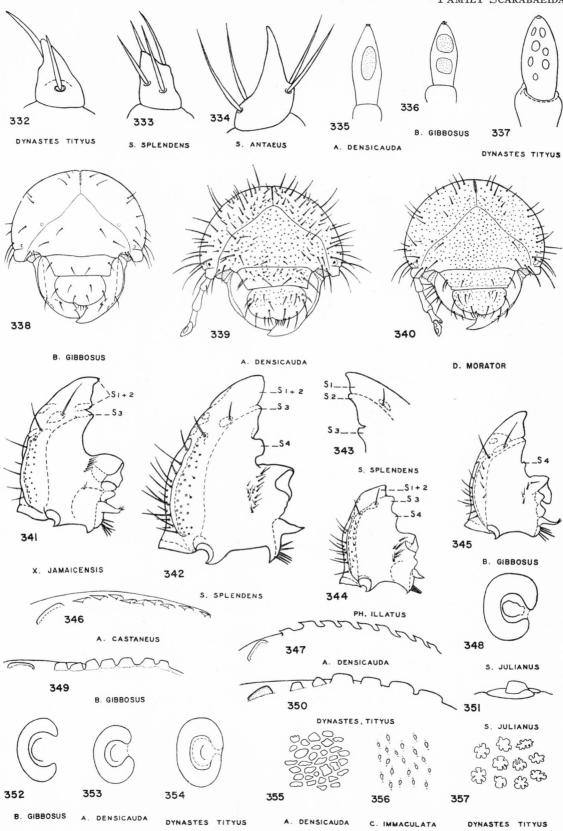

332 DYNASTES TITYUS

333 S. SPLENDENS

334 S. ANTAEUS

335 A. DENSICAUDA

336 B. GIBBOSUS

337 DYNASTES TITYUS

338 B. GIBBOSUS

339 A. DENSICAUDA

340 D. MORATOR

341 X. JAMAICENSIS

342 S. SPLENDENS

343 S. SPLENDENS

344 PH. ILLATUS

345 B. GIBBOSUS

346 A. CASTANEUS

347 A. DENSICAUDA

348 S. JULIANUS

349 B. GIBBOSUS

350 DYNASTES, TITYUS

351 S. JULIANUS

352 B. GIBBOSUS

353 A. DENSICAUDA

354 DYNASTES TITYUS

355 A. DENSICAUDA

356 C. IMMACULATA

357 DYNASTES TITYUS

Plate XXIX

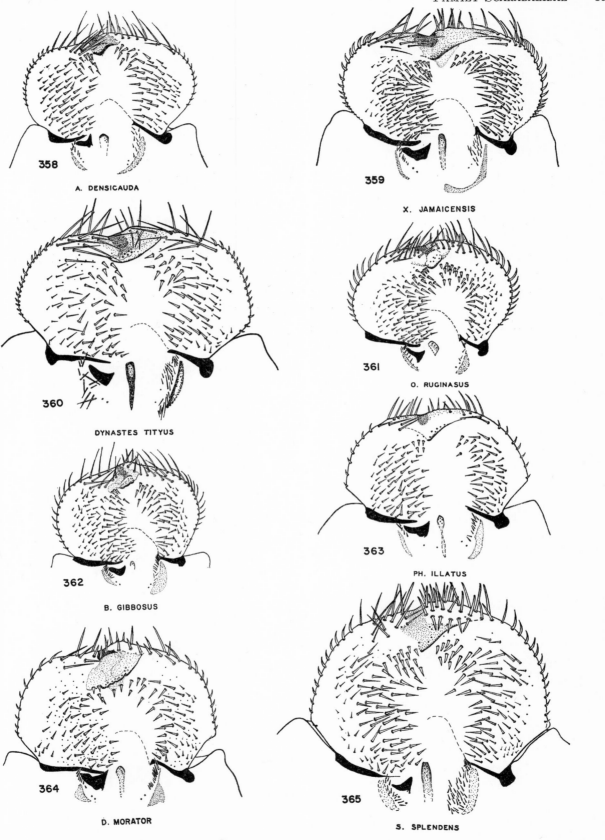

358   A. DENSICAUDA

359   X. JAMAICENSIS

360   DYNASTES TITYUS

361   O. RUGINASUS

362   B. GIBBOSUS

363   PH. ILLATUS

364   D. MORATOR

365   S. SPLENDENS

Plate XXX

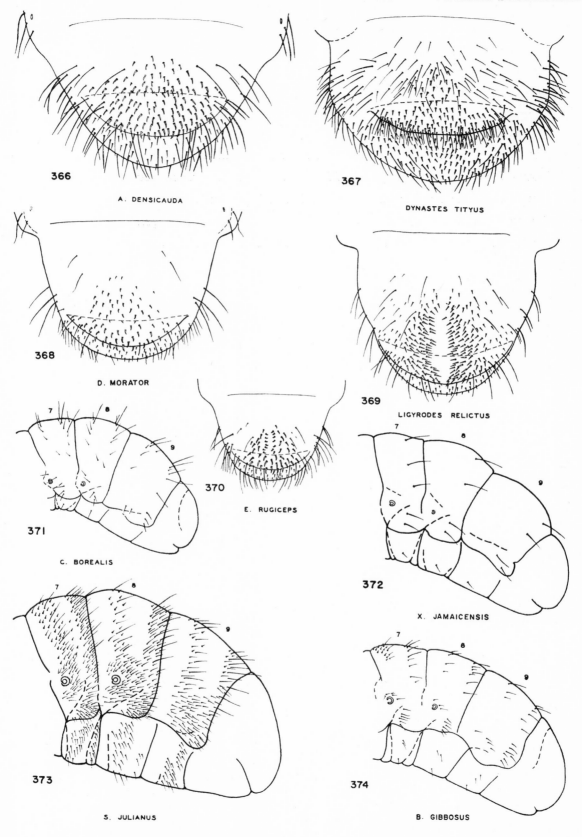

366   A. DENSICAUDA

367   DYNASTES TITYUS

368   D. MORATOR

369   LIGYRODES RELICTUS

370   E. RUGICEPS

371   C. BOREALIS

372   X. JAMAICENSIS

373   S. JULIANUS

374   B. GIBBOSUS

Plate XXXI

FIGURE 375. *Ancognatha manca* Lec. Head.

FIGURE 376. *Ancognatha manca* Lec. Epipharynx.

FIGURE 377. *Ancognatha manca* Lec. Left mandible, dorsal view.

FIGURE 378. *Phileurus didymus* (Linn.). Left mandible, dorsal view.

FIGURE 379. *Ancognatha manca* Lec. Left maxilla, dorsal view.

FIGURE 380. *Ancognatha manca* Lec. Venter of last abdominal segment.

FIGURE 381. *Ancognatha manca* Lec. Left mandible, ventral view.

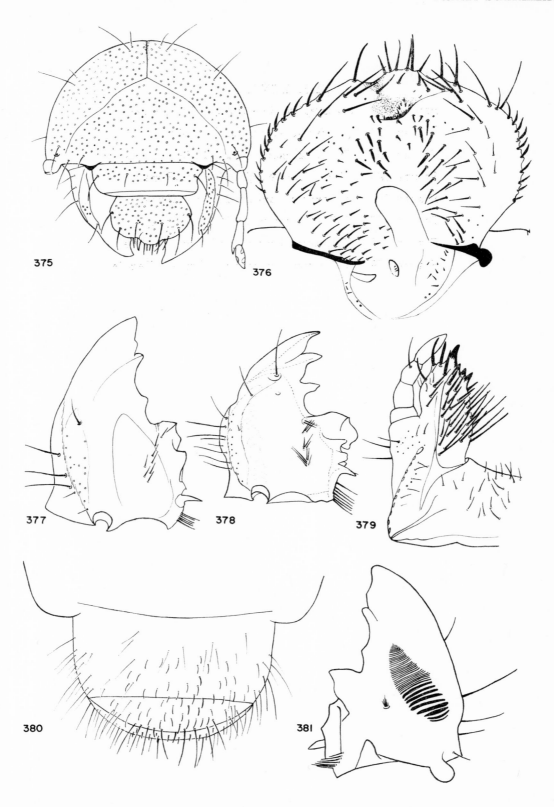

375

376

377

378

379

380

381

Plate XXXII

FIGURE 382.   *Cyclocephala immaculata* Oliv. Female.

FIGURE 383.   *Dyscinetus morator* (Fab.). Male.

FIGURE 384.   *Euetheola rugiceps* Lec. Male.

FIGURE 385.   *Cheiroplatys pyriformis* Lec. Female.

FIGURE 386.   *Aphonus castaneus* (Melsh.). Female.

FIGURE 387.   *Phileurus castaneus* Hald. Female.

FIGURE 388.   *Xyloryctes jamaicensis* (Drury). Male.

FIGURE 389.   *Xyloryctes jamaicensis* (Drury). Female.

Plate XXXIII

Figure 390. *Dynastes tityus* (L.). Female.

Figure 391. *Dynastes tityus* (L.). Male.

Figure 392. *Strategus julianus* Burm. Male.

Figure 393. *Xyloryctes jamaicensis* (Drury). Male.

Figure 394. *Strategus antaeus* (Fab.). Male.

Figure 395. *Strategus antaeus* (Fab.). Female.

Figure 396. *Ligyrodes relictus* (Say.). Female.

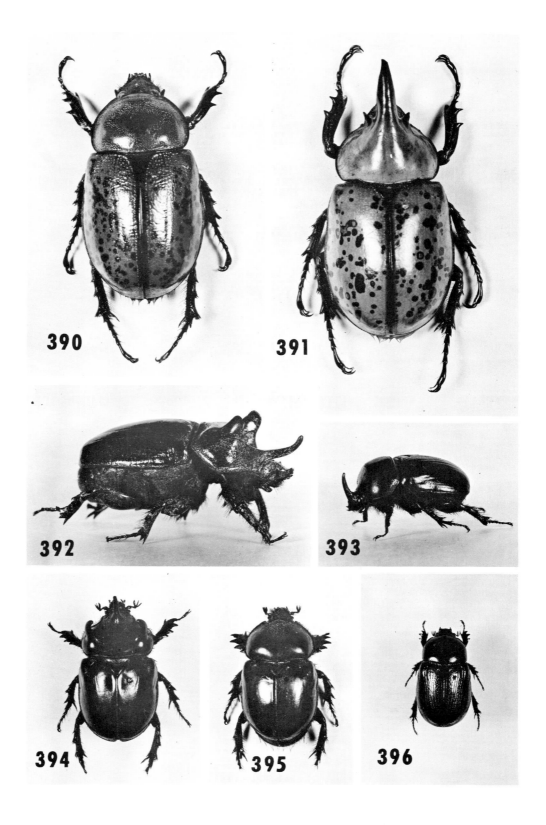

Subfamily Cetoniinae

The scarabaeid subfamily Cetoniinae includes over 2,500 species of small to very large, mostly bright-colored or highly variegated beetles. A majority of the species are tropical or subtropical, and, according to Casey (1915), the subfamily occurs in Africa in greatest profusion and variety. In America the group is represented, north of Mexico, by about 120 species belonging to 19 genera (Leng, 1920).

Many of the North American species of Cetoniinae are brightly colored, while others are dull black or have black and brown markings. The adults are subquadrate in shape, flattened dorsally. Their mouth parts are poorly developed. The name "flower beetles" is often given to the group since many species (such as those belonging to the genera *Trichiotinus, Trigonopeltastes,* and *Valgus*) frequent blossoms. Other species, such as the green June beetle, *Cotinis nitida* (L.), cause considerable damage by feeding on the juices of ripening grapes, apples, peaches, figs, and corn in the milk stage. Most of the species are diurnal.

The food of cetoniid larvae consists of organic matter in the soil, decaying wood or trash, and other debris accumulated in the hollows of trees and elsewhere. Larvae of the genera *Euphoria* and *Cotinis,* all of which crawl on their backs, feed in the soil on organic matter and are usually restricted to poorly drained areas or the vicinity of straw stacks, manure piles, or refuse heaps. *Cotinis nitida* larvae frequently cause severe damage by disturbing the soil of tobacco beds, lawns, and golf fairways and greens.

Trichiotinus larvae feed on decaying wood, often just under the bark. Larvae of the genus *Osmoderma* feed in decaying wood or on the accumulated wood fragments and other debris in cavities in fallen logs or standing trees. Both adults and larvae of the genus *Cremastocheilus* are found associated with ants (Wheeler, 1908); the larvae probably feed on vegetable debris. *Valgus* larvae feed on the wooden walls of termite burrows in standing dead trees or in logs.

Larvae of a number of North American species of Cetoniinae have been briefly described in the American literature (Böving and Craighead, 1931; Davis, 1916; Davis and Luginbill, 1921; Hayes, 1925, 1928, and 1929; and Sim, 1934). A key to six genera was published by Hayes in 1928 and 1929 and a key to groups of genera by Böving and Craighead in 1931. The material presented here, which includes 20 species belonging to 9 genera, is largely reprinted from my 1945 paper. All of the species treated by previous writers are included except the rare species, *Stephanucha pilipennis* Kr., partially described by Hayes (1928 and 1929) from a cast skin. The larva of *Valgus californicus* Horn is described for the first time.

Cetoniid larvae may be distinguished from other scarabaeid larvae by the following combination of characters: Labrum symmetrical. Mandibles each with a ventral, oval stridulatory structure consisting of transverse ridges (absent in some species of *Valgus*). Maxillary stridulatory teeth with anteriorly directed points (poorly developed in some species of *Valgus*). Epipharynx with a single nesium (sclerotized plate absent). Plegmata and proplegmata absent. Haptomerum of epipharynx frequently with a conspicuous, transverse, curved row of stout setae. Lacinia of each maxilla with a single distal uncus or with 2 unequal distal unci fused at their bases. Ninth and tenth abdominal segments in some genera fused dorsally. Palidia present or absent. Anal slit transverse, often slightly curved.

Key to Genera of the Subfamily Cetoniinae
Based on Characters of Third-Stage Larvae

1. Labrum trilobed; clithra present (Figs. 425 and 427). Distal segment of antenna with 2 or more dorsal sensory spots (Figs. 401 and 402). Ninth and tenth abdominal segments fused dorsally (Fig. 438) 2
   Labrum bilobed or entire (Figs. 406 and 409); clithra absent (Figs. 428 and 429). Distal segment of antenna with 1 dorsal sensory spot (Figs. 400, 405, and 409). Ninth and tenth abdominal segments not fused dorsally (Fig. 440) 9

2. Claws cylindrical, rounded apically, and bearing 7 or more setae (CL, Fig. 397). Venter of last abdominal segment with or without longitudinal palidia (Figs. 430 to 435 and 439). (Larvae crawl on their backs.) 3
   Claws subconical or falcate, each bearing 2 setae (Figs. 398 and 399). Venter of last abdominal segment without longitudinal palidia (Figs. 436 and 437). (Larvae not back-crawlers.) 8

3. Raster with each longitudinal palidium consisting of 2 or more irregular rows of pali (Figs. 430 and 431). Distal segment of antenna with 4 or more dorsal sensory spots (Fig. 402) and with 5 to 13 ventral sensory spots ....*Cotinis*
   Raster with or without longitudinal palidia; palidia monostichous if present (Figs. 432, 434, 435, and 439). Distal segment of antenna with 2 to 4 dorsal sensory spots (Fig. 401) and 2 to 4 ventral sensory spots 4

4. Raster with palidia 5
   Raster without palidia 6

5. Palidia elliptical ....*Gymnetis sallei* Schaum
   Palidia parallel or diverging posteriorly .... most *Euphoria*

6. Exterior and anterior frontal setae absent ....
   ....*Euphoria inda* (Linn.)
   Exterior and anterior frontal setae present 7

7. With long, anterior frontal setae ....
.... *Euphoriaspis hirtipes* (Horn)[29]
   With 2 or 3 very short anterior frontal setae ....
.... *Gymnetis cretacia* Lec.[29]

8. Claws short, subconical (Fig. 398) ....*Osmoderma*
   Claws falcate (Fig. 399) ....*Cremastocheilus*

9. Distal margin of labrum emarginate (Fig. 406). Haptomeral region of epipharynx without a circlet of
   stout spine-like setae interrupted by a tooth-like
   process (Figs. 426 and 429). Raster without short,
   spine-like setae (Fig. 436) ....*Valgus*
   Distal margin of labrum broadly rounded (Fig. 409).
   Haptomeral region of epipharynx with a circlet of
   stout, spine-like setae interrupted by a tooth-like
   process (HP, Fig. 428). Raster with several to
   many short, spine-like setae (Fig. 437) 10

10. Scissorial area of left mandible with 2 teeth, S_1 and
    S_2 fused (Fig. 412). Lacinia of maxilla with a
    single terminal uncus (LU, Fig. 420)......*Trigonopeltastes*
    Scissorial area of left mandible with 3 teeth, S_1 and
    S_2 not fused (Fig. 419). Lacinia of maxilla with 2
    terminal unci fused at their bases (Fig. 424) 11

11. Frons with an anterior, semicircular sunken area.
    Labrum without a median, transverse, emarginate
    protuberance (Fig. 405) ....*Gnorimella*
    Anterior part of frons convex. Labrum with a me-
    dian, transverse, emarginate protuberance (Fig.
    409) ....*Trichiotinus*

TRIBE Gymnetini

Genus *Gymnetis* MacLeay
Gymnetis sallei Schaum., Third-Stage Larva
(Figs. 414, 427, and 435)

This description is based on the following material
and reprinted from my 1945c bulletin:

> Three third-stage larvae collected December 31,
> 1919, at Blue Mott, Victoria County, Texas, by J.
> D. Mitchell, in rotten moss in a hollow anachua
> tree. (Loaned by the U. S. National Museum.)

Maximum width of head capsule of third-stage larva
4.06 to 4.34 mm. Surface of cranium smooth, light
reddish-brown. Frons with a median, longitudinal depression extending anteriorly from the epicranial stem.
Frons with a single, posterior frontal seta and a single
anterior angle seta on each side; other frontal setae
absent. Dorsoepicranial setae consisting of 1 large and
2 or 3 smaller setae on each side. Labrum trilobed;
clithra present. Haptomeral region of epipharynx with
a slightly curved, transverse row of conical setae (Fig.
427). Left mandible with 2 scissorial teeth posterior to
the scissorial notch (S_3 and S_4 present, Fig. 414). Maxillary stridulatory area consisting of a row of 3 to 5

---
[29] Larvae of these two species were received too late for their
descriptions to be included.

curved teeth with anteriorly projecting points and an
anterior truncate process about the size of the base of a
stridulatory tooth. Lacinia of maxilla with 2 terminal
unci, fused basally; dorsal uncus much the larger. Last
antennal segment with 2 dorsal sensory spots.

Thoracic spiracles .56 to .62 mm long and .42 to .46
mm wide. Respiratory plate with a maximum of about
35 small, irregularly shaped "holes" along any diameter. "Holes" not in definite rows. Distance between
the two lobes of the respiratory plate much to very
much less than the dorsoventral diameter of the bulla.
Spiracles of abdominal segments 1 to 8 similar in size.

Dorsa of abdominal segments 7 and 8 each divided
into 2 annulets, a prescutum, and a scutum. Each annulet
covered with short, stiff setae and fringed posteriorly
with long setae. Abdominal segments 9 and 10 fused
dorsally and with a dense covering of setae. Venter of
tenth abdominal segment posteriorly with paired palidia
and a pair of tegilla (Fig. 435). Each palidium set with
15 to 18 rather stout, moderately long, somewhat compressed pali in a single, slightly irregular, longitudinal,
curved row. Pali separated from each other by a space
equal to or somewhat larger than the width of a palus
at its base. Septula elongate-obovate to subelliptical.
The paired tegilla consist of numerous backward-pro-
jecting, sharp-pointed, slightly curved setae. Among
the preseptular, tegillar setae are numerous long setae.
Central transverse area of lower anal lip with 10 to 20
long, cylindrical setae. Claws cylindrical, each rounded
apically and bearing 10 to 12 setae.

Gymnetis sallei Schaum. is a moderately large southwestern species known to occur in Texas, Arizona, and
Mexico (Leng, 1920). Adult black with mustard-yellow
markings. Elytra and pronotum velvety-black with irregular yellow margins. Dorsal surface of head almost
entirely yellow. Venter and legs shiny black with scattered yellow markings. Shape very similar to *Cotinis*.
Adult 21 mm long and 12.5 mm wide. Little is known
about the habits of the larvae. Those described in this
paper were found at Blue Mott, Texas, in rotten moss
in a hollow anachua tree.

Genus *Cotinis* Burmeister

Key to Larvae of the Species of *Cotinis*

1. Raster with inner row of each palidium set with pali
   only slightly larger than those in the outer row
   (Fig. 430) ....*Cotinis nitida* (Linn.)
   Raster with inner row of each palidium having 7 to 10
   pali which are much stouter and larger than the
   other pali (Fig. 431) ....*Cotinis texana* Casey

Cotinis nitida (Linn.), Third-Stage Larva
(Figs. 397, 402, 403, 407, 408, 421, 425, and 430)

This description is based on the following material and reprinted from my 1945c bulletin:

Six third-stage larvae collected November 9, 1942, at Lexington, Kentucky, by P. O. Ritcher, from soil near a manure pile.

Four cast skins of third-stage larvae reared to the adult stage. Larvae collected May 26, 1943, at Malaga, Kentucky, by P. O. Ritcher, from pasture soil, behind the plow.

Maximum width of head capsule of third-stage larva 5.2 to 5.95 mm. Surface of cranium fairly smooth, reddish, with reddish-black clypeus, labrum, and mandibles. Frons with a shallow, median, longitudinal depression extending anteriorly from the epicranial stem. Frons (Fig. 403) bearing on each side a single (rarely 2) posterior frontal seta, a very small exterior frontal seta, a very small anterior frontal seta, and a single anterior angle seta. Dorsoepicranial setae 2 to 9 on each side, 2 or 3 large, the rest very small. Anterior margin of labrum trilobed; each lobe only slightly pigmented and bearing a number of long setae. Clithra present. Haptomeral region of epipharynx with a curved transverse row of moderately long, fairly stout setae, caudad of which are a number of similarly shaped setae (Fig. 425). Mandibles with 2 scissorial teeth posterior to the scissorial notch (Figs. 407 and 408). Maxillary stridulatory area (Fig. 421) consisting of 7 to 10 rather stout stridulatory teeth and an anterior truncate process. Stridulatory teeth with anteriorly directed points (Fig. 421). Lacinia of maxilla with 2 terminal unci, fused basally; the dorsal uncus much the larger. Last antennal segment with 3 to 7 dorsal sensory spots (Fig. 402).

Thoracic spiracles .74 to .88 mm long and .58 to .66 mm wide. Respiratory plate with a maximum of about 40 very small, irregularly margined "holes" along any diameter. Distance between the two lobes of the respiratory plate much less than the dorsoventral diameter of the bulla. Spiracles of abdominal segments similar in size.

Dorsa of abdominal segments 7 and 8 each divided into 2 annulets, each having a rather dense, uniform covering of fine setae, among which, across the central part of each annulet, is a sparse, transverse row of long setae. Dorsa of abdominal segments 9 and 10 fused. Venter of abdominal segment 10 with palidia. Each palidium consisting of 2 somewhat irregular, rather sparsely set, longitudinal rows of short, stout, compressed, falcate setae (Fig. 430). Palidia set closely together with pali mesally directed. Septula long and very narrow. Palidia surrounded laterally and anteriorly by a teges of numerous short, fine setae, among which are interspersed a few longer setae. Claws cylindrical, each rounded apically and bearing 11 or 12 terminal setae (Fig. 397).

Cotinis nitida (Linn.), commonly called the green June beetle or "June bug," is a moderately large species, very common in most of the southeastern states. Its range is from Connecticut to Kansas and southward to Texas and Florida (Chittenden and Fink, 1922). Adult with velvety-green and tan elytra and pronotum. Head, venter, and legs shiny, with green and purple iridescence. Adult 15 to 22 mm long and 9 to 11.5 mm wide.

Adults appear the last of June or early in July and have been taken at Lexington, Kentucky, as late as September 15. The beetles often cause considerable damage by feeding on ripening fruits, such as apples, peaches, nectarines, figs, blackberries, and grapes. They also feed on tomatoes, corn in the milk stage, and the sap of oak and other trees. The larvae feed on organic matter in the soil. They are often pests in tobacco plant beds (Jewett, 1943), where they injure the small plants by disturbing the soil or covering them with dirt. Favorite breeding places are in barnyards, in poorly drained soil, and near manure piles and straw stacks. They are seldom abundant in pastureland, but sometimes become troublesome in lawns or on golf courses where they push up unsightly mounds of dirt. The species has a one-year life cycle.

Cotinis texana Casey, Third-Stage Larva
(Fig. 431)

This description is based on the following material and reprinted from my 1945c bulletin:

Three third-stage larvae (and 3 second-stage larvae) collected October 26, 1933, at Tucson, Arizona, by L. P. Wehrle, from soil.

Four third-stage larvae collected September 13, 1943, at Tucson, Arizona, by C. T. Vorhies, from garden soil.

Three third-stage larvae collected February 20, 1934, at Tucson, Arizona, by K. B. Cowart.

One third-stage larva collected April 9, 1931, at Tucson, Arizona, by L. P. Wehrle.

Maximum width of head capsule of third-stage larva 5.6 to 5.95 mm. Surface of cranium slightly roughened, dark red to reddish-black. Frons with a shallow, median, longitudinal depression extending anteriorly from the epicranial stem. Frons bearing on each side a single, large posterior frontal seta, a very small exterior frontal seta, a very small anterior frontal seta, and a single, large anterior seta. Dorsoepicranial setae consisting of a patch of 10 to 17 setae of which 3 or 4 are large. Anterior margin of labrum trilobed; each lobe only slightly pigmented and bearing a number of long setae. Clithra present. Haptomeral region of epipharynx with a curved, transverse row of moderately long, fairly stout setae, caudad of which are a number of similarly shaped setae. Mandibles with 2 scissorial teeth posterior to the scissorial notch. Maxillary stridulatory area consisting of 5 to 9 rather stout stridulatory teeth and an anterior truncate process. Stridulatory teeth with anteriorly directed points. Lacinia of maxilla with 2 terminal unci, fused at their bases; dorsal uncus much the larger. Last antennal segment with 3 to 7 dorsal sensory spots.

Thoracic spiracles .76 to .94 mm long and .58 to .7 mm wide. Respiratory plate with a maximum of 40 to 50 very small, irregularly margined "holes" along any diameter. Distance between the two lobes of the respiratory plate much less than the dorsoventral diameter of the bulla. Spiracles of abdominal segments 1 to 8 similar in size.

Dorsa of abdominal segments 7 and 8 each with 2 annulets. Each annulet with a dense covering of fine setae among which is a transverse, sparse row of longer setae. Dorsa of abdominal segments 9 and 10 fused. Venter of abdominal segment 10 with 2 palidia. Each palidium consists of 2 or 3 somewhat irregular, sparsely set rows of stout, falcate, compressed setae (Fig. 431). Inner row of each palidium with 7 to 10 much stouter and larger pali. Septula oblong (noticeably wider than in *Cotinis nitida*). Caudo-laterad of each palidium is a number of rather short, straight, spine-like setae. Laterad of these setae and laterad and cephalad of the palidia, the venter of abdominal segment 10 is covered with many slender setae. Claws cylindrical, each rounded apically and bearing 10 or 11 terminal setae.

Cotinis texana Casey is a large southwestern species occurring in Texas and Arizona. In Arizona it is found along the Gila River and all its tributaries, and along the upper tributaries of the Bill Williams River (Nichol, 1935). Adult with dull-green elytra having tan margins; pronotum dull-green. Head, venter, and legs reddish-brown with greenish iridescence. Adult 22 to 26 mm long and 12 to 15.5 mm wide.

Adults injure soft-skinned fruits, such as figs, peaches, and grapes, during the latter half of July and in August and September. The worst injury occurs in the valleys and on the mesas between 1,000 and 3,500 feet in elevation (Nichol, 1935). The larvae feed on organic matter in the soil, and the most favorable breeding ground is a moist, but not wet, sandy clay covered by a 3- to 5-inch layer of rich organic matter. Favorite breeding grounds are corrals, manure piles, and stack bottoms.

TRIBE Cetoniini

Genus *Euphoria* Burmeister

Key to Larvae of the Species of *Euphoria*

1. Raster without palidia (Fig. 433)........*Euphoria inda* (Linn.)
   Raster with definite palidia (Figs. 432, 434, and 439)............ 2

2. Palidia with 9 to 12 sparsely set pali (Fig. 432). Septula narrow (5 to 9 times as long as its greatest width). Spiracular areas of abdominal segments 1 to 8 with 3 to 6 setae. Pleural lobes of same segments with 5 to 7 setae....................*Euphoria herbacea* Oliv.
   Palidia with 11 to 16 rather closely set pali. Septula broad, length less than 5 times its greatest width. Spiracular areas of abdominal segments 1 to 8 with 8 to 14 setae. Pleural lobes of same segments with 12 or more setae 3

3. Septula at least 4 times as long as its greatest width, very narrow anteriorly (Fig. 439). Pedal areas of abdominal segments 1 to 8 with 1 seta on each side, or none ....*Euphoria fulgida* (Fab.)
   Septula very broad, length less than 3 times its width (Fig. 434); usually broadly rounded at its cephalic end (palidia joined anteriorly). Pedal areas of abdominal segments 1 to 8 with 2 large setae and from none to 4 small setae on each side..............................
   .... *Euphoria sepulchralis* (Fab.)

Euphoria fulgida (Fab.), Third-Stage Larva
(Fig. 439)

This description is based on the following material and reprinted from my 1945c bulletin:

> One third-stage larva reared by William P. Hayes from an egg laid by a segregated female at Manhattan, Kansas, in 1921. Loaned by William P. Hayes of the University of Illinois.

Maximum width of head capsule of third-stage larva 3.9 mm. Surface of head smooth, faintly reticulate, yellow-brown. Frons with a shallow, median, longitudinal depression, forked anteriorly, extending forward from the epicranial stem. Frons with a single, large, posterior frontal seta and a single, large, anterior angle seta; other frontal setae absent. One large dorsoepicranial seta on each side. Labrum trilobed; clithra present. Haptomerum of epipharynx with a curved, transverse, closely set row of spine-like setae. Left mandible with 2 scissorial teeth anterior to the scissorial notch and 2 scissorial teeth posterior to the same notch. Lacinia of maxilla with 2 terminal unci which are fused at their bases.

Maxillary stridulatory area consisting of a row of 6 to 7 curved, sharp-pointed teeth and a small, distal, conical process. Stridulatory teeth with anteriorly projecting points. Last antennal segment with 3 to 4 dorsal sensory spots.

Thoracic spiracles about .5 mm long and .34 mm wide. Respiratory plate with a maximum of about 30 "holes" along any diameter. Distance between the two lobes of the respiratory plate much less than the dorsoventral diameter of the bulla. Spiracles of abdominal segments similar in size.

Spiracular areas of abdominal segments 1 to 8 each bearing 8 to 14 setae; pleural lobes of same segments with 12 to 17 setae. Venters of abdominal segments 1 to 9 each with a short, median, transverse group of setae confined to the central half of each eusternum; each group of setae containing 4 to 14 long setae. Pedal areas of abdominal segments 1 to 9 with 1 or no seta on each side, anterior to which are several short setae. Dorsa of abdominal segments 7 and 8 each divided into 2 annulets. Each annulet with a narrow, transverse band composed of 2 or 3 irregular rows of short, stiff setae and a caudal, sparsely set row of very long setae. Dorsa of abdominal segments 9 and 10 fused. Venter of abdominal segment 10 with paired palidia and a pair of tegilla (Fig. 439). Each palidium consists of a single, rather closely set row of 11 to 15 rather stout, blunt, slightly curved setae. The two palidia are fairly straight and slightly diverging posteriorly. Septula fairly narrow (four times as long as its greatest width). Tegilla composed of a number of rather short setae interspersed with a number of very long setae, united anterior to the palidia. Central transverse area of lower anal lip with about 25 long setae. Claws cylindrical, rounded apically, each with 10 or 11 setae borne on the sides and near the end of the claw.

Euphoria fulgida (Fab.) is a rather uncommon greenish species of no economic importance. According to Hayes (1925), it is found throughout most of the United States east of the Rocky Mountains and also occurs in Arizona, New Mexico, and Canada. Elytra, scutellum, and pronotum of beetle are light tan to light reddish-brown with bright green-coppery iridescence. Scutellum and pronotum sometimes bright green. Venter green with white patches on the sides of the abdominal segments. Legs tan to red-brown with greenish markings. Unlike the genus *Cotinis,* the scutellum is exposed in this and other species of *Euphoria.* Adult 15 to 17.5 mm long and 8 to 10 mm wide.

Adults in flight in Kansas from late May through July, feeding on sap exuding from wounds of trees or frequenting plants such as thistles (Hayes, 1925). Little is known of the food habits of the larvae. Hayes (1925) reared a number of them on manure. He found that the winter was passed in the prepupal stage.

Euphoria herbacea (Oliv.), Third-Stage Larva
(Figs. 417 and 432)

This description is based on the following material and reprinted from my 1945c bulletin:

One cast-skin of third-stage larva reared to the pupal stage. Larva (in cell) collected August 23, 1939, at Minorsville, Kentucky, by P. O. Ritcher, from soil of a woodland pasture.

One cast skin of third-stage larva reared to the adult stage. Larva (in cell) found November 20, 1936, at Versailles, Kentucky, by P. O. Ritcher, at a depth of 3 inches in the soil of a bluegrass pasture.

One third-stage larva (also 7 first-stage and 3 second-stage larvae) reared by P. O. Ritcher from eggs laid by isolated females, summer of 1943, at Lexington, Kentucky.

Maximum width of head capsule of third-stage larva about 3.9 mm. Surface of head smooth, faintly reticulate, light yellow-brown. Frons with a shallow, median, longitudinal depression which is forked anteriorly and extends forward from the epicranial stem. Frons with a single, large, posterior frontal seta and a single, large, anterior angle seta; other frontal setae absent. Dorsoepicranial setae 2 to 4 on each side, of which one on each side is quite large. Labrum trilobed; clithra present. Haptomeral region of epipharynx with a curved, transverse, closely set row of spine-like setae. Left mandible with 2 scissorial teeth anterior to the scissorial notch and 2 scissorial teeth posterior to the same notch. Lacinia of maxilla with 2 terminal unci, fused at their bases; dorsal uncus much the larger. Maxillary stridulatory area consisting of a row of 7 to 9 stridulatory teeth and a small, distal, conical process (Fig. 417). Stridulatory teeth with anteriorly projecting points. Last antennal segment with 2 to 4 dorsal sensory spots. Ocelli present.

Thoracic spiracles .5 to .54 mm long and .36 to .42 mm wide. Respiratory plate with a maximum of about 32 "holes" along any diameter. Distance between the two lobes of the respiratory plate much less than the dorsoventral diameter of the bulla. Spiracles of abdominal segments 1 to 7 similar in size, those of abdominal segment 8 slightly smaller.

Spiracular areas of abdominal segments 1 to 8 bearing 3 to 6 setae; pleural lobes of same segments with 5 to 7 setae. Venters of abdominal segments 1 to 8 each with a long, transverse row of 10 to 14 long setae and a few or no very short setae on each eusternum. Pedal areas of the same segments with 2 or 3 long setae on each side. Dorsa of abdominal segments 7 and 8 each with 2 annulets. Each annulet with a sparsely set, narrow, transverse band of setae composed of scattered, very short, stiff setae and a posterior, sparsely set row of long setae. Dorsa of abdominal segments 9 and 10 fused. Venter of abdominal segment 10 with paired palidia and a pair of tegilla (Fig. 432). Each palidium consisting of a single, longitudinal, very sparsely set row of 9 to 12 rather short pali with incurved tips. Palidia nearly parallel. Septula narrow (5 to 9 times as long as its greatest width). Tegilla composed of numerous rather short, flattened setae interspersed with a few very long setae; tegilla united anterior to the palidia. Central, transverse area of lower anal lip with about 17 very long setae. Claws cylindrical, rounded apically, and each bearing 7 to 9 terminal setae.

Euphoria herbacea (Oliv.) is a rather uncommon dull-colored species of no economic importance. According to Leng, it occurs from New York to Maryland. It also occurs in Indiana and Kentucky. Elytra and pronotum of beetle dull purplish-green; elytra often with scattered white patches and margined with white posteriorly. Head a shiny green or purplish-green with white hairs. Venter shiny green with a metallic luster. Venter and legs with numerous white hairs. Adult 12 to 16 mm long and 7 to 9 mm wide.

Adults in flight at Lexington, Kentucky, from the middle of June through July, frequenting flowers, such as yarrow and milkweed, or feeding on ripe fruits, such as blackberries and drop peaches, and on sap exuding from wounds on small limbs of *Crataegus*. Beetles are attracted to fermenting sugar-solution baits. Larvae feed in the soil. Winter is passed in the prepupal stage.

Euphoria inda (L.), Third-Stage Larva
(Figs. 413, 415, 433, 442, 448, and 449)

This description is based on the following material and reprinted from my 1945c bulletin:

Two third-stage larvae collected July 8, 1933, at Gays Mills, Wisconsin, by P. O. Ritcher, in rich soil beneath strawy litter. Loaned by T. R. Chamberlin, formerly with the Federal White Grub Laboratory.

Two third-stage larvae collected August 5, 1901, at Springfield, Massachusetts, by G. Dimmock. Said to be feeding on roots of *Akebia quinata*. (Loaned by the U. S. National Museum.)

Ten third-stage larvae and cast skins of 11 third-stage larvae reared to the adult stage. This material is from a lot of 7 second-stage and 53 third-stage larvae collected June 23 ,1944, on the Western Kentucky Substation Farm at Princeton, by W. D. Armstrong and P. O. Ritcher, in and beneath moist, decaying hay at the base of a small, conical, three-year-old stack.

Maximum width of head capsule of third-stage larva 3.7 to 4.0 mm. Surface of head smooth and reticulate, yellow-brown. Frons with a median, longitudinal depression, forked anteriorly, extending forward from the epicranial stem. Frons with a single posterior frontal seta and a single anterior angle seta; other frontal setae absent. Dorsoepicranial setae consisting of 1 large and 3 to 5 very small setae, on each side. Labrum trilobed, clithra present. Haptomeral region of epipharynx with a curved transverse row of rather short, spine-like setae. Left mandible (Fig. 415) with 2 scissorial teeth anterior to the scissorial notch and 2 scissorial teeth posterior to the same notch. Lacinia of maxilla with 2 terminal unci which are fused at their bases. Maxillary stridulatory area consists of a row of 8 to 10 teeth and a small, distal, conical process. Stridulatory teeth with anteriorly projecting points. Last antennal segment with 2 or 3 dorsal sensory spots (Fig. 448).

Thoracic spiracles .45 to .55 mm long and .29 to .38 mm wide. Respiratory plate with a maximum of about 32 "holes" along any diameter. Distance between the two lobes of the respiratory plate much less than the dorsoventral diameter of the bulla. Spiracles of abdominal segments 1 to 8 similar in size.

Dorsa of abdominal segments 7 and 8 each with 2 rather sparse, transverse patches of setae, each patch consisting of 2 or 3 sparse, irregular rows of short, stout setae and a few longer setae. Each caudal patch with several very long setae among the short, stout setae. Dorsa of abdominal segments 9 and 10 fused (Fig. 442).

Spiracular areas of abdominal segments 1 to 8 bear 9 to 13 setae; pleural lobes of same segments with 12 to 15 long setae (and several very short setae besides). Venters of abdominal segments 1 to 8 each with a long transverse patch of setae on each eusternum, each patch consisting of 14 to 25 long setae and a number of smaller setae. Pedal areas of same segments with 2 (rarely 1 or 3) long setae and 1 to 3 very short setae. Raster without palidia (Fig. 433). Teges consisting of numerous caudally directed, rather short, flattened setae, laterally interspersed with several very long, cylindrical setae. Lower anal lip with scattered, short, flattened setae similar to those of the teges, and among these are 20 to 30 long, cylindrical setae. Posterior to these long setae and bordering the anal slit are about 5 irregular rows of fairly short, sharp-pointed setae. Claws cylindrical, rounded apically, and each bearing about 10 setae (Fig. 449).

Euphoria inda is a fairly large cetoniid occurring from Connecticut to Florida, and westward to Oregon and Arizona. It is rather uncommon in Kentucky. Adult has cinnamon-colored elytra with irregular longitudinal rows of small black spots, many of which are rectangular. Head and pronotum densely hairy, pronotum black with vague cinnamon markings and frequently margined with cinnamon. Venter and legs reddish-brown, bearing numerous whitish hairs. Adults 12 to 16 mm long and 8 to 10 mm wide.

Larvae feed in rich soil, such as at the edges of old hay or straw stacks or in decaying vegetation or manure. According to Lugger (1899), larvae were very abundant in a low spot where spoiled melons, potatoes, and other refuse had been dumped. At Gays Mills, Wisconsin, the writer found that pupation began early in July at a depth of 2 to 5 inches, within oval earthen cells formed by the larvae. Adults were in flight from late August through September. In Wisconsin and Kentucky, a number of reports have been received of adults injuring corn in the milk stage. The adults also feed on flowers; ripe fruits, such as apples, pears, and peaches (Bruner, 1891); and on sap exuding from tree wounds. They are attracted to fermenting sugar baits. Winter is passed in the adult stage, and the adults are again in flight during the first warm days. The species has a one-year life cycle.

Euphoria sepulchralis (Fab.), Third-Stage Larva (Fig 434)

This description is based on the following material and reprinted from my 1945c bulletin:

Six third-stage larvae reared by P. O. Ritcher, from eggs laid by segregated females, summer of 1943, at Lexington, Kentucky.

One third-stage larva and one cast skin associated with an adult reared from a second third-stage larva. Both larvae collected August 9, 1943, at Lexington, Kentucky, by P. O. Ritcher, under a piece of dead bluegrass sod lying in a corn patch.

Maximum width of head capsule of third-stage larva 2.8 to 3 mm. Surface of head smooth, faintly reticulate, light yellow-brown. Frons with a shallow, median, longitudinal depression, forked anteriorly, extending forward from the epicranial stem. Frons with a single posterior frontal seta and a single anterior angle seta; other frontal setae absent. Dorsoepicranial setae 1 or 2 on each side. Labrum trilobed; clithra present. Haptomeral region of epipharynx with a curved, transverse, closely set row of short, spine-like setae. Left mandible with 2 scissorial teeth anterior to the scissorial notch and 2 scissorial teeth posterior to the same notch. Lacinia of maxilla with 2 terminal unci which are fused at their bases; dorsal uncus much the larger. Maxillary stridulatory area consisting of a row of 5 to 6 curved teeth and a small, distal, conical process. Stridulatory teeth

with anteriorly projecting points. Last antennal segment with 2 dorsal sensory spots.

Thoracic spiracles .36 to .4 mm long and .24 to .3 mm wide. Respiratory plate with a maximum of 22 to 24 "holes" along any diameter. Distance between the two lobes of the respiratory plate somewhat to much less than the dorsoventral diameter of the bulla. Thoracic spiracles larger than abdominal spiracles. Spiracles of abdominal segments 1 to 7 similar in size, those of abdominal segment 8 slightly smaller.

Dorsa of abdominal segments 7 and 8 each with 2 transverse bands of stiff setae, each band consisting of 2 to 4 sparse, irregular rows of short setae and a caudal fringe of long setae. Dorsa of abdominal segments 9 and 10 fused.

Spiracular areas of abdominal segments 1 to 8 each bearing 9 to 13 setae; pleural lobes of same segments with 12 to 22 setae. Venters of abdominal segments 1 to 9 each with a long transverse band of setae; each group of setae containing 12 to 17 long setae anterior to which are several short setae. Pedal areas of abdominal segments 1 to 9 each with 2 long setae and from none to 4 small setae. Venter of tenth abdominal segment with paired palidia and a pair of tegilla (Fig. 434). Each palidium consisting of a single close-set row of 12 to 16 rather stout, blunt, moderately long, depressed pali. The two palidia join anteriorly and are parallel posteriorly or slightly diverging. Septula rather broad (length less than three times its width), usually rounded anteriorly. Tegilla composed of short-to-long, straight, sharp-pointed setae, united anterior to the palidia. Central, transverse area of lower anal lip with 15 to 21 long, cylindrical setae. Claws cylindrical, rounded apically, and each bearing 6 to 9 terminal setae.

Euphoria sepulchralis is a common, fairly small, dull-colored species. It is found over the southern states as far north as Indiana and as far west as Kansas and Texas (Hayes, 1925 and Leng, 1920). It is quite common in Kentucky. Adult with dark reddish-brown to reddish-black elytra bearing a few inconspicuous white spots and several irregular transverse white lines. Head, pronotum, venter, and legs reddish-black to blackish with faint purple iridescence and covered with numerous whitish hairs. Adult 9 to 12 mm long and 5.5 to 7.5 mm wide.

Larvae feed in rich soil, such as beneath dead sod or under manure (Hayes, 1925). Pupation occurs in August in Kentucky. Adults frequently found injuring ears of corn in the milk stage. Adults also found on the flowers of apple, thistles, mock orange, milkweed, yarrow, and field daisies, and feeding on ripe fruits, such as apples and watermelons. Adults are in flight during August and September into October, and again in the spring. The species has a one-year life cycle.

Genus *Gnorimella* Casey

Gnorimella maculosa (Knoch), Third-Stage Larva
(Figs. 405, 437, and 440)

This description is based on the following material and reprinted from my 1945c bulletin:

Two third-stage larvae collected November 3, 1910, at Plummer's Island, Maryland, by H. S. Barber, in a hollow *Cercis* trunk. (Loaned by the U. S. National Museum.)

Maximum width of head capsule of third-stage larva 3.5 to 3.8 mm. Surface of cranium reticulate, light reddish-brown. Frons (Fig. 405) posteriorly with a slight, median, longitudinal depression extending anteriorly from the epicranial stem; anteriorly with a half-moon shaped, flattened area. Frons bearing on each side a single posterior frontal seta, a single exterior frontal seta, 1 or 2 minute anterior frontal setae, and a single anterior angle seta. Dorsoepicranial setae consist of 3 setae on each side. Labrum rather truncate apically; clithra absent. Haptomeral region of epipharynx with a curved transverse row of setae interrupted by a tooth-like haptomeral process. Left mandible bearing 2 sharp teeth anterior to the scissorial notch and a single sharp tooth posterior to the same notch; right mandible with a single sharp tooth anterior to the scissorial notch. Maxillary stridulatory area consisting of a row of 3 to 4 curved teeth with anteriorly projecting points, and an anterior truncate process. Lacinia of maxilla with 2 terminal unci, fused at their bases; the dorsal uncus much the larger. Last antennal segment with a single, oval, dorsal sensory spot.

Thoracic spiracles .46 to .54 mm long and .33 to .36 mm wide. Respiratory plate with a maximum of about 18 irregularly shaped "holes" along any diameter. Distance between the two lobes of the respiratory plate very much less than the dorsoventral diameter of the bulla. Spiracles of abdominal segments 1 to 8 similar in size.

Dorsum of abdominal segment 7 divided into 2 annulets, each bearing a transverse row of long setae. Anterior annulet with 1 or 2 rows of short, stout setae cephalad of the row of long setae; posterior annulet with scattered, short setae cephalad of the row of long setae. Dorsa of abdominal segments 8 and 9 each with a caudal, transverse row of long setae. Cephalad of each transverse row are a few scattered, short, stouter setae. Dorsa of abdominal segments 9 and 10 not fused (Fig. 440). Venter of abdominal segment 10 without palidia (Fig. 437). Teges consisting of scattered, very short, spine-like setae, among which are interspersed a number of very long setae. Lower anal lip fringed caudally

with about 15 long setae. Claws falcate, sharp-pointed, each bearing 2 proximal setae.

Gnorimella maculosa (Knoch) is a rather uncommon, medium-sized species ranging from Connecticut to Florida and westward to Indiana (Leng, 1920). Adult with light red-brown to red-brown elytra spotted with black. Head and pronotum black and hairy with dull yellow spots and a suggestion of a median yellow stripe. Venter and legs black. Abdominal segments with a yellow spot on each side; pygidium mostly yellow. Adult 12.5 to 15.5 mm long and 7 to 8 mm wide. Very little is known about the habits of the larvae. The two larvae upon which the description above is based were collected in a hollow in the trunk of a redbud (*Cercis*) tree.

TRIBE Cremastocheilini

Genus *Cremastocheilus* Knoch

Cremastocheilus pugetanus Casey, Third-Stage Larva
(Figs. 399, 404, 416, and 418)

This description is based on the following material and reprinted from my 1945c bulletin:

Five third-stage (and one second-stage) larvae collected June 27, 1939, at Dayton, Washington, by S. H. Lyman, from ant nests. (Loaned by the U. S. National Museum.)

Maximum width of head capsule of third-stage larva 2.9 to 3.1 mm. Surface of cranium slightly roughened, reticulate, yellow-brown. Frons with median, longitudinal depression extending anteriorly from the epicranial stem. Frons (Fig. 404) bearing on each side a single posterior frontal seta; a single small exterior frontal seta, a single anterior frontal seta, and a single anterior angle seta. Dorsoepicranial setae, 1 or 2 on each side. Labrum trilobed; clithra present. Haptomeral region of epipharynx with a curved, transverse row of stout, spine-like setae. Left mandible (Fig. 416) with 2 scissorial teeth anterior to the scissorial notch and 2 scissorial teeth posterior to the same notch. Lacinia of maxilla with 2 terminal unci, fused at their bases. Maxillary stridulatory area (Fig. 418) consisting of 4 to 6 curved teeth with anteriorly projecting points and an anterior conical process. Last antennal segment with 2 dorsal sensory spots.

Thoracic spiracles .26 to .28 mm long and .16 to .2 mm wide. Respiratory plate with a maximum of about 14 elongate-oval "holes" along any diameter. Distance between the two lobes of the respiratory plate more than to slightly less than the dorsoventral diameter of the bulla. Spiracles of abdominal segments 1 to 7 similar in size, those of abdominal segment 8 distinctly smaller.

Dorsum of abdominal segment 7 divided into 2 annulets, each bearing a transverse patch of short setae and a posterior, transverse row of long setae. Dorsum of abdominal segment 8 covered with fairly short setae and with 2 widely separated rows of long setae. Dorsa of abdominal segments 9 and 10 fused. Venter of abdominal segment 10 without palidia. Teges consisting of 25 to 30 scattered, short, flattened setae, laterad and cephalad of which are found many long and short, slender setae. Barbulae dense. Lower anal lip with a few short setae similar to those of the teges, and a dense caudal fringe of about 75 long setae. Claws (Fig. 399) falcate, sharp-pointed, each bearing 2 proximal setae.

Cremastocheilus pugetanus is a rather small, myrmecophilous species occurring in Washington, Idaho, and Utah. Adult elongate, dull black with coarsely pitted pronotum and elytra. Adult 11.5 to 13 mm long and 5.5 to 6 mm wide. The adults are found in the nests of ants of the genus *Formica*. On June 6, 1939, at Dayton, Washington, S. H. Lyman collected a pair of beetles from the nest of *Formica obscuripes* Ford. On June 27 of the same year he collected 1 second-stage and 5 third-stage larvae from ant nests in the same locality.

According to Wheeler (1908), the larval and pupal stages of *Cremastocheilus* are probably passed in the vegetable debris of ant nests. Recent knowledge of the relationships between *Cremastocheilus* adults and ants of several genera was summarized by Cazier and Statham (1962). In the Willamette Valley of Oregon, larvae of *C. armatus* Walker have been found in numbers in rich garden soil but have never been found in ant nests west of the Cascades. East of the Cascade Mountains, both larvae and adults of this species are common in *Formica* nests.

TRIBE Trichiini

Genus *Osmoderma* Serv.

Key to Larvae of the Species of *Osmoderma*

1. Dorsa of abdominal segments 7 and 8 each with 2 annulets, each bearing a transverse patch of fairly short setae and a posterior, sparsely set row of longer setae (Fig. 438). Spiracles of abdominal segments 7 and 8 larger than those of abdominal segments 1 to 6 .... *Osmoderma eremicola* Knoch.
   Dorsa of abdominal segments 7 and 8 each with 2 widely separated, sparsely set rows of long setae but with only a few short setae anterior to each row. Spiracles of abdominal segments similar in size ....*Osmoderma scabra* (Beauv.)

Osmoderma eremicola Knoch., Third-Stage Larva
(Figs. 398, 401, 422, 438, 441, and 445)

This description is based on the following material and reprinted from my 1945c bulletin:

Ten third-stage larvae and 10 cast skins of third-stage larvae reared to the adult stage. All 20 larvae collected December 31, 1942, at Lexington, Kentucky, by P. O. Ritcher, from a large cavity in a soft maple tree.

Three third-stage larvae and 1 cast skin of a third-stage larva reared to the adult stage. The 4 larvae collected March 25, 1939, in Clark County, Kentucky, by P. O. Ritcher, from a large hollow in an ash tree.

Maximum width of head capsule of third-stage larva 5.8 to 6.7 mm. Surface of cranium slightly roughened, light reddish-brown with darker-colored clypeus and labrum. Frontal sutures sinuate. Frons bears on each side 1 to 3 posterior frontal setae, 1 or 2 exterior frontal setae, 1 or 2 anterior frontal setae (if 2, one very small), and a single anterior angle seta. Dorsoepicranial setae 4 to 11 on each side, of which 2 or 3 are large and the rest very small. Labrum trilobed, clithra present. Haptomeral region of epipharynx with a curved zygum and a curved transverse row of short, spine-like setae (Fig. 441). Left mandible (Fig. 445) with 4 scissorial teeth, right mandible with 3. Maxillary stridulatory area (Fig. 422) consisting of a row of 4 to 6 (rarely 7 or 8) sharp, curved teeth and an anterior truncate process. Stridulatory teeth with anteriorly projecting points. Lacinia of maxilla with 2 terminal unci, fused basally; dorsal uncus much the larger. Cardo with 3 sclerites; anterior sclerite with a patch of setae. Last antennal segment (Fig. 401) with 2 (rarely 3 or 4) dorsal sensory spots. Ocelli present.

Thoracic spiracles .94 to 1.1 mm long and .72 to .84 mm wide. Respiratory plate with a maximum of 40 to 45 very small, irregularly margined "holes" along any diameter. "Holes" not in definite rows. Distance between the two lobes of the respiratory plate much less than the dorsoventral diameter of the bulla. Spiracles of abdominal segments 1 to 6 similar in size, those of abdominal segments 7 and 8 slightly larger.

Dorsum of abdominal segment 7 with 2 annulets, each with a transverse patch of fairly short setae and a posterior, sparsely set row of longer setae. Dorsum of abdominal segment 8 with 2 transverse, sparsely set patches of short setae, each with a posterior row of long setae; the 2 patches of setae sometimes confluent. Dorsa of abdominal segments 9 and 10 fused (Fig. 438). Venter of abdominal segment 10 with a broad teges of very short, slightly curved, spine-like setae, anteriorly and laterally interspersed with a number of long, slender setae. Palidia absent. Lower anal lip anteriorly with a transverse patch of spine-like setae similar to those of the teges; mesally with a transverse band of long and short, slender setae; posteriorly with a sparsely set patch of short, slender setae. Claws (Fig. 398) very short, subconical, each bearing 2 distal setae.

Osmoderma eremicola is a large, broad, wood-inhabiting species commonly called the smooth *Osmoderma.* According to Hoffman (1939), it occurs along the southern border of Canada from Maine westward to Montana and southward to northern Oklahoma, northern Arkansas, northern Georgia, and northern South Carolina. It is very common in Kentucky. Adult 24 to 31 mm long and 13 to 17 mm wide; dark, shiny red-brown with smooth elytra.

The larvae feed on decaying wood and are most commonly found in the accumulated wood debris in large cavities in standing trees. They are sometimes found in rotten logs. Hoffman (1939) states that larvae have been reported by various writers from the decaying wood of beech, hickory, elm, apple, cherry, oak, maple, and cottonwood trees. The writer has collected larvae from soft maple, ash, hackberry, and elm. Full-grown larvae overwinter within oval cocoons formed of wood fragments or wood loam. They pupate in the late spring. Adults in flight from late June until the middle of August; sometimes taken at lights or in bait traps containing fermenting sugar solution. Hoffman (1939) and Sweetman and Hatch (1927) found evidence that the species has a three-year life cycle in Minnesota. Observations at Lexington, Kentucky, suggest a two-year cycle in that latitude.

Osmoderma scabra (Beauv.), Third-Stage Larva

This description is based on the following material and reprinted from my 1945c bulletin:

> One third-stage larva and 5 cast skins of third-stage larvae reared to the adult stage. All 6 larvae collected April 21, 1940, at Oak Ridge, Kentucky, by H. Dorman, from rotten wood.
>
> One third-stage larva and 1 cast skin of a third-stage larva associated with a pupa. The 2 larvae were collected December 27, 1935, at Easton, Massachusetts, by T. H. Jones from the decayed stump of an apple tree. Determined by C. H. Hoffman.
>
> One third-stage larva collected May 6, 1933, at Taylor's Falls, Minnesota, by C. H. Hoffman, from a decayed black birch tree.

Maximum width of head capsule of third-stage larva 5.1 to 5.8 mm. Surface of cranium slightly roughened, reddish-brown. Frontal suture sinuate. Frons bearing on each side a single posterior frontal seta, a single exterior frontal seta, 1 to 2 anterior frontal setae, and a single anterior angle seta. One or 2 large dorsoepicranial setae on each side and several very small setae. Labrum trilobed, clithra present. Haptomeral region of

epipharynx with a curved, transverse row of short, spine-like setae. Left mandible with 4 scissorial teeth, right mandible with 3. Maxillary stridulatory area consisting of a row of 7 to 10 (rarely 6) sharp, curved teeth and an anterior truncate process. Stridulatory teeth with anteriorly projecting points. Lacinia of maxilla with 2 terminal unci, fused basally; dorsal uncus much the larger. Anterior sclerite of cardo with a patch of setae. Last antennal segment with 2 or 3 dorsal sensory spots. Ocelli present.

Thoracic spiracles .64 to .73 mm long and .46 to .52 mm wide. Respiratory plate with a maximum of about 40 very small, irregularly margined "holes" along any diameter. "Holes" not in definite rows. Distance between the two lobes of the respiratory plate much less than the dorsoventral diameter of the bulla. Spiracles of abdominal segments similar in size.

Dorsum of abdominal segment 7 with 2 annulets, each bearing a transverse row of fairly long setae anterior to each of which are a few short setae. Dorsum of abdominal segments 8 with 2 widely separated, sparsely set rows of fairly long setae anterior to each of which are a very few short setae. Dorsa of abdominal segments 9 and 10 fused. Venter of abdominal segment 10 with a teges of very short, spine-like setae, extending from the lower anal lip to the caudal margin of abdominal segment 9, and interspersed anteriorly and laterally with a number of long setae. Palidia absent. Claws very short, subconical, each bearing 2 distal setae.

Osmoderma scabra is a fairly large, wood-inhabiting species commonly called the rough *Osmoderma.* According to Hoffman (1939), it occurs in southeastern Canada, from Maine to Minnesota, and southward to Kansas, Tennessee, northern Georgia, and northwestern South Carolina. It is rare in Kentucky. Adult with rugose elytra, blackish and shiny with a coppery luster.

Head and pronotum blackish with metallic luster; venter and legs blackish. Adult 19 to 22 mm long and 9.5 to 10 mm wide.

The larvae have been found feeding in the decaying wood of apple, cherry, beech, sweet gum, hickory, poplar, willow, sycamore, sassafras, maple, oak, chestnut, and birch (Hoffman, 1939). According to the same writer, the adults are in flight in June, July, and August.

Genus *Trigonopeltastes* Burm.

Trigonopeltastes delta (Forst.), Third-Stage Larva
(Figs. 412 and 420)

This description is based on the following material and reprinted from my 1945c bulletin:

Two cast skins of 2 third-stage larvae collected February 26, 1919, about 18 miles southwest of Paradise Key, Florida, by H. S. Barber, in trash in bracts of living palm (Sabal?). Adults emerged about April 7, 1919. (Loaned by the U. S. National Museum.)

Maximum width of head capsule of third-stage larva about 3.7 mm. Surface of cranium smooth. Frons bearing on each side 2 posterior frontal setae, a single exterior frontal seta, a single anterior frontal seta, and a single anterior angle seta. Labrum broadly rounded apically, clithra absent. Labrum anteriorly with a slightly raised, bilobed elevation, posteriorly with a transverse row of 8 setae. Haptomeral region of epipharynx with a curved, transverse row of setae interrupted by a tooth-like process. Left mandible (Fig. 412) with an oblique blade-like portion anterior to the scissorial notch and a single tooth posterior to the notch. Maxillary stridulatory area consisting of a row of 3 to 4 curved teeth with anteriorly projecting points and an anterior conical process. Lacinia of maxilla with a single very small terminal uncus (Fig. 420). Last antennal segment with a single oval, dorsal sensory spot. Ocelli present.

Thoracic spiracle about .29 mm long and .18 mm wide. Distance between the two lobes of the respiratory plate slightly less than the dorsoventral diameter of the bulla. Spiracles of abdominal segments 1 to 8 similar in size.

Dorsa of abdominal segments 9 and 10 not fused. Venter of abdominal segment 10 without palidia. Teges consisting of numerous rather short, slightly curved, flattened setae. Claws falcate, sharp-pointed, each bearing 2 setae.

Trigonopeltastes delta is a small distinct species occurring in Kentucky, North Carolina, Florida, and Texas. The species is rather common in North Carolina and is in flight from May to August (Brimley, 1938).

T. delta is fairly common in some localities in western and southeastern Kentucky. On June 19 and 23, 1944, the writer collected over 50 adults from the flowers of slender mountain mint, *Pycnanthemum flexuosum* (Walt.) BSP., along the margin of a woods near Princeton, Kentucky. A number of adults were also taken June 28, 1945, by Lee H. Townsend and the writer, on flowers in Whitley County, west of Corbin, Kentucky. Most of the beetles were on slender mountain mint, a few were on flowers of Queen Anne's lace, and one was taken on ox-eye daisy. The species has also been taken at Smithland, Nortonville, Central City, Mt. Vernon, and Cumberland Falls, Kentucky.

Adult with dull orange elytra and legs. Pronotum margined with white and with a whitish delta on a dull

black background. Venter whitish. Adult 8.5 to 9 mm long and 4.5 mm wide. The description of the larva given above is based on the cast skins of 2 larvae found by Barber in trash in the bracts of a palm. Dozier (1920) reared adults from larvae found in an oak stump at Gainesville, Florida.

Genus *Trichiotinus* Casey

Surface of cranium smooth, faintly reticulate, yellow-brown to light reddish-brown. Frontal sutures fairly straight. Frons (Fig. 409) bearing on each side 2 (rarely 1) posterior frontal setae, a single exterior frontal seta, 1 or 2 anterior frontal setae, and a single anterior angle seta. One or 2 dorsoepicranial setae on each side. Labrum symmetrical, broadly rounded apically, clithra absent. Labrum with a median, transverse, bilobed ridge (Fig. 409). Haptomeral region of epipharynx (Fig. 428) with a curved zygum and a curved transverse row of blunt, spine-like setae interrupted mesally by a tooth-like process. Left mandible (Fig. 419) with 2 scissorial teeth anterior to the scissorial notch and a single scissorial tooth posterior to the notch. Maxillary stridulatory area (Fig. 423) consisting of a row of 2 to 6 sharp, slightly curved teeth and an anterior truncate process. Stridulatory teeth with anteriorly projecting points. Lacinia with 2 terminal unci which are fused; the dorsal uncus much the larger (Fig. 424). Cardo with 3 sclerites. Last antennal segment (Fig. 400) with a single, oval, dorsal sensory spot. Ocelli present.

Dorsa of abdominal segments 9 and 10 not fused. Venter of abdominal segment 10 without palidia. Raster consisting of a few to many, very short, stout setae and a few to many long, acicular setae. Lower anal lip clothed with setae similar in shape to those of the raster. Claws sharp-pointed, each bearing 2 proximal setae.

Key to Larvae of the Species of *Trichiotinus*

1. Spiracles of abdominal segment 8 much smaller than those of abdominal segments 1 to 7. Raster with more than 50 very small, stout setae among which and anterior to which are only a very few long, acicular setae ....*T. lunulatus* (Fab.)

   Spiracles of abdominal segments 1 to 8 similar in size. Raster with 40 or fewer, very small, stout setae among which, or anterior to which, are many long, acicular setae 2

2. Raster with less than 10 very short setae, all of them spine-like and borne close to the lower anal lip posterior to the acicular setae ....*T. affinis* (G. and P.)

   Raster with 15 or more very short setae but all of them not always spine-like. Short setae scattered among the long, acicular setae 3

3. Maxillary stridulatory area with 4 or fewer sharp teeth .... *T. assimilis* (Kirby)
   Maxillary stridulatory area with 5 to 6 sharp teeth (rarely 4 on one maxilla) 4

4. Raster with 25 to 40 very short, slightly hooked, spine-like setae which are scattered among many long, acicular setae. Each lateral, pedal area on the sterna of abdominal segments 3 to 6 inclusive usually with 3 to 5 setae (never with 2, rarely with 6 or 7) ......
   .... *T. piger* (Fab.)
   Raster with from 25 to 35 short setae; about 10 or 12 of them slightly hooked and spine-like, the rest acicular in shape. Each lateral, pedal area on the sterna of abdominal segments 3 to 6 inclusive usually with 2 long setae (a few pedal areas with only 1 or with 3 or 4 setae) .... *T. bibens* (Fab.)

Trichiotinus affinis (G. and P.), Third-Stage Larva

This description is based on the following material and reprinted from my 1945c bulletin:

One third-stage larva and cast skin of one third-stage larva associated with a reared adult male. The 2 larvae collected April 19, 1938, at Greenbelt, Maryland, by William H. Anderson, in dry, decaying oak, No. L. 264. (Loaned by the U. S. National Museum.)

Maximum width of head capsule of third-stage larva 2.66 mm. Maxillary stridulatory area with 4 sharp, anteriorly directed teeth and an anterior truncate process. Thoracic spiracle .255 mm long and .18 mm wide. Respiratory plate with a maximum of about 18 "holes" along any diameter. "Holes" of irregular shapes, separated from each other by a distance much less than the diameter of a "hole." Distance between the lobes of the respiratory plate much less than the dorsoventral diameter of the bulla. Spiracles of abdominal segments 1 to 8 similar in size.

Dorsa of abdominal segments 7 and 8 each with 2 annulets. Each annulet clothed with 3 or 4 irregular, transverse rows of short setae and a caudal transverse row of long, slender setae. Dorsum of abdominal segment 9 similarly clothed but without distinct annulets. Pedal areas of abdominal segments 1 to 8 each with 1 to 3 setae on each side. Raster consisting of about 8 very short, stout setae adjacent to the lower anal lip; anteriorly with numerous long setae (about 35). Lower anal lip with an anterior, irregular, transverse row of about 8 very short, stout setae; posteriorly with about 30 long and short, slender setae.

Trichiotinus affinis is a small species occurring from New Hampshire to northern Illinois and southward to northern Alabama and northern South Carolina. There is a single specimen in the collection of the Kentucky Department of Entomology and Botany collected by H. Garman, June 16, 1892, near Pineville, Kentucky, along Clear Creek. On June 28, 1945, Lee H. Townsend and the writer collected a female from the flowers of Queen Anne's lace, west of Corbin in Whitley County, Kentucky. Elytra of adult black with orange-brown discal area. Black marginal portion of each elytron interrupted by 2 short, transverse white bars. Head, pronotum, and pygidium black with metallic luster; hairy. Pygidium also with a white patch on each side. Venter and legs black, covered with rather long, whitish hairs. Adults 8 to 9.5 mm long and 4 to 5.5 mm wide.

Larvae or pupae of this species have been collected from the decaying wood of aspen, oak, and hickory (Hoffman, 1935). According to the same writer, adults are found on many species of flowers, including raspberry, blackberry, daisy, and *Spiraea latifolia.*

Trichiotinus assimiliis (Kirby), Third-Stage Larva

This description is based on the following material and reprinted from my 1945c bulletin:

Two third-stage larvae collected March 10, 1898, at Springfield, Massachusetts, by Dimmock, from decaying maple bark. No. 1281. See Hoffman (1935), page 138. (Loaned by the U. S. National Museum.)

Cast skin of 2 third-stage larvae collected and reared by C. H. Hoffman. No. F-50 and F-51. (Loaned by the U. S. National Museum.)

Cast skin of one third-stage larva collected May 27, 1934, at Itasca Park, Minnesota, by C. H. Hoffman, from a basswood stump. No. F-42. Associated with a pupa. Donated by C. H. Hoffman.

Maximum width of head capsule of third-stage larva 3.0 to 3.4 mm. Maxillary stridulatory area with 2 to 4 sharp, anteriorly directed teeth and an anterior truncate process. Thoracic spiracle .29 to .36 mm long and .21 to .25 mm wide. Respiratory plate with a maximum of 18 to 21 "holes" along any diameter. "Holes" of oval to elongate-oval irregular shapes, separated by distances much less than the diameters of the "holes." Distance between the two lobes of the respiratory plate much less than the dorsoventral diameter of the bulla. Spiracles of abdominal segments 1 to 8 similar in size.

Dorsa of abdominal segments 7 and 8 each with 2 annulets. Each annulet with a patch of about 4 irregular, transverse rows of short setae and a posterior, transverse row of long setae. Dorsum of abdominal segment 9 similarly clothed but not divided into annulets. Pedal areas of abdominal segments 1 to 8 with 2 to 6 setae on each side. Raster with a teges of 18 to 25 very short, stout setae, anterior to which are borne and among which are interspersed many long setae. Lower anal lip anteriorly with 12 to 14 very short, stout setae, posteriorly with a fringe of 25 to 32 long and rather short, slender setae.

Trichiotinus assimilis is a northern species occurring in Canada from Nova Scotia and New Brunswick westward to British Columbia. In the United States its range extends from the east coast as far west as Oregon, Idaho, and Utah and as far south as northern New Mexico, northern Kansas, Iowa, northern Illinois, and northern Pennsylvania (Hoffman, 1935). Elytra of adult usually black with a small orange-brown discal area. On many specimens each elytron bears 2 lateral, transverse white bars extending more than half way across and has one posterior, longitudinal white line along each inner margin. Head and pronotum black, covered with short, yellowish hairs. Pygidium black, with a large white patch on each side, densely clothed with whitish or yellowish hairs. Legs and venter black, clothed with numerous, long, whitish hairs. Adult 9 to 12 mm long and 4.5 to 6.5 mm wide.

The larvae of this species have been found in decaying maple bark, in the decaying wood of white birch, and in oak and basswood stumps (Hoffman, 1935). According to the same writer, the adults are found on wild and cultivated roses and a number of other flowers.

Trichiotinus bibens (F.), Third-Stage Larva

This description is based on the following material and reprinted from my 1945c bulletin:

Four cast skins of third-stage larvae, collected May 19, 1919, at Falls Church, Virginia, by A. D. Hopkins. U. S. No. 100878. (Loaned by the U. S. National Museum.)

Four third-stage larvae and 5 cast skins of third-stage larvae, the latter associated with 4 pupae and a reared female. Larvae and prepupae collected March 14, 1945, north of Mt. Vernon, Kentucky, by H. H. Jewett and the writer, from beneath the bark of an oak log.

Maximum width of head capsule of third-stage larva 3.2 to 3.3 mm. Maxillary stridulatory area with 4 to 6 sharp, anteriorly directed teeth and an anterior truncate process. Thoracic spiracle .35 to .36 mm long and .247 to .29 mm wide. Respiratory plate with a maximum of 20 to 25 "holes" along any diameter. "Holes" of oval irregular shapes, separated by a distance less than the diameter of a "hole." Distance between the two lobes of the respiratory plate much less than the dorsoventral diameter of the bulla. Spiracles of abdominal segments 1 to 8 similar in size.

Dorsa of abdominal segments 7 and 8 each with 2 annulets. Each annulet covered with a transverse patch of setae consisting of 3 or 4 irregular rows of short setae and a caudal, transverse row of long setae. Dorsum of abdominal segment 9 covered with short setae among which are an anterior transverse row of long setae and a posterior row of long setae. Pedal areas of abdominal segments 3 to 6 inclusive each usually with 2 long setae (a few pedal areas with only 1 or 3 or 4 setae). Raster consisting of about 12 very short, spine-like setae located anterior to the lower anal lip and a broad patch of acicular setae ranging in length from short to very long. Lower anal lip anteriorly with 19 to 30 very short, spine-like setae and posteriorly with a transverse patch of long, acicular setae.

Trichiotinus bibens occurs from southeastern New York to southern Michigan, westward to central Illinois, and southward to northern Alabama and northern Georgia (Hoffman, 1935). Elytra of adult bright green or shiny orange to orange-brown with a metallic luster. The head, pronotum, pygidium, legs, and venter are metallic green. Pygidium also frequently with a white patch on each side. Elytra, pronotum, and head covered with short, yellow hairs. Venter and legs with long, whitish hairs. Adult 11.5 to 12.5 mm long and 5.5 to 6 mm wide.

The larvae feed in the wood of fallen trees. Hoffman (1935) mentions finding adults in decaying chestnut logs. On May 27, 1943, the writer collected one adult in Breathitt County, Kentucky, from beneath the bark of a fallen white oak. On March 14, 1945, north of Mt. Vernon, Kentucky, H. H. Jewett and the writer collected 11 full-grown larvae and prepupae from cells under the bark of an oak log lying in a hillside woodland. According to Hoffman (1935), adults have been collected on the flowers of *Ceanothus*, blackberry, *Viburnum pubescens*, and different species of *Cornus*. Lee H. Townsend and the writer found a female on the flowers of Queen Anne's lace, on June 28, 1945, several miles west of Corbin in Whitley County, Kentucky.

Trichiotinus lunulatus (F.), Third-Stage Larva

This description is based on the following material and reprinted from my 1945c bulletin:

Two third-stage larvae collected April 15, 1932, at Mobile, Alabama, by H. P. Loding, in dead oak limbs. (Loaned by the U. S. National Museum.)

Maximum width of head capsule of third-stage larva 3.57 mm. Maxillary stridulatory area with 2 to 3 sharp, anteriorly directed teeth and an anterior, subconical process. Thoracic spiracle .27 mm long and .19 mm wide. Respiratory plate with a maximum of 15 "holes" along any diameter. "Holes" of elongate, irregular shapes, separated by distances less than the diameter of a "hole." Distance between the two lobes of the respiratory plate only slightly less than the dorsoventral diameter of the bulla. Spiracles of abdominal segments 8 very much smaller than those of abdominal segments 1 to 7.

Dorsa of abdominal segments 7 and 8 each with 2 annulets. Vestiture of each annulet consisting of 1 to 2 sparse, irregular, transverse rows of short setae and a caudal, transverse row of long setae. Dorsum of abdominal segment 9 similarly clothed but not divided into annulets. Pedal areas of abdominal segments 1 to 8 each with 1 to 3 setae on each side. Raster consisting of a large patch of about 52 very short, stout setae, interspersed with a few (about 20) long setae. Lower anal lip anteriorly with about 20 short, stout setae and posteriorly with a row of 10 long setae.

Trichiotinus lunulatus is a southeastern species which is very variable in color. It occurs throughout the southeastern states south of a line running from southern Virginia through western North Carolina, northern Louisiana, and northeastern Texas (Hoffman, 1935). Elytra orange-brown, bright green, or purplish with a metallic luster. Each elytron with 2 short, lateral, transverse white lines. Head and pronotum metallic green or purplish with a rather sparse vestiture of short, gray hairs. Pygidium metallic green or purple, with a whitish patch on each side. Frequently the 2 patches are united anteriorly. Vestiture on pygidium sparse. Legs and venter green or purplish, bearing a number of long, whitish hairs. Adult 10.5 to 11.5 mm long and 5 to 6.5 mm wide.

Little is known of the biology of *T. lunulatus* (Hoffman, 1935). The 2 larvae upon which the description is based were collected from dead oak limbs.

Trichiotinus piger (F.), Third-Stage Larva
(Figs. 400, 409, 419, 423, 424, and 428)

This description is based on the following material and reprinted from my 1945c bulletin:

Two third-stage larvae collected October 5, 1901, at Wilbraham, Massachusetts, by Dimmock, under a chestnut log. No. 2016. See Hoffman, (1935), page 138, lines 7 and 8. (Loaned by the U. S. National Museum.)

Five cast skins of larvae collected and reared to the adult stage by C. H. Hoffman. No. T-2. (Loaned by the U. S. National Museum.)

Maximum width of head capsule of third-stage larva 3.1 to 3.4 mm. Maxillary stridulatory area (Fig. 423) with 4 to 6 sharp, anteriorly directed teeth and an anterior truncate process. Thoracic spiracle .26 to .37 mm long and .21 to .28 mm wide. Respiratory plate with a maximum of 18 to 22 "holes" along any diameter. "Holes" of oval, irregular shapes, separated by less than the diameter of a "hole." Distance between the two lobes of the respiratory plate much less than the dorsoventral diameter of the bulla. Spiracles of abdominal segments 1 to 8 similar in size.

Dorsa of abdominal segments 7 and 8 each with 2 annulets. Each annulet covered with a transverse patch of setae consisting of 3 or 4 irregular rows of short setae and a caudal transverse row of long setae. Dorsum of abdominal segment 9 covered with short setae among which are an anterior transverse row of long setae and a caudal double row of long setae. Pedal areas of abdominal segments 1 to 8 each with 3 to 7 setae on each side. Raster consisting of a broad teges of 25 to 40 scattered, very short, spine-like setae among which are interspersed many long, acicular setae. Lower anal lip anteriorly with 10 to 25 very short, stout setae and posteriorly with a transverse patch of long setae.

Trichiotinus piger is a hairy cetoniid of little or no economic importance. It occurs throughout the eastern half of the United States (Hoffman, 1935), and is fairly common in Kentucky. Elytra of adult with rather short, yellowish hairs; red-brown to blackish with 2 transverse, white, marginal bars on each elytron. Head and pronotum densely hairy, blackish with a metallic luster. Venter and legs red-brown. Venter and femora of legs with a dense covering of whitish hairs. Pygidium with a white patch on each side. Adult 9.5 to 12 mm long and 4.5 to 6.5 mm wide.

Larvae found in decayed oak, poplar, maple, chestnut, and cherry, preferring old stumps and fallen branches that have seasoned 3 or 4 years (Hoffman, 1935). According to Hoffman, larvae in stumps are usually found tunnelling in wood just under the bark and a few inches below the soil surface. On November 14, 1942, near Urbana, Illinois, the writer collected 3 mature larvae from the central cavity of a small, dry oak stump. Adults in flight in Kentucky from late May through July, often frequenting rose blossoms. A few specimens have been collected on *Aesculus parviflora*. Hoffman (1935) found them in the blossoms of rose, iris, peony, mock orange, dogbane, dogwood, lilac, and *Sorbaria*. The species has a one-year life cycle.

Genus *Valgus* Scriba

Key to Larvae of the Species of *Valgus*

1. Frons of head with only a few pits.........*V. seticollis* (Beauv.)
   Frons of head with 30 or more pits (Fig. 406) 2
2. Frons with pits which are much larger than those on the epicranium. Raster with a sparse, transverse patch of 30 to 40 slender setae, anterior to the lower anal lip ....*V. californicus* Horn
   Frons with pits which are similar in size to those on the epicranium. Raster with a sparse, transverse patch of 10 or fewer slender setae, anterior to the lower anal lip ....*V. canaliculatus* (Fab.)

Valgus californicus Horn, Third-Stage Larva

This description is based on the following material:

Three third-stage larvae collected July 27, 1948, from a termite's nest, at Snowline Camp, Eldorado County, California, by J. W. MacSwain.

Maximum width of head capsule of third-stage larva 1.1 to 1.2 mm.[30] Head yellow-brown, epicranium with small pits; frons with 70 to 80 very large, prominent, round pits. Frons bearing, on each side, 2 posterior frontal setae, 1 exterior frontal seta, 1 or 2 anterior frontal setae, and 1 or 2 anterior angle setae. Labrum slightly emarginate distally. Haptomerum of epipharynx poorly developed. Each chaetoparia with 11 to 15 scattered setae. Pedium vague. Tormae poorly developed, knob-like, not extended mesally. Ventral stridulatory area of mandibles consisting of 3 broad, vague ridges. Maxillary stridulatory area without sharp teeth but with a round anterior process. Last segment of antenna with a single, oval, dorsal sensory spot.

Thoracic spiracles .11 mm long and .066 wide. Respiratory plate with a maximum of 8 to 10 elongate "holes" along any diameter. "Holes" not in definite rows. Arms of respiratory plate not constricted.

Dorsa of abdominal segments 7 to 9 not divided into annulets. Each dorsum with 2 widely separated, transverse rows of long slender setae anterior to which are a very few short, stout setae. Dorsa of abdominal segments 9 and 10 not fused. Dorsum of abdominal segment 10 slightly swollen, divided into 2 cushion-like areas by a slight median groove. Venter of abdominal segment 10 without palidia. Raster consisting of a sparse, transverse patch of 30 to 40 long and short, straight, slender setae, anterior to the lower anal lip. Lower anal lip with about 30 long and short slender setae. Claws falcate, sharp-pointed, each bearing 2 setae.

Valgus californicus is mentioned by Banks and Snyder (1920) as being an inquiline in the nest of the termite *Zootermopsis angusticollis* Hagen in the Pacific coast region. It was taken by Snyder at Little Bear Lake, California, at an elevation of 5,000 feet.

Concerning this species, Hinton (1930) has written, "while collecting near Riverton, El Dorado County, California, during August 1930, I was attracted to three fire-killed sugar pines which were badly infested with termites. Beneath the loose bark and at the base of the trees in the damp, rotten wood and termite castings all stages of this insect were to be found. The pupae

were lying in small hollows which looked as if they had been made out of the castings."

Linsley and Ross (1940) found *V. californicus* in several localities in the Transition Zone in the San Jacinto Mountains, California, in association with *Zootermopsis* sp. under bark of *Pinus ponderosa* stumps. These authors refer to this beetle as rather uncommon. Linsley and Michener (1943) took adults in several localities in Shasta County, California, in the nests of *Zootermopsis nevadensis* (Hagen).

Dr. E. C. Van Dyke (personal letter, February 1945) gave the writer the following notes on the distribution of *Valgus californicus*: "The species is found throughout the southern Cascades and Sierra Nevada Mountains, as well as in the San Jacinto Mountains of southern California and the Coast Range (Santa Cruz) Mountains. Specimens are to be found in the collection of the California Academy of Sciences from the following counties in California: Trinity, Siskiyou, Placer, Merced, Lassen, Alpine, Tuolumne, Calaveras, Tulare, Santa Cruz, Riverside; and Klamath County, Oregon." The writer has also seen a specimen from Jackson County, Oregon.

Valgus canaliculatus (Fab.), Third-Stage Larva[31]
(Figs. 406, 411, and 426)

This description is based on the following material and reprinted from my 1945c bulletin:

One hundred fifty-five third-stage larvae, 6 prepupae, 6 pupae, and 36 newly transformed adults (a number of these reared and associated with cast larval skins). These specimens were collected July 13, 15, and 28, 1944, at Lexington, Kentucky, by Lee H. Townsend and the writer, from the galleries of termites in a soft-maple log.

Maximum width of head capsule of third-stage larva 1.2 to 1.4 mm. Surface of cranium prominently pitted, light red-brown (Fig. 406). Frons bearing on each side 2 long, posterior frontal setae, 1 or 2 exterior frontal setae, a very small anterior frontal seta, and a single anterior angle seta. Two to 4 dorsoepicranial setae on each side. Labrum bilobed distally; clithra absent. Haptomerum of epipharynx poorly developed, lacking a distinct curved row of setae (Fig. 426). Pedium indistinct. Mandibles with the ventral, oval stridulatory area consisting of 3 to 4 short, broad ridges (Fig. 411). Maxilla with the uncus of the lacinia represented by a small, indistinct elevation. Maxillary stridulatory area consists of an anterior subconical process, posterior to

---

[30] There is a possibility that these may be second, not third instars.

[31] The larvae described as this species by Böving and Craighead (1931) belong to the species *V. seticollis* (Beauv.).

which are 1 or 2 sharp teeth with anteriorly directed points, surrounded by a group of 5 to 14 minute teeth. Last antennal segment with a single, oval, dorsal sensory spot.

Thoracic spiracles .21 to .247 mm long and .15 to .18 mm wide. Respiratory plate with a maximum of 15 to 19 oval to elongate-oval "holes" along any diameter. Distance between the two lobes of the respiratory plate much less than the dorsoventral diameter of the bulla. Thoracic spiracles much larger than the abdominal spiracles. Spiracles of abdominal segment 1 considerably larger than those of abdominal segments 2 to 8.

Dorsa of abdominal segments 7 to 9 not divided into annulets. Each dorsum with 2 widely separated, narrow, transverse patches of setae, each consisting of a posterior, rather sparsely set row of long, slender setae anterior to which are a few scattered, short, stout setae. Dorsa of abdominal segments 9 and 10 not fused. Dorsum of abdominal segment 10 slightly swollen, divided into 2 oval, cushion-like areas by a slight median groove. Venter of abdominal segment 10 without palidia. Setation of raster limited to a few (10 or fewer) scattered, slender setae; short, spine-like setae absent. Lower anal lip fringed posteriorly with 20 to 31 long and short, slender setae. Claws falcate, sharp-pointed, each bearing 2 setae.

Valgus canaliculatus (Fab.) is a very small, termitophilous cetoniid occurring in eastern United States. Adult with reddish-brown head, pronotum, and elytra, these mottled with irregular patches of black and white squamae. Legs and venter of thorax reddish-brown, covered with white squamae. Venter of male abdomen with a very dense patch of yellow to orange squamae. Female with a prominent spine-like, pygidial process about 1.3 mm long. Adults 5.1 to 6.1 mm long and 2.7 to 3.2 mm wide (not including pygidial process of female).

Adult males are found from late May through early June feeding on the nectar of mock orange (*Philadelphus*) blossoms. At Lexington, Kentucky, during July 1944, larvae were found in considerable numbers in two decaying soft maple logs, feeding on the walls of termite galleries. Presence of the larvae could be detected by the accumulation of many small pellets of frass in the termite burrows. Pupation began about July 10, in small, oval cells 9.5 to 11.5 mm long and 3.5 to 5.5 mm wide, constructed of wood fragments or dirt adhering to the wood. By August 14 almost all the larvae had pupated. The first adult emerged about July 20 and the last about August 25. Of 36 newly transformed adults, 21 were females. The winter is passed in the adult stage.

Valgus seticollis (Beauv.), Third-Stage Larva
(Figs. 410, 429, and 436)

This description is based on the following material and reprinted from my 1945c bulletin:

Three third-stage larvae, associated with 2 adults. No collection data. Determined by E. A. Chapin, February 1944. (Loaned by the U. S. National Museum.)

Three third-stage larvae, associated with an adult. Collected August 9, 1903, in Rock Creek Park, D. C., by W. V. Warner, from a stump infested with termites.

Six third-stage larvae reared by Lee H. Townsend, from eggs laid by segregated females. The adults were collected March 14, 1945, near Mt. Vernon, Kentucky, by H. H. Jewett and the writer, from termite galleries in a beech log.

Maximum width of head capsule of third-stage larva 1.4 to 1.55 mm. Surface of cranium smooth, light yellow-brown. Frons bearing on each side 2 to 4 posterior frontal setae, 2 exterior frontal setae, a single anterior frontal seta, and a single anterior angle seta. Two to 3 dorsoepicranial setae on each side. Labrum bilobed distally, clithra absent. Haptomeral region of epipharynx (Fig. 429) poorly developed, lacking a distinct curved row of setae. Pedium indistinct. Mandibles without a distinct ventral, oval stridulatory area. Mandibles (Fig. 410) with a single scissorial tooth posterior to the scissorial notch. Maxilla with the unci of the lacinia represented by a small, indistinct elevation. Maxillary stridulatory area with one or more indistinct elevations and an anterior subconical process. Last antennal segment with a single oval, dorsal sensory spot.

Thoracic spiracles .22 to .23 mm long and .16 to .18 mm wide. Respiratory plate with a maximum of about 14 oval "holes" along any diameter. Distance between the two lobes of the respiratory plate somewhat to much less than the dorsoventral diameter of the bulla. Thoracic spiracles several times larger than the abdominal spiracles. Spiracles of abdominal segments 2 to 8 similar in size, those of the first abdominal segment larger.

Dorsa of abdominal segments 7 to 9 each with 2 annulets. Each anterior annulet with a transverse patch of short, stout setae and a caudal, transverse row of long, slender setae; each posterior annulet with a few short, stout setae and a caudal, transverse row of long setae. Dorsa of abdominal segments 9 and 10 not fused. Dorsum of abdominal segment 10 slightly swollen, divided into 2 oval, cushion-like areas by a slight median groove. Venter of abdominal segment 10 without palidia

(Fig. 436). Setation of raster limited to about 10 to 15 scattered, slender setae; short, spine-like setae absent. Lower anal lip with about 30 longer, slender setae. Claws falcate, sharp-pointed, each bearing 2 setae.

Valgus seticollis (Beauv.) is a small, termitophilous species occurring in eastern United States. Adult male with reddish-brown head, pronotum, and elytra, these mottled with scattered patches of white and black squamae. Legs and venter reddish-brown, covered with white squamae. Venter of abdomen of both sexes with a sparse covering of squamae. Female blackish, without a pygidial process. Adult 6.5 to 6.8 mm long and 3 to 3.2 mm wide. Adults and larvae are found in wood infested with termites.

Plate XXXIV

FIGURE 397. *Cotinis nitida* (Linn.). Distal portion of mesothoracic leg.

FIGURE 398. *Osmoderma eremicola* Knoch. Distal portion of metathoracic leg.

FIGURE 399. *Cremastocheilus pugetanus* Casey. Distal portion of mesothoracic leg.

FIGURE 400. *Trichiotinus piger* (Fab.). Last antennal segment, dorsal view.

FIGURE 401. *Osmoderma eremicola* Knoch. Last antennal segment, dorsal view.

FIGURE 402. *Cotinis nitida* (Linn.). Last antennal segment, dorsal view.

FIGURE 403. *Cotinis nitida* (Linn.). Head, dorsal view.

FIGURE 404. *Cremastocheilus pugetanus* Casey. Head, dorsal view.

FIGURE 405. *Gnorimella maculosa* (Knoch). Head, dorsal view.

FIGURE 406. *Valgus canaliculatus* (Fab.). Head, dorsal view.

FIGURE 407. *Cotinis nitida* (Linn.). Left mandible, dorsal view.

FIGURE 408. *Cotinis nitida* (Linn.). Right mandible, ventral view.

FIGURE 409. *Trichiotinus piger* (Fab.). Head, dorsal view.

FIGURE 410. *Valgus seticollis* (Beauv.). Left mandible, dorsal view.

FIGURE 411. *Valgus canaliculatus* (Fab.). Left mandible, ventral view.

FIGURE 412. *Trigonopeltastes delta* (Forst.). Left mandible, dorsal view.

FIGURE 413. *Euphoria inda* (Linn.). Labium and ental view of left maxilla.

FIGURE 414. *Gymnetis sallei* Schaum. Left mandible, dorsal view.

FIGURE 415. *Euphoria inda* (Linn.). Left mandible, dorsal view.

FIGURE 416. *Cremastocheilus pugetanus* Casey. Left mandible, dorsal view.

FIGURE 417. *Euphoria herbacea* Oliv. Maxillary stridulatory teeth.

FIGURE 418. *Cremastocheilus pugetanus* Casey. Maxillary stridulatory teeth.

FIGURE 419. *Trichiotinus piger* (Fab.). Left mandible, dorsal view.

FIGURE 420. *Trigonopeltastes delta* (Forst.). Inner distal portion of right maxilla, showing unci.

FIGURE 421. *Cotinis nitida* (Linn.). Maxillary stridulatory teeth.

FIGURE 422. *Osmoderma eremicola* Knoch. Maxillary stridulatory teeth.

FIGURE 423. *Trichiotinus piger* (Fab.). Maxillary stridulatory teeth.

FIGURE 424. *Trichiotinus piger* (Fab.). Inner distal portion of right maxilla, showing unci.

Symbols Used

AA—Seta of anterior angle of frons
AFS—Anterior frontal seta
CL—Claw
DES—Dorsoepicranial setae
DSS—Dorsal sensory spots
E—Epicranium
EFS—Exterior frontal seta
ES—Epicranial stem
F—Frons
FS—Frontal suture

GL—Glossa
GU—Uncus of galea
HP—Haptomeral process
L—Labium
LP—Labial palpus
LT—Laeotorma
LU—Uncus of lacinia
MP—Maxillary palpus
MS—Macrosensillae
O—Ocellus
PC—Preclypeus

PFS—Posterior frontal seta
PSC—Postclypeus
S_{1-4}—Scissorial teeth
SD—Maxillary stridulatory area
SDT—Process anterior to maxillary stridulatory teeth
SN—Scissorial notch
STA—Stridulatory area of mandible
TP—Truncate process of hypopharynx

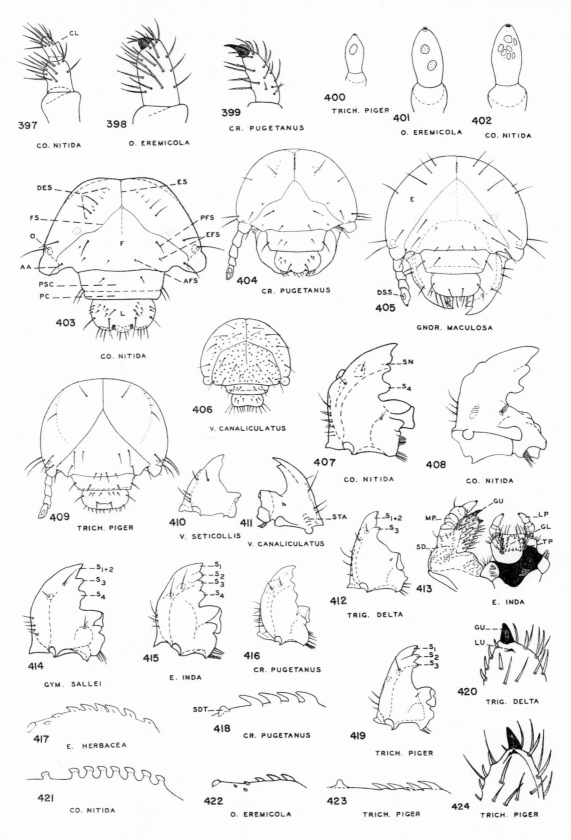

397 CO. NITIDA
398 O. EREMICOLA
399 CR. PUGETANUS
400 TRICH. PIGER
401 O. EREMICOLA
402 CO. NITIDA
403 CO. NITIDA
404 CR. PUGETANUS
405 GNOR. MACULOSA
406 V. CANALICULATUS
407 CO. NITIDA
408 CO. NITIDA
409 TRICH. PIGER
410 V. SETICOLLIS
411 V. CANALICULATUS
412 TRIG. DELTA
413 E. INDA
414 GYM. SALLEI
415 E. INDA
416 CR. PUGETANUS
417 E. HERBACEA
418 CR. PUGETANUS
419 TRICH. PIGER
420 TRIG. DELTA
421 CO. NITIDA
422 O. EREMICOLA
423 TRICH. PIGER
424 TRICH. PIGER

Plate XXXV

FIGURE 425. *Cotinis nitida* (Linn.). Epipharynx.

FIGURE 426. *Valgus canaliculatus* (Fab.). Epipharynx.

FIGURE 427. *Gymnetis sallei* Schaum. Epipharynx.

FIGURE 428. *Trichiotinus piger* (Fab.). Epipharynx.

FIGURE 429. *Valgus seticollis* (Beauv.). Epipharynx.

FIGURE 430. *Cotinis nitida* (Linn.). Central portion of raster.

FIGURE 431. *Cotinis texana* Casey. Central portion of raster.

FIGURE 432. *Euphoria herbacea* Oliv. Central portion of raster.

FIGURE 433. *Euphoria inda* (Linn.). Central portion of raster.

FIGURE 434. *Euphoria sepulchralis* (Fab.). Central portion of raster.

FIGURE 435. *Gymnetis sallei* Schaum. Central portion of raster.

FIGURE 436. *Valgus seticollis* (Beauv.). Venter of last abdominal segment.

FIGURE 437. *Gnorimella maculosa* (Knoch). Venter of last abdominal segment.

FIGURE 438. *Osmoderma eremicola* Knoch. Lateral view of eighth to tenth abdominal segments.

FIGURE 439. *Euphoria fulgida* (Fab.). Central portion of raster.

FIGURE 440. *Gnorimella maculosa* (Knoch). Lateral view of eighth to tenth abdominal segments.

Symbols Used

ACP—Acanthoparia
ASL—Anal slit
C—Campus
CI—Clithrum
CPA—Chaetoparia
DX—Dexipohoba

EUS—Eusternum
H—Haptomerum
HL—Haptolachus
LAL—Lower anal lip
PE—Pedium
PEA—Pedal area

PLA—Palidium
PLL—Pleural lobe
SC—Sense cone
SE—Septula
T—Teges

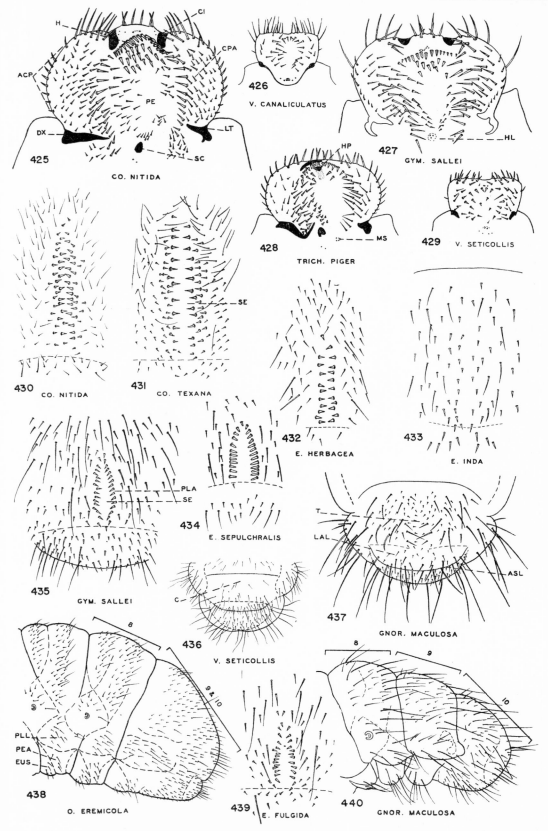

425 CO. NITIDA

426 V. CANALICULATUS

427 GYM. SALLEI

428 TRICH. PIGER

429 V. SETICOLLIS

430 CO. NITIDA

431 CO. TEXANA

432 E. HERBACEA

433 E. INDA

434 E. SEPULCHRALIS

435 GYM. SALLEI

436 V. SETICOLLIS

437 GNOR. MACULOSA

438 O. EREMICOLA

439 E. FULGIDA

440 GNOR. MACULOSA

Plate XXXVI

Figure 441.   *Osmoderma eremicola* Knoch. Epipharynx.

Figure 442.   *Euphoria inda* (Linn.). Entire larva (setae omitted); left lateral view.

Figure. 443.   *Crematocheilus armatus* Walker. Epipharynx.

Figure 444.   *Genuchinus ineptus* (Horn). Epipharynx.

Figure 445.   *Osmoderma eremicola* Knoch. Left mandible, dorsal view.

Figure 446.   *Cremastocheilus armatus* Walker. Left mandible, ventral view.

Figure 447.   *Cremastocheilus armatus* Walker. Last antennal segment, dorsal view.

Figure 448.   *Euphoria inda* (Linn.). Last antennal segment, dorsal view.

Figure 449.   *Euphoria inda* (Linn.). Distal portion of leg.

Figure 450.   *Cremastocheilus armatus* Walker. Labium and right maxilla, ental view.

Figure 451.   *Cremastocheilus armatus* Walker. Venter of last two abdominal segments.

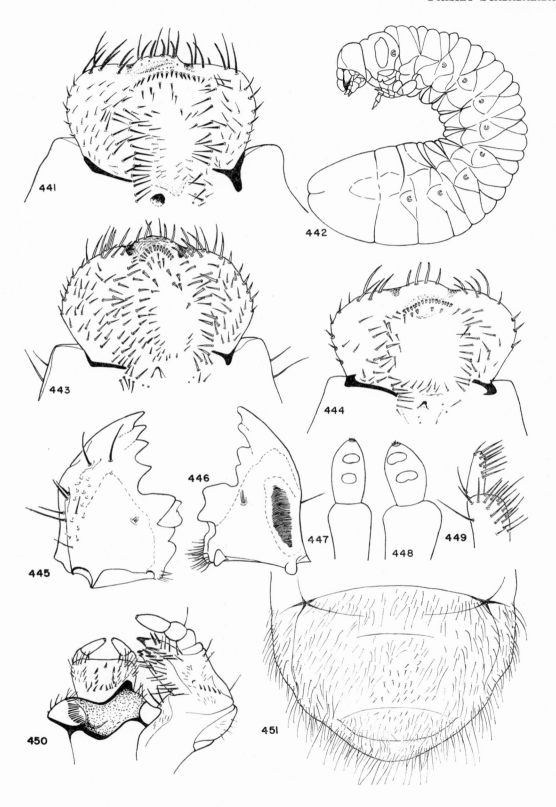

441

442

443

444

445

446

447

448

449

450

451

Family Lucanidae

The Lucanidae, commonly called stag beetles, are usually associated with decaying wood. Some 40 species belonging to 9 genera occur in the United States. Adults range in size from small to very large (genus *Lucanus*), are elongate in shape, and in some cases possess greatly elongated mandibles. The adults are usually black or castaneous in color, but one species, *Platycerus oregonensis* Westw., has bluish elytra.

Lucanidae breed in old stumps and in decaying roots and logs of both coniferous and deciduous trees, where all stages may be found. Little is known about the habits of the adults. Adults of several species of *Lucanus* are frequently attracted to lights. Adults of other species are found most commonly in or under decaying wood or seen resting on foliage of nearby shrubs. Larvae of one species of *Lucanus* are known to feed on live roots (Shenefelt and Simkover, 1950, and Milne, 1933).

Larvae of Lucanidae are C-shaped and whitish or bluish in color. They also resemble larvae of Scarabaeidae in possessing cribriform spiracles. The antennae, epipharynges, and maxillae of lucanid larvae show morphological similarities to those of the less specialized Scarabaeidae. On the other hand, mandibles and hypopharynges of lucanid larvae are similar to those of the higher Scarabaeidae. In general, based on larval characters, the Lucanidae are much more closely related to the Scarabaeidae than are the Passalidae.

Hayes (1929) described the epipharynx of *Dorcus* sp. and of *Sinodendron rugosum* Mann. and gave a key to five genera of Lucanidae using characters of the number of antennal segments and of the last abdominal segment. Böving and Craighead (1930), in their key to families and subfamilies of Scarabaeoidea, gave characters for separating the subfamilies and known genera of Lucanidae occurring in the United States, including *Nicagus*. Van Emden (1935) published a detailed study of lucanid larvae from the world standpoint, including keys. In 1941, in his paper on British beetles, he gave a key for separating three species belonging to three genera occurring in the United States. Benesh, the leading authority on adult lucanids, described the larva of *Platycerus oregonensis* Westw. in 1946. Peterson (1951) figured the larva of *Ceruchus piceus* Web., its mouthparts, the mesothoracic and metathoracic legs, the caudal end of the body, and an abdominal spiracle.

Larvae of the family Lucanidae may be characterized as follows: Antenna 3- or 4-segmented; last segment greatly reduced in size. Epipharynx with united tormae and usually with three nesia. Mandibles with a ventral process. Maxilla with galea and lacinia distinctly separate. Maxillary stridulatory teeth usually absent (present in *Platycerus*). Cribriform spiracles present. Abdominal terga usually not plicate. Anal opening Y-shaped or longitudinal, lying between two fleshy lobes. Legs well developed, not reduced in size, and possessing stridulatory organs on the mesothoracic and metathoracic legs.

Key to Tribes and Genera of the Family Lucanidae Based on Characters of the Larvae

1. Thoracic spiracles with concavities of respiratory plates facing cephalad (Fig. 453). Trochanters of metathoracic leg each with a stridulatory area consisting of a single longitudinal row of very short transverse ridges (Figs. 472, 476, and 478). Inner margin of left mandible, distad of the molar areas, set with one or more teeth (Figs. 457 and 459). Raster with 2 patches of spine-like setae (Figs. 480, 482, and 483) 2

   Thoracic spiracles with concavities of respiratory plates facing caudoventrad or cephaloventrad. Trochanters of metathoracic leg each with a stridulatory area consisting of a patch of small teeth or granules (Figs. 471 and 475). Inner margin of left mandible, distad of the molar areas, smooth, not toothed (Fig. 455 and 460). Raster without spine-like setae (Figs. 479 and 481) 4

2. Trochanter of metathoracic leg with a rather long, pointed, apical lobe (Fig. 472). Claws with 4 to 6 setae (Fig. 470)...................subfamily **Lucaninae**, *Lucanus*
   page 188

   Trochanter of metathoracic leg swollen apically but without a strongly projecting lobe (Figs. 476 and 478). Claws with 2 setae 3

3. Coxae of mesothoracic legs each with a stridulatory area consisting of a slightly curved, longitudinal row of subconical teeth posterior to which is a patch of small granules (Fig. 474) ....
   ....subfamily **Dorcinae**, *Dorcus*
   page 188

   Coxae of mesothoracic legs each with a stridulatory area consisting of several oblique rows of subconical granules (Fig. 477)....................................
   ....subfamily **Aesalinae** (part), *Platycerus*
   page 187

Genus *Ceruchus* MacLeay

Ceruchus piceus Web., Third-Stage Larva
(Figs. 460, 461, 467, 481, and 485)

Description based on the following material:

Ten third-stage larvae associated with 3 adults, collected March 28, 1945, from decaying wood, in Brownfield Woods, near Urbana, Illinois, by P. O. Ritcher.

Maximum width of head capsule 2.97 to 3.24 mm. Surface of head (Fig. 461) smooth, straw colored, with reddish-brown anterior portion of frons, clypeus, and labrum. Primary frontal setae, on each side, consisting of 1 anterior frontal seta, 1 anterior angle seta, 1 exterior frontal seta, and an oblique pair of posterior frontal setae along the frontal suture. Epicranial setae consisting of 5 or 6 setae, on each side, not in a definite row. Labrum symmetrical, feebly trilobed. Pedium of epipharynx (Fig. 467) surrounded by 3 phobae. Anterior phoba consisting of a curved, closely set row of 10 to 12 truncate processes. Pedium bounded, on each side, by a phoba consisting of a single row of setae-like filaments. Chaetoparia bare except for a single seta in each caudolateral angle. Haptolachus with 3 nesia, the left one less sclerotized than the other two. Tormae united, symmetrical, and with a prominent triangular epitorma. Scissorial area of left mandible (Fig. 460) with 3 teeth, 2 anterior to and 1 posterior to the scissorial notch. Right mandible with scissorial area consisting of 2 teeth. Inner face of mandibles, between scissorial and molar areas, without additional teeth. Lacinia with 2 terminal, unequal unci fused at their bases; ventral uncus the smaller. Antenna 3-segmented. Last antennal segment much reduced in size. Apex of second antennal segment with a round sensory area laterad of the base of the third segment.

Spiracles with concavities of respiratory plates facing cephaloventrally. Respiratory plate of thoracic spiracle reniform. Spiracles on abdominal segments 1 to 4 progressively smaller in size, those on abdominal segments 5 to 8 all small and similar in size. Prothorax without projecting anterior processes. Abdominal terga not plicate. Dorsa of abdominal segments 1 to 6 each with a broad, transverse band of short, stout setae fringed posteriorly with a few long, slender setae. Venter of last abdominal segment (Fig. 481), anterior to lower anal lobes, with a median, bulbous swelling flanked

on each side with 4 to 6 long, curved setae. Anal opening Y-shaped with arms and stem of Y of about equal length. Lateral anal lobes setose (Fig. 485), upper anal lobe with a few, small, scattered setae. Upper and lateral anal lobes similar in size, without oval, pad-like areas on lateral anal lobes. Legs well developed, 4-segmented. Coxa of mesothoracic leg with a stridulatory area consisting of a long, oval patch of short, diagonal, truncate ridges (Fig. 473). Trochanter of metathoracic leg with a stridulatory area consisting of a dense, elongate patch of granules (Fig. 475). Claws longest on prothoracic legs, quite short on metathoracic legs. Each claw bearing 2 setae.

Ceruchus piceus Web. is a small lucanid occurring from New York westward to Illinois and southward as far as the District of Columbia. The adult is reddish-brown to black and from 10 to 15 mm long. The elytra are striated with finely punctured intervals. The mandibles of the male are as long as, or longer than, the pronotum, and each has 2 prominent teeth on the inner margin, one larger and median and the other near the base. Hoffman (1937) obtained larvae from a decayed elm log, Norway pine, and a decayed birch stump. Felt (1906) found larvae in beech, chestnut, willow, birch, and black cherry. This species is also reported from oak (Blatchley, 1910).

Genus *Sinodendron* Hellw.

Sinodendron rugosum Mann., Third-Stage Larva
(Figs. 452, 455, 458, 462, 469, 471, 479, and 487)

This description is based on the following material:

Eleven third-stage larvae collected May 17, 1962, from decaying ash and maple logs, 2 miles west of Dead Indian Soda Springs, Jackson County, Oregon, by P. O. Ritcher.

Maximum width of head capsule 3.52 to 3.74 mm. Surface of head smooth, yellow-brown, with straw-colored frons and reddish-brown clypeus and labrum. Primary frontal setae, on each side, consisting of 1 (rarely 2) anterior frontal seta, 1 anterior angle seta, 1 exterior frontal seta, and an oblique, elongate patch of 6 to 8 setae close to the frontal suture (Fig. 458). Labrum symmetrical with almost truncate anterior margin. Pedium of epipharynx (Fig. 462) surrounded by a spiculate phoba. Chaetoparia with an irregular row of 5 to 7 setae near each lateral margin. Haptolachus with 3 nesia, the left one smaller than the other two. Tormae united, nearly symmetrical, and with a prominent triangular epitorma. Scissorial area of left mandible (Fig. 455) with 3 teeth; scissorial area of right mandible with 2 teeth. Inner face of mandibles, between scissorial area and molar area, without additional teeth. Lacinia with a single, large falcate uncus. Antenna 4-segmented, the first segment very short and partly fused to the base of the second segment. Apex of third segment with a small, conical sensory appendage. Last antennal segment much reduced in size.

Thoracic spiracle with reniform respiratory plate, concavity facing caudoventrad. Spiracles on abdominal segments 3 to 8 much smaller than those on abdominal segments 1 and 2. Prothorax, on each side, with a projecting anterior process (Fig. 452). Dorsum of first abdominal segment with a transverse patch of very short, stout setae. Second abdominal segment with a similar transverse patch of short, stout setae which surrounds a median bare area. Abdominal segments 3 to 9 each with 2 dorsal folds, a wide scutum, and a narrow scutellum. Each scutum covered with very short, stout setae; each scutellum with a narrow, transverse patch of similar setae and with a few longer, slender setae along the posterior margin of the patch. Venter of last abdominal segment (Fig 479) with about 25 scattered, long and short setae, on each side, anterior to the lower anal lobes. Anal slit longitudinal, located between the 2 bulbous lower anal lobes (Fig. 487). Lower anal lobes each with from 12 to 23 short, scattered setae and each with an oval area that is bare of setae. Upper anal lobe very small. Coxa of mesothoracic leg (Fig. 469) with a stridulatory area consisting of an oval patch of short, truncate teeth arranged in about 10 oblique rows. Trochanter of metathoracic leg (Fig. 471) with a stridulatory area consisting of an elongate patch of small, subconical teeth. Claws well developed with large bases and short apices; each basal part with 2 setae.

Sinodendron rugosum Mann., known as the rugose stag beetle, is a very common western species occurring from California into British Columbia (Essig, 1958) and found in a few localities in Idaho (Barr, 1957). The adult is dark reddish-black to black, with cylindrical body, and is 10 to 17 mm long. The head, pronotum, and elytra are punctate with especially coarse punctures on the elytra. Both sexes have a median horn on the head, but that of the male is much more prominent. The male also has an anterior median process on the cephalic margin of the pronotum. This species breeds in wet, rotten wood (especially logs), and the larvae have been found in oak, alder, willow, cherry, (Fender, 1948), poplar, cottonwood, water birch (Barr, 1957), and California laurel (Essig, 1958). I have collected larvae, in Oregon, from alder, ash, maple, and oak. Frequently, adults in pupal cells are found associated with larvae.

Genus *Platycerus* Geoffrey

Platycerus oregonensis Westwood, Third-Stage Larva
(Figs. 456, 457, 463, 468, 477, 478, 483, and 486)

This description is based on the following material:

Eleven third-stage larvae, associated with two adults, collected May 17, 1962, from decaying ash and maple logs 2 miles west of Dead Indian Soda Springs, Jackson County, Oregon, by P. O. Ritcher.

Maximum width of head capsule 3.0 to 3.2 mm. Surface of head straw colored, with yellow-brown clypeus and light reddish-brown labrum. Primary frontal setae, on each side, consisting of 1 anterior frontal seta, 1 anterior angle seta, 1 very long exterior frontal seta, and a widely separated, oblique pair of posterior frontal setae along the frontal suture (Fig. 456). Epicranial setae, on each side, 4 to 7 in number, in a definite row. Labrum symmetrical with truncate apical margin. Pedium of epipharynx (Fig. 463) with a phoba on each side; phobal filaments rather short and with blunt tips. Chaetopariae each with 11 to 15 setae. Tormae united, nearly symmetrical, and with a prominent elongate epitorma. Haptolachus with 3 nesia, the left one very small. Median nesium with a large posterior extension. Scissorial area of left mandible (Fig. 457) with 3 teeth; scissorial area of right mandible with 2 teeth. Inner face of left mandible with a small tooth. Maxilla (Fig. 468) with a patch of about 60 small, sharp stridulatory teeth on stipes. Lacinia with a single falcate uncus. Antenna 3-segmented, last segment much reduced in size. Apex of second antennal segment with an elongate sensory spot.

Thoracic spiracle reniform with concavity of respiratory plate facing cephalad; abdominal spiracles with concavities of respiratory plates facing cephaloventrad. Spiracles on abdominal segments 1 to 3 much larger than those on abdominal segments 4 to 8. Prothorax without anterior processes. Abdominal terga not plicate. Dorsum of first abdominal segment with a narrow, transverse band of short, stiff setae fringed posteriorly with a row of long, slender setae. Dorsa of abdominal segments 2 to 6 each with a broad, transverse band of short, stiff setae fringed posteriorly with a row of long, slender setae. Venter of last abdominal segment (Fig. 483), anterior to lower anal lobes, with 2 large patches of intermixed long and very short setae. Posterior setae, on each side, arranged obliquely with tips directed caudomesad. Anal slit Y-shaped with arms of Y very short. Lateral anal lobes bulbous and each with an oval bare area (Fig. 486). Lateral anal lobes each with from 4 to 6 setae. Upper anal lobe small, triangular in shape. Coxa of mesothoracic leg (Fig. 477) with a stridulatory area consisting of 8 regular rows of subconical granules. Trochanter of metathoracic leg (Fig. 478) with a stridulatory area consisting of a single, long row of short, transverse ridges. Claws well developed, falcate, each bearing 2 setae.

Platycerus oregonensis Westwood, known as the Oregon stag beetle, is a common western species occurring from California into British Columbia. The adult has a black head and thorax and dark-bluish elytra, and ranges from 10 to 15 mm in length. The elytra are smooth with rows of shallow punctures. In Oregon, this species occurs on eastern slopes of the Coast Range, along streams, and in Willamette Valley woodlots (Fender, 1948). It has also been found in southwestern Oregon. The species breeds in rotten wood of ash, alder, and maple in Oregon and has been reported infesting California holly, madrone, live oak, blue gum, and red alder in California (Benesh, 1946). Adults fly in the heat of the day (Essig, 1958).

Genus *Dorcus* MacLeay

Dorcus parallelus (Say), Third-Stage Larva
(Figs. 453, 459, 464, 474, 476, and 480)

This description is based on the following material:

Eight third-stage larvae, associated with one male adult, collected November 30, 1942, at Lexington, Kentucky, from the decayed base and roots of a soft maple tree, by P. O. Ritcher. No. 42-6C. Adult identified by B. Benesh.

Maximum width of head capsule 5.72 to 6.05 mm. Surface of head smooth, yellow-brown, with light reddish-yellow clypeus and labrum. Primary frontal setae, on each side, consisting of 2 anterior frontal setae, 1 anterior angle seta, 1 exterior frontal seta, and from 4 to 6 posterior frontal setae along the frontal suture. Labrum wider than long, symmetrical, broadly rounded apically. Haptomerum of epipharynx (Fig. 464) with a curved, transverse group of 15 to 17 stout setae. Chaetopariae well developed, each covered with from 25 to 36 sharp setae. Pedium with a small dexiophoba. Tormae asymmetrical, with pternotorma present on right side only; tormae with a long, narrow epitorma. Haptolachus with 3 nesia but with the left nesium poorly developed. Scissorial area of left mandible with 3 teeth (Fig. 459). Right mandible with scissorial area having 2 teeth. Inner face of left mandible, between scissorial and molar areas, with 2 small additional teeth; inner face of right mandible with 1 additional tooth. Maxilla with 1 falcate uncus on lacinia. Stipes without maxillary stridulatory teeth but with a prominent triangular projection. Antenna distinctly 4-segmented; last segment much reduced in size. Third segment with an elongate, apical sensory spot.

Spiracles C-shaped, with concavities of respiratory plates facing cephalad (Fig. 453). Thoracic spiracle with arms of respiratory plate somewhat constricted. Abdominal spiracles progressively smaller in size posteriorly. Prothorax with projecting anterior processes. Abdominal terga not plicate. Dorsa of abdominal seg-

ments 1 to 6 each covered with a broad, transverse patch of very short setae and with a few longer setae along the posterior edge of the patch. Dorsa of abdominal segments 7 to 9 sparsely covered with short setae and with a few longer setae intermixed, especially along the posterior margin of each segment. Venter of last abdominal segment (Fig. 480) with a raster consisting of 2 dense patches of caudomesally directed, spine-like setae. Anal opening Y-shaped; stem of Y much longer than arms. Each lower anal lobe with an oval, whitish lobe which is bare of setae. Coxa of mesothoracic leg (Fig. 476) with a stridulatory area consisting of a slightly curved, longitudinal row of distinctly separate, subconical teeth. Posterior to this row of teeth is a patch of minute granules. Trochanter of metathoracic leg (Fig. 474) with a stridulatory area consisting of a longitudinal row of short, transverse ridges, which are slightly wider apically. Trochanter of metathoracic leg swollen apically but without a strongly projecting lobe. Claws well developed on all legs, each with 2 setae.

Dorcus parallelus (Say) is a rather common eastern stag beetle found from Connecticut to South Carolina and ranging westward to Indiana and Tennessee (Leng, 1920). The adult is dark reddish-brown to black with deeply striated elytra and is 15 to 26 mm long (Blatchley, 1910). The mandible of the male is shorter than the length of the pronotum and has a prominent tooth. Larvae of this species are found in decaying roots and stumps of oak, linden, and maple trees (Dillon and Dillon, 1961).

Genus *Lucanus* Scop.

Larvae of this genus (including species formerly placed in the genus *Pseudolucanus* Hope) may be distinguished by the following characters: Haptomerum of epipharynx with a semicircular group of stout setae. Phobae absent. Chaetoparia well developed. Haptolachus with 3 nesia; left nesium poorly developed. Inner face of left mandible, between scissorial and molar areas, with 2 blunt teeth. Antenna 4-segmented. Spiracles C-shaped with concavities of respiratory plates facing cephalad. Venter of last abdominal segment with 2 dense patches of stiff setae which are directed caudomesad. Coxa of mesothoracic leg with a stridulatory area consisting of a long, slightly curved, dense row of narrow striae and a posterior patch of granulae. Trochanter of metathoracic leg with a long, straight, single row of transverse striae. Distal end of trochanter with a projecting acute lobe. Claws with 4 to 7 setae.

Larvae of *Lucanus elaphus* Fab., *L. capreolus* Linn., *L. mazama* Lec., and *L. placidus* Say., are so similar in morphological characters that I have not been able to make a satisfactory key for their separation. Some differences have been noted in the comparative lengths of the stiff setae on the raster.

Lucanus placidus Say., Third-Stage Larva
(Figs. 465, 470, 472, and 482)

This description is based on the following material:
Four third-stage larvae collected September 3 to 4, 1947, from soil, at the Griffith State Nursery, Wisconsin Rapids, Wisconsin, by Ray D. Shenefelt.

Maximum width of head capsule 7.7 to 8.58 mm. Head reddish-brown. Primary frontal setae, on each side, consisting of 2 anterior frontal setae, 1 anterior angle seta, 2 exterior frontal setae, and 1 or 2 posterior frontal setae. Labrum symmetrical, broadly rounded apically. Haptomerum of epipharynx (Fig. 465) with a semicircular group of about 30 to 35 stout setae which adjoin a double row of similar setae extending along the right side of the pedium. Phobae absent. Chaetopariae well developed, each with about 50 to 70 long, sharp setae. Tormae almost symmetrical, without pterno-tormae, but with a long, narrow epitorma. Scissorial area of left mandible with 3 teeth, that of right mandible with 2 teeth. Inner face of left mandible, between scissorial and molar areas, with 2 blunt teeth; inner face of right mandible with 1 small tooth. Lacinia of maxilla with one falcate uncus. Antenna 4-segmented. Last antennal segment much reduced in size. Apex of third antennal segment with an elongate sensory area.

Spiracles C-shaped, with concavities of respiratory plates facing cephalad. Thoracic spiracle with arms of respiratory plate slightly constricted. Abdominal spiracles on segments 1 to 3 similar in size, those on segments 4 to 8 progressively smaller. Dorsa of abdominal segments 1 to 6 covered with many very short, stiff setae. Dorsa of abdominal segments 7 to 9 with scattered, short, stiff setae and with a few long, slender setae posteriorly. Venter of last abdominal segment largely covered by 2 dense patches of intermixed long and very short, stiff setae which are directed caudomesad. Anal slit longitudinal. Lateral anal lobes with 2 bare, oval lobes. Upper anal lobe very small, triangular. Coxa of mesothoracic leg (Fig. 470) with a stridulatory area consisting of a long, slightly curved row of narrow striae and a posterior patch of minute granules. Trochanter of metathoracic leg (Fig. 472) with a long, straight, single row of transverse striae. Distal end of tranchanter with a projecting acute lobe. Claws with 4 to 7 setae.

Lucanus placidus (Say) is a rather large, locally common stag beetle which is known to occur in many of the north-central and eastern states and in Kansas, Alabama, Arkansas, New Mexico, Oklahoma, and Ontario (Milne, 1933). The adult is a dark red-brown in color and 24 to 31 mm long, with elytra which are rather smooth and without striae. The mandibles of the male are no longer than the pronotum and each has several teeth on its inner margin. According to Hoffman (1937), the larvae are found in decayed oak logs and in roots of old oak stumps. Shenefelt and Simkover (1950), however, reported that larvae of this species were found damaging roots of conifers in forest-tree nurseries in Wisconsin. According to Milne (1933), the larvae feed on the live roots of oak, Virginia creeper, Boston ivy, various shrubs, flowers, and grass.

Plate XXXVII

Figure 452.   *Sinodendron rugosum* Mann. Head, thorax, and part of first abdominal segment; left lateral view.

Figure 453.   *Dorcus parallelus* (Say). Entire larva, left lateral view (setae omitted).

Figure 454.   *Lucanus placidus* (Say). Head.

Figure 455.   *Sinodendron rugosum* Mann. Left mandible, dorsal view.

Figure 456.   *Platycerus oregonensis* Westw. Head.

Figure 457.   *Platycerus oregonensis* Westw. Left mandible, dorsal view.

Figure 458.   *Sinodendron rugosum* Mann. Head.

Figure 459.   *Dorcus parallelus* (Say). Left mandible, dorsal view.

Figure 460.   *Ceruchus piceus* Web. Left mandible, dorsal view.

Figure 461.   *Ceruchus piceus* Web. Head.

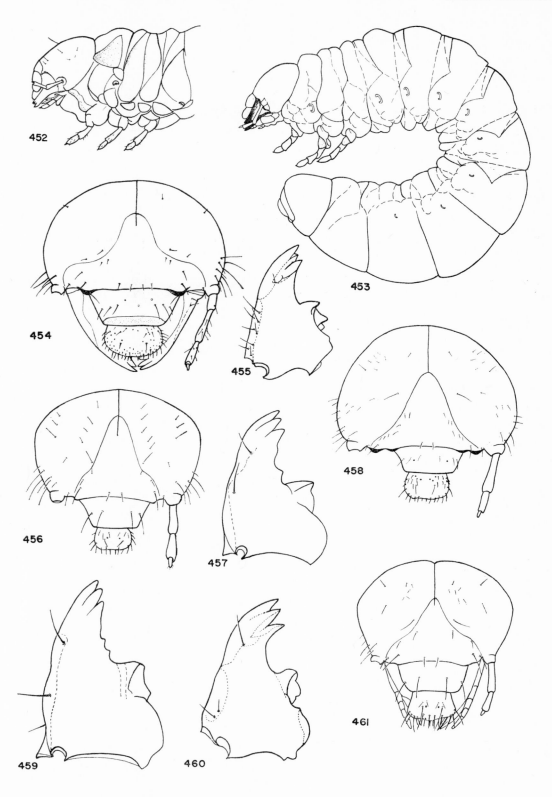

Plate XXXVIII

FIGURE 462.   *Sinodendron rugosum* Mann. Epipharynx.

FIGURE 463.   *Platycerus oregonensis* Westw. Epipharynx.

FIGURE 464.   *Dorcus parallelus* (Say). Epipharynx.

FIGURE 465.   *Lucanus placidus* Say. Epipharynx.

FIGURE 466.   *Lucanus capreolus* (L.). Left maxilla, dorsal view.

FIGURE 467.   *Ceruchus piceus* Web. Epipharynx.

FIGURE 468.   *Platycerus oregonensis* Westw. Right maxilla and hypopharynx.

Symbols Used

CPA—Chaetoparia
DPH—Dexiophoba
ETA—Epitorma
LPH—Laeophoba

NI-3—Nisia
P—Pedium
PH—Phoba

PPH—Protophoba
PTT—Pternotorma
SD—Stridulatory teeth

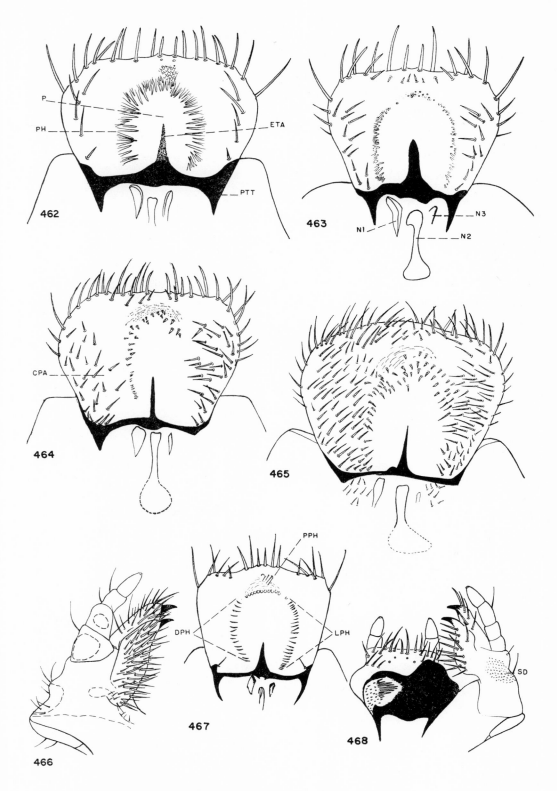

Plate XXXIX

FIGURE 469.   *Sinodendron rugosum* Mann. Right mesothoracic leg.

FIGURE 470.   *Lucanus placidus* Say. Right mesothoracic leg.

FIGURE 471.   *Sinodendron rugosum* Mann. Left metathoracic leg.

FIGURE 472.   *Lucanus placidus* Say. Left metathoracic leg.

FIGURE 473.   *Ceruchus striatus* Lec. Right mesothoracic leg.

FIGURE 474.   *Dorcus parallelus* (Say). Right mesothoracic leg.

FIGURE 475.   *Ceruchus striatus* Lec. Left mesothoracic leg.

FIGURE 476.   *Dorcus parallelus* (Say). Left metathoracic leg.

FIGURE 477.   *Platycerus oregonensis* Westw. Left mesothoracic leg.

FIGURE 478.   *Platycerus oregonensis* Westw. Right metathoracic leg.

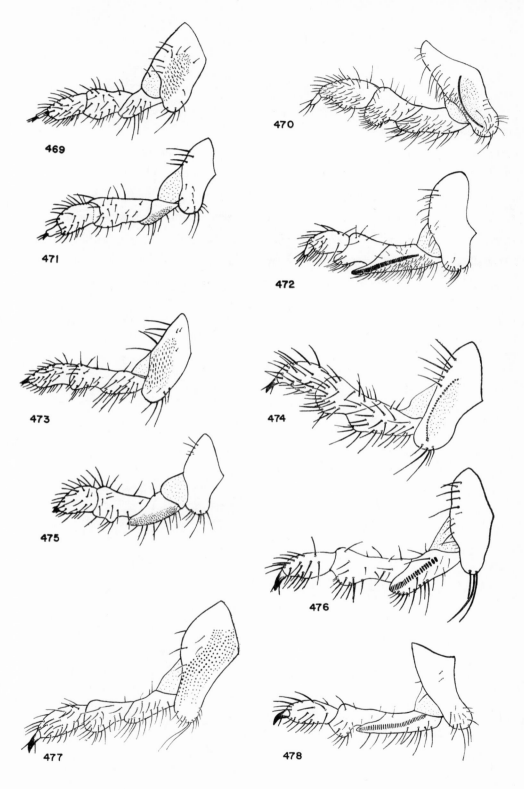

469

470

471

472

473

474

475

476

477

478

Plate XL

Figure 479. *Sinodendron rugosum* Mann. Venter of last abdominal segment.

Figure 480. *Dorcus parallelus* (Say). Venter of last abdominal segment.

Figure 481. *Ceruchus piceus* Web. Venter of last abdominal segment.

Figure 482. *Lucanus placidus* Say. Venter of last abdominal segment.

Figure 483. *Platycerus oregonensis* Westw. Venter of last abdominal segment.

Figure 484. *Lucanus mazama* (Lec.). Caudal view of last abdominal segment.

Figure 485. *Ceruchus piceus* Web. Caudal view of last abdominal segment.

Figure 486. *Platycerus oregonensis* Westw. Caudal view of last abdominal segment.

Figure 487. *Sinodendron rugosum* Mann. Caudal view of last abdominal segment.

Symbols Used

AO—Anal opening   LAL—Lateral anal lobe   UAL—Upper anal lobe
AP—Anal pad

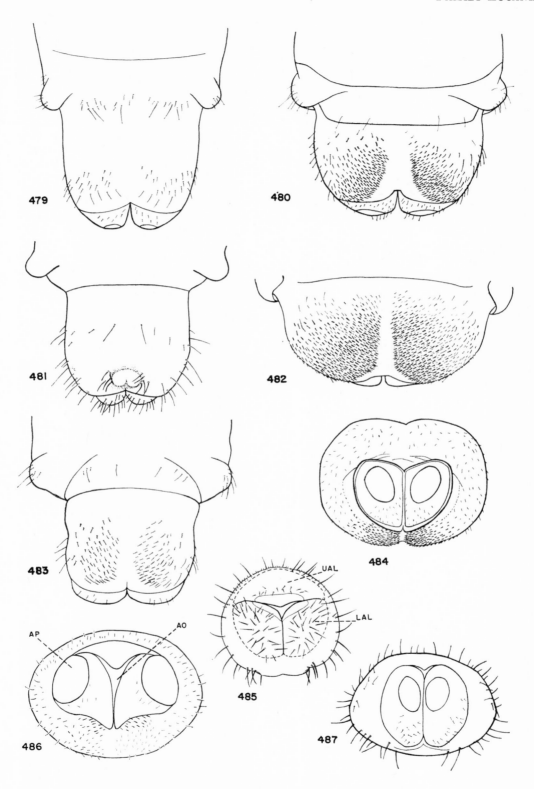

Family Passalidae

The Passalidae are a group of social beetles which inhabit decaying stumps and logs in moist, warm climates. Of the approximately 150 species found in North America (Blackwelder, 1944), only 3 have been recorded from the United States (Blackwelder, 1939). These are *Popillius disjunctus* Ill., which is very common in the eastern United States, *Passalus interruptus* Linn., and *Passalus punctiger* St. Farg. and Serv. The latter two are South American and Central American species which occur as far north as Texas.

All stages of Passalidae are found in galleries excavated by the adults. The larvae feed on wood pulp prepared by the adults and packed along the galleries (Gray, 1946). Both adults and larvae have stridulatory organs which produce audible sounds. Ohaus (1900, 1909) has published on the habits of South American species.

Hayes (1929) compared the morphology of *Popillius disjunctus* Illiger (under the name *Passalus cornutus*) with that of other lamellicorn larvae and published numerous figures showing structural details. Böving and Craighead (1931) included the same species in their key. Many of the oriental species have been described by Gravely (1916 and 1919) who points out that passalid larvae are all much alike. He states "In Oriental forms specific differences are found in the arrangement and nature of the large hairs, and in the manner in which, if at all, pile is developed on the body." (Gravely, 1916.) The various species also vary in regard to whether the lower anal lobe is entire or divided into two lobes. The North American larvae I have examined all have two anal lobes.

Larvae of Passalidae may be distinguished as follows: Antenna 2-segmented. Mandible tridentate at apex; without a ventral process. Maxilla with galea and lacinia distinctly separate. Dorsal surface of stipes with a patch of conical stridulatory teeth. Maxillary palpus 3-segmented. Epipharynx symmetrical; tormae not united. Hypopharynx fleshy, without a heavily sclerotized sclerome. Prothoracic spiracle with concavity of respiratory plate facing anteriorly; abdominal spiracles with concavities of respiratory plates facing posteriorly. Body orthosomatic; dorsa of thoracic and abdominal segments not plicate. Anal slit transverse. Prothoracic and mesothoracic legs 4-segmented, each with a long, curved claw. Metathoracic legs reduced to an unsegmented stub with several apical teeth which rub against a striated stridulatory area on the coxa of the mesothoracic leg.

Key to Third-Stage Larvae of Several Genera and Species of Passalidae

1. With several long, prominent setae anterior to prothoracic spiracle (Fig. 488). Maxilla with uncus of lacinia bifid (Fig. 494) 2
   Without long, prominent setae anterior to prothoracic spiracle. Maxilla with uncus of lacinia entire (Fig. 489) 3
2. Dorsa of abdominal segments 1 to 5 with a single long seta on each side (Fig. 488) ....*Popillius disjunctus* Illiger
   Dorsa of abdominal segments 1 to 5 with a pair of long setae on each side ....*Spurius* sp.
3. Dorsa of abdominal segments 1 to 5 with a transverse pair of long setae on each side. Maxillary stridulatory area consisting of a patch of 30 to 45 teeth ....*Passalus interstitialis* (Esch.)
   Dorsa of abdominal segments 1 to 5 with a single long seta on each side. Maxillary stridulatory area consisting of a patch of 10 to 12 teeth....*Paxillus leachi* Mel.

Genus *Popillius* Kaup

Popillius disjunctus Illiger, Third-Stage Larva
(Figs. 488 and 490 to 496)

This description is based on the following material:
Eleven third-stage larva collected in 1944 at Lexington, Kentucky, in rotting logs, by P. O. Ritcher.

Two third-stage larvae collected in 1959 at Rocky Mount, North Carolina, in a rotting log.

Maximum width of head capsule from 4.5 to 4.7 mm. Head (Fig. 493) straw colored with smooth surface, covered with scattered, short setae interspersed with minute sensilla. Entire larva ranging from 32 to 39 mm in length. Frons lacking primary setae. Epipharynx (Fig. 490) with clithra. Haptomerum poorly developed, with a transverse patch of about 20 short setae; heli absent, chaetoparia well developed, covered with long chaetae. Six sensilla located anterior to the crepis. Lacinia (Fig. 494) with a bifurcate apical uncus.

Maxillary stridulatory area with a patch of 16 to 20 conical teeth borne on the stipes (Fig. 494).

Prothorax with setation limited to a curved longitudinal row of 4 or 5 long setae, located anterior to the prothoracic spiracle. Spiracles reniform (Fig. 488). Respiratory plates not surrounding bullae; spiracles gradually decreasing in size posteriorly. Dorsum of each abdominal segment with a sparse covering of very short setae; dorsa of abdominal segments 1 to 5 or 1 to 6 also with a long seta on each side. Dorsum of abdominal segment 9 with 2 transverse pairs of long setae on each side. Venter of abdominal segment 9 with a pair of long setae on each side. Dorsum of abdominal segment 10 with a sparse, transverse row of 4 long setae on each side. Raster consisting of a rather dense patch of short, caudally directed setae, which covers most of the ventral surface of the last abdominal segment; posteriorly among these setae is a sparse, transverse row of 8 longer setae (Fig. 495). Lower anal lip bare, with a median cleft. Upper anal lip also bare.

Adults of *Popillius disjunctus,* often called "betsy bugs" or the horned passalus, are common in eastern United States where they may be found in colonies in decaying hardwood logs and stumps together with their larvae. The adults are large, elongate, shining, black beetles with a curved horn on the head.

According to Gray's excellent paper on the life history (1946), there is one generation a year. He states that eggs are bright red when first laid and that there are three larval instars. The bluish larvae are nourished on wood pulp prepared for them by the adults. Both larvae and adults possess stridulatory organs which produce audible sound, but their exact function and whether or not they are used for communication is not known.

Genus *Spurius* Kaup

Spurius sp., Third-Stage Larva
(Figs. 497, 499, and 501)

This description is based on the following material:
Seven third-stage larvae, found with 2 pupae and 2 adults, August 1, 1952, 32 miles east of San Christobel, Chiapas, Mexico, at 7,500 feet elevation, in a log.

Maximum width of head capsule from 3.3 to 3.5 mm. Head (Fig. 497) straw colored with smooth surface. Entire larva ranging from 30 to 33 mm in length. Frons with primary setae; with 1 or 2 anterior frontal setae, a patch of 9 to 15 exterior frontal setae, and 1 or 2 posterior frontal setae, on each side. Epipharynx (Fig. 499) very similar to that of *Popillius,* each chaetoparia with a sparse covering of 18 to 21 chaetae. Lacinia with a bifurcate apical uncus of which the dorsal tooth is much the larger. Maxillary stridulatory area with a patch of 15 to 19 conical teeth.

Prothorax with a transverse row of 3 or 4 very long setae anterior to the thoracic spiracle. Spiracles of abdominal segments 2 to 8 much smaller than those on the thoracic and first abdominal segments, decreasing in size posteriorly. Respiratory plates of spiracles not surrounding bullae; surface of plates appearing as a series of transverse, serrated slits.

Dorsa of abdominal segments 1 to 8 with a sparse covering of short, inconspicuous setae; dorsa of abdominal segments 1 to 6 with a transverse pair of long, prominent setae. Dorsum of abdominal segment 9 with a long seta toward each side. Venter of abdominal segment 9 also with a long seta on each side. Dorsum of abdominal segment 9 (Fig. 501) with a transverse row of 3 long setae on each side of the median line. Pleural lobes on thoracic and abdominal segments with numerous (30-50) short setae. Raster with a sparse covering of short, slender, caudally directed setae interspersed posteriorly with a transverse, slightly curved row of 6 long, curved, cylindrical setae. Lower anal lobes bare, distinctly cleft.

Genus *Paxillus* MacLeay

Paxillus leachi Mel., Third-Stage Larva
(Figs. 489 and 500)

This description is based on the following material:
Three third-stage larvae, found with 5 pupae and 2 adults, October 31, 1955, at Musawas, Nicaragua (Waspuc River), under bark, by Boris Malkin.

Maximum width of head capsule from 3.1 to 3.2 mm. Head yellow-brown, smooth; frons and epicranium covered with small, irregularly scattered setae. Epipharynx very similar to that of *Popillius* and *Spurius.* Lacinia of maxilla with a single apical uncus (Fig. 489). Maxillary stridulatory teeth consisting of a patch of 10 to 12 conical teeth on the stipes. Long setae absent from the area anterior to the prothoracic spiracle. Spiracles on abdominal segments 2 to 8 similar in size; those on first abdominal segment somewhat larger. Respiratory plates with closely set, transverse rows of "holes." Thoracic spiracle with approximately 50 rows of 15 to 22 "holes."

Dorsa of abdominal segments 1 to 7 each with a long seta toward each side. Long setae absent on the dorsum of abdominal segment 9; a single long seta on the venter toward each side. Venter of last abdominal segment with scattered, very short setae and posteriorly with a transverse row of 5 very large setae on each side which extends onto the dorsum. Lower anal lobes bare, prominent, ovoid in shape.

Genus *Passalus* Fabricius

Passalus interstitialis (Esch.), Third-Stage Larva
(Fig. 498)

This description is based on the following material:

One second-stage larva and 2 third-stage larvae, found with 1 pupa and several adults, July 10, 1918, at "Bracho" Papaya Plantation, Canal Zone, Panama, by H. F. Dietz and J. Zetck, in a rotten log. Adults determined by E. A. Chapin.

Maximum width of head capsule from 6.8 to 7 mm. Head yellow-brown with smooth surface. Entire larvae about 52 mm in length. Frons with a few primary setae but with small setae irregularly scattered over surface; 2 or 3 rather small exterior frontal setae. Epipharynx (Fig. 498) similar to that of *Popillius*. Chaetopariae well developed with 30 to 38 mesally directed chaetae on each side. Between the chaetopariae, anterior to the pedium, is a patch of about 20 short, stout setae. Tormae inconspicuous, confined to the posterior lateral margins of the epipharynx, unbranched. Lacinia with a single uncus. Maxillary stridulatory teeth with a patch of 30 to 45 spine-like, conical teeth.

Prothorax bare of large setae anterior to the thoracic spiracle. Abdominal spiracles decreasing in size posteriorly. Respiratory plates reniform, not surrounding bullae; surface of plates transversely ribbed with transverse rows of closely set, subquadrate to circular holes between the "ribs."

Dorsa and pleura of abdominal segments with a sparse covering of very small, short setae. Dorsa of abdominal segments 1 to 5 typically with a transverse pair of long, prominent setae, on each side; dorsa of abdominal segments 6 and 7 with a single large seta on each side. Venter of abdominal segment 9 also with one long seta, on each side. Venter of abdominal segment 10 sparsely and irregularly set with small setae; also on each side, with a transverse row of 5 or 6 long, prominent setae which extend onto the dorsum. Lower anal lobes bare, distinctly separated.

Plate XLI

FIGURE 488.   *Popillius disjunctus* Illiger. Third-stage larva, left lateral view.

FIGURE 489.   *Paxillus leachi* Mel. Left maxilla, dorsal view.

FIGURE 490.   *Popillius disjunctus* Illiger. Epipharynx.

FIGURE 491.   *Popillius disjunctus* Illiger. Left mandible, dorsal view.

FIGURE 492.   *Popillius disjunctus* Illiger. Left mandible, ventral view.

FIGURE 493.   *Popillius disjunctus* Illiger. Head.

FIGURE 494.   *Popillius disjunctus* Illiger. Left maxilla and labium, dorsal view.

FIGURE 495.   *Popillius disjunctus* Illiger. Venter of last two abdominal segments.

Symbols Used

CL—Clithrum
CPA—Chaetoparia
DX—Dexiotorma

G—Galea
LA—Lacinia
LAL—Lower anal lobe

LT—Laeotorma
SD—Stridulatory area
SP—Sclerotized plate

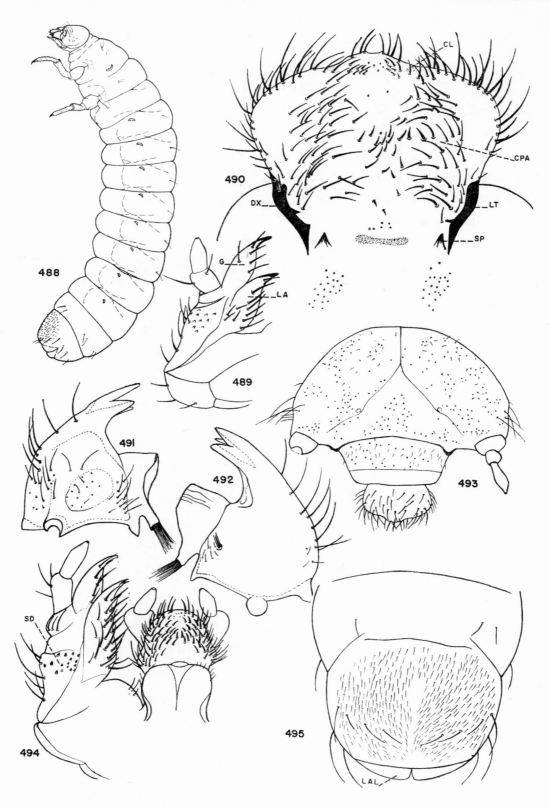

Plate XLII

FIGURE 496. *Popillius disjunctus* Illiger. Left mesothoracic and metathoracic legs.

FIGURE 497. *Spurius* sp. Head.

FIGURE 498. *Passalus interstitialis* (Esch.). Epipharynx.

FIGURE 499. *Spurius* sp. Epipharynx.

FIGURE 500. *Paxillus leachi* Mel. Venter of last abdominal segment.

FIGURE 501. *Spurius* sp. Venter of last two abdominal segments.

Symbols Used

A—Antenna
E—Epicranium
F—Frons

FS—Frontal suture
LAL—Lower anal lobe
ML—Metathoracic leg

PFS—Posterior frontal setae
SD—Stridulatory area
TT—Tibiotarsus

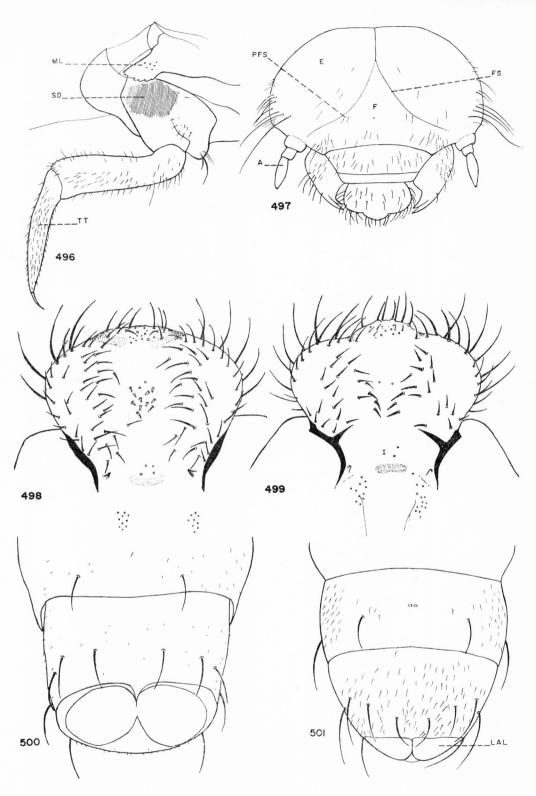

496

497

498

499

500

501

Bibliography

Allen, W. W.
1959. Strawberry pests in California. Calif. Agr. Expt. Sta. Cir. 484, pages 30-32.

Anderson, W. H.
1936. A comparative study of the labium of coleopterous larvae. Smithsn. Inst. Misc. Collection, *95*(13):1-20.

Anonymous.
1961. The European chafer . . . how we fight it. U. S. Dept. Agr. PA-455, 6 pp.

Arrow, G. J.
1904. Sound-production in the lamellicorn beetles. Trans. Ent. Soc., London, 1904:709-750.

1906. A contribution to the classification of the Coleopterous family Passalidae. Trans. Ent. Soc., London, 1906: 441-469.

1937. Systematic notes on beetles of the subfamily Dynastinae with descriptions of a few species in the British Museum collection (Coleoptera). Trans. Royal Ent. Soc., London, *86*(3):35-57.

1951. *Horned Beetles*. W. Junk Publishers, The Hague, 154 pp.

Baerg, W. J.
1942. Rough-headed corn-stalk beetle. Ark. Agr. Expt. Sta. Bull. 415, 22 pp.

Baerg, W. J., and C. E. Palm.
1932. Rearing the rough-headed corn-stalk beetle. Jour. Econ. Ent., *25*(2):207-212.

Banks, N., and T. E. Snyder.
1920. A revision of the nearctic termites. Bull. U. S. Nat. Mus. *108*:1-211.

Barr, W. F.
1957. Notes on the occurrence of *Sinodendron rugosum* Mannerheim in Idaho. Pan. Pac. Ent., *33*(2):86.

Benesh, B.
1937. Some notes on boreal American Dorcinae. Trans. Amer. Ent. Soc., *63*:1-16.

1942. Some notes on nearctic stagbeetles, with description of a new species of *Platycerus* from the Pacific Northwest (Coleoptera:Lucanidae). Ent. News, *56*(8):221-223.

1946. A systematic revision of the holarctic genus *Platycerus* Geoffry (Coleoptera:Lucanidae). Trans. Amer. Ent. Soc., *72*: 139-202.

1956. Some notes on boreal American stagbeetles (Coleoptera:Lucanidae). Ent. News, *67*(2):43-46.

Blackwelder, R. E.
1939. *Fourth Supplement to the Leng Catalogue of Coleoptera of America, North of Mexico, 1933 to 1938 (inclusive)*. John D. Sherman, Jr., Mt. Vernon, N. Y., 146 pp.

1944. Check list of the Coleopterous insects of Mexico, Central America, the West Indies, and South America, Part 2. Bull. U. S. Nat. Mus., *185*:189-341.

Blackwelder, R. E., and R. M. Blackwelder.
1948. *Fifth Supplement 1939 to 1947 (inclusive) to the Leng Catalogue of Coleoptera of America, North of Mexico*. John D. Sherman, Jr., Mount Vernon, N. Y., 87 pp.

Blanchard, F.
1888-1890. Some account of our species of *Geotrupes*. Psyche, *5*:103-110.

Blatchley, W. S.
1910. *An Illustrative Descriptive Catalogue of the Coleoptera or Beetles (exclusive of the Rhynchophora) Known to Occur in Indiana*. Nature Publishing Co., Indianapolis, Indiana, 1,386 pp.

Boucomont, A.
1902. Coleoptera Lamellicornia: Fam. Geotrupidae. Genera Insectorum, Fasc., *7*:1-20.

1912. Scarabaeidae: Taurocerastinae, Geotrupinae. Coleopterorum Catalogus, *46*:1-47. Edited by S. Schenkling. W. Junk Publishers, Berlin.

Böving, A. G.
1921. The larva of *Popillia japonica* Newman and a closely related undetermined ruteline larva. A systematic and morphological study. Proc. Ent. Soc. Wash., *23*(3): 51-62.

1936. Description of the larva of *Plectris aliena* Chapin and explanation of new terms applied to the epipharynx and raster. Proc. Ent. Soc. Wash., *38*(8):169-185.

1937. Keys to the larvae of four groups and forty-three species of the genus *Phyllophaga*. U. S. Dept. Agr., Bur. of Ent. and Plant Quar., *E-417*:1-8.

1939. Descriptions of the three larval instars of the Japanese beetle. *Popillia japonica* Newm. (Coleoptera, Scarabaeidae). Proc. Ent. Soc. Wash., *41*(6):183-191.

1942a. Descriptions of the larvae of some West Indian melolonthinae beetles and a key to the known larvae of the tribe. Proc. U. S. Nat. Mus., *92*(3146):167-176.

1942b. Description of the third-stage larva of *Amphimallon majalis* (Razoumowski). Proc. Ent. Soc. Wash., *44* (6):111-121.

1942c. A classification of larvae and adults of the genus *Phyllophaga*. Mem. Ent. Soc. Wash., *2*:1-95.

1945. Description of the larva and pupa of the scarab beetle *Anclyonycha mindanaona* (Brenske). Jour. Wash. Acad. Sci., *35*(1):13-15.

Böving, A. G., and Craighead, F. C.
1931. An illustrated synopsis of the principal larva forms of the order Coleoptera. Ent. Amer., *11*(N.S.)(1-4):1-351.

Boyer, B. L.
1940. A revision of the species of *Hoplia* occurring in America north of Mexico (Coleoptera: Scarabaeidae). Microent., *5*(1):1-31.

Brimley, C. S.
1938. *The Insects of North Carolina*. North Carolina Dept. Agr., 560 pp.

1942. *Supplement to Insects of North Carolina*. North Carolina Dept. Agr., 39 pp.

Britton, W. E.
1920. Check-list of the insects of Connecticut. Conn. State Geo. and Nat. Hist. Survey Bull. 31, 397 pp.

Brown, W. J.
1928. Two new species of *Bolbocerosoma* with notes on the habits and genitalia of other species. Can. Ent., *60*:192-196.
1931. Revision of the North American Aegialiinae. Can. Ent., *63*(1) :9-19, (2) :42-49.
1940. Some new and poorly known species of Coleoptera. Can. Ent., *72*(9) :182-188. (Review of genus *Polyphylla*.)

Bruner, L.
1891. Report of the Entomologist. Nebr. State Bd. Agr. Ann. Rpt., pp. 240-309. (The Indian cetonia (*Euphoria inda* Linn.), pp. 279-80.)

Burmeister, H.
1848. Die Entwichkelungeschichte der Gattung *Deltochilum* Esch. Zeitung für Zoologie, Zootomie und Palaeozoologie, *1*(17) :133-136 and *1*(18) :141-144. (Published by D'Alton and Burmeister.)

Butt, F. H.
1944. External morphology of *Amphimallon majalis* (Razoumowski) (Coleoptera, the European chafer). Cornell Univ. Agr. Expt. Sta. Mem. 266, 18 pp.

Carillo, J. L. S., and W. W. Gibson.
1960. Repaso de las especies mexicanas del género *Macrodactylus* (Coleoptera, Scarabaeidae), con observaciones biológicas de algunas especies. Sec. de Agric. and Ganad. Oficina de Est. Espec. Foll. Tecn. 39, 102 pages.

Carne, P. B.
1950. The morphology of the immature stages of *Aphodius howitti* Hope (Coleoptera, Scarabaeidae, Aphodiinae). Proc. Linn. Soc. N.S.W., *75*(3 and 4) :158-166.
1951. Preservation techniques for scarabaeid and other insects' larvae. Proc. Linn. Soc. N. S. W., *76*:26-30.

Cartwright, O. L.
1948a. The American species of *Pleurophorus* (Coleoptera: Scarabaeidae). Trans. Amer. Ent. Soc., *74*(3 and 4) : 131-145.
1948b. *Ataenius strigatus* (Say) and allied species in the United States (Coleoptera:Scarabaeidae). Trans. Amer. Ent. Soc., *74*(3 and 4) :147-153.
1949. The egg-ball of *Deltochilum gibbosum* (Fab.). Coleopt. Bull., *3*(3) :38.
1952. *Aphotaenius,* a new genus of dung beetle (Coleoptera: Scarabaeidae). Proc. U. S. Nat. Mus., *102*(3295) :181-184.
1953. Scarabaeid beetles of the genus *Bradycinetulus* and closely related genera in the United States. Proc. U. S. Nat. Mus., *103*(3318) :95-120.
1955. Scarab beetles of the genus *Psammodius* in the western hemisphere. Proc. U. S. Nat. Mus., *104*(3344) :413-462.
1959. Scarab beetles of the genus *Bothynus* in the United States (Coleoptera:Scarabaeidae). Proc. U. S. Nat. Mus., *108*(3409) :515-541.

Casey, T. L.
1915. *Memoirs on the Coleoptera.* Vol. 6, I. A review of the American species of Rutelinae, Dynastinae and Cetoniinae. New Era Printing Co., Lancaster, Pa., 394 pp.

Cazier, Mont A.
1937. A new species of *Valgus* and a new generic record for Mexico (Coleoptera—Scarabaeidae). Pan-Pac. Ent., *13*(4) :190-192.
1937. A revision of the Pachydemini of North America (Coleoptera, Scarabaeidae). Pomona College Jour. of Ent. and Zool., *29*:73-87.

1938. A generic revision of the North American Cremastocheilini with description of a new species (Coleoptera—Scarabaeidae). Bull. South. Calif. Acad. Sci., *37* (2) :83-87.
1939. Revision of Phileurini of America north of Mexico. Bull. South. Calif. Acad. Sci., *38*(3) :169-171.
1940. The species of *Polyphylla* in North America, North of Mexico (Coleoptera:Scarabaeidae). Ent. News, *51* (5) :134-139.
1953. A review of the scarab genus *Acoma* (Coleoptera, Scarabaeidae). Amer. Mus. Novitates, No. 1624, 13 pp.
1961. A new species of myrmecophilous Scarabaeidae, with notes on other species (Coleoptera, Cremastocheilini). Amer. Mus. Novitates, No. 2033, 12 pp.

Cazier, M. A., and McClay, A. T.
1943. A revision of the genus *Ceononycha* (Coleoptera, Scarabaeidae). Amer. Mus. Novitiates, No. 1239, 27 pp.

Cazier, M. A., and M. Statham.
1962. The behavior and habits of the myremecophilous scarab *Cremastocheilus stathamae* Cazier with notes on other species (Coleoptera:Scarabaeidae). Jour. N. Y. Ent. Soc., *70*(3) :125-149.

Chamberlin, T. R., and C. L. Fluke.
1947. White grubs in cereal and forage crops and their control. Wisc. Agr. Expt. Sta. Res. Bull. 159, 15 pp.

Chapin, E. A.
1934. An apparently new scarab beetle (Coleoptera) now established at Charleston, South Carolina. Proc. Biol. Soc. Wash., *47*:33-36.
1938. The nomenclature and taxonomy of the genera of the scarabaeid subfamily Glaphyrinae. Proc. Biol. Soc. Wash., *51*:79-86.

Chittenden, F. H.
1899. Notes on the rhinoceros beetle. (*Dynastes tityus* Linn.) U. S. Dept. Agr., Div. Ent. Bull. (N.S.), *38*:29-30.

Chittenden, F. H., and D. E. Fink.
1922. The green June beetle. U. S. Dept. Agr. Bull. 891, 52 pp.

Chittenden, F .H., and A. L. Quaintance.
1916. The rose-chafer: A destructive garden and vineyard pest. U. S. Dept. Agr. Farmers' Bull. 721, 6 pp.

Cody, F. P., and I. E. Gray.
1938. The changes in the central nervous system during the life history of the beetle, *Passalus cornutus* Fabricius. Jour. Morph., *63*:503-521.

Comstock, J. H.
1950. *An Introduction to Entomology.* Comstock Publishing Co., Inc., Ithaca, N. Y., 1,064 pp.

Cooper, R. H.
1938. Tumble-bugs. Can. Ent., *70*(8) :155-157.

Corbett, G. H., and N. C. E. Miller.
1939. The identification of grubs from rubber estates. Fed. Malay States Dept. Agr. (Bull.), Sci. Ser. 22, 7 pp.

Crowson, R. A.
1955. *The Natural Classification of the Families of Coleoptera.* Nathaniel Lloyd and Co., Ltd., London, 187 pp.

Davis, J. J.
1916. A progress report on white grub investigations. Jour. Econ. Ent., *9*(2) :261-281.

Davis, J. J., and Philip Luginbill.
1921. The green June beetle or fig eater. North Carolina Agr. Expt. Sta. Bull. 242, 35 pp.

Davis, A. C.
1934. A revision of the genus *Pleocoma*. Bull. South. Calif. Acad. Sci., *33*(3) :123-130. Continued in *34*(1) :4-36.

Dawson, R. W., and McColloch, J. W.
1924. New species of *Bolbocerosoma*. Can. Ent., *56*(1) :9-15.

DeHann, W.
1836. Mémoires sur les métamorphoses des coleoptères. Nouv. Ann. Mus. Nat. Hist., *4*:125.

Denier, P.
1936. Estado actual de mis conocimientos acera del "champi" (*Trox suberosus* F.). Mem. Com. cent. Invest. Langosta, pp. 203-216.

Dillon, E. S., and L. S. Dillon.
1961. *A Manual of Common Beetles of Eastern North America.* Row Peterson and Co., Evanston, Ill., 884 pp.

Downes, W.
1928. On the occurrence of *Aphodius pardalis* Lec., as a pest of lawns in British Columbia. Rept. Ent. Soc. Ontario, *58*:59-61.

Downes, W., and Andison, H.
1941. Notes on the life history of the June beetle *Polyphylla perversa* Casey. Proc. Ent. Soc. Brit. Columbia, *37*: 5-8.

Dozier, H. L.
1920. An ecological study of hammock and piney woods insects in Florida. Ann. Ent. Soc. Amer., *13*(4):325-380.

Edwards, E. E.
1930. On the morphology of the larva of *Dorcus parallelopipedus.* Jour. Linn. Soc., London, Zool., *37*:93-108.

Ellertson, F. E.
1956. *Pleocoma oregonensis* Leach as a pest in sweet cherry orchards. Jour. Econ. Ent., *49*(3):431.

Ellertson, F. E., and P. O. Ritcher.
1959. Biology of rain beetles, *Pleocoma* spp., associated with fruit trees in Wasco and Hood River counties. Ore. Agr. Expt. Sta. Tech. Bull. 44, 42 pp.

Emden, F. van.
1935. Die Gattungsunterschiede der Hirschkäferlarven, ein Beitrag zum natürlichen System der Familie (Col. Lucan.). Stettin Ent. Zeit., *96*:178-200.
1941. Larvae of British beetles. 11: A key to the British lamellicornia larvae. Ent. Monthly Mag., *77*:117-127, 181-192.
1948. A *Trox* larva feeding on locust eggs in Somalia. Proc. R. Ent. Soc., London (B), *17*(11-12):145-148.
1952. The larvae of *Dendezia* and *Figulus,* with notes on some other larvae of Lucanidae. Rev. Zool. Bot. Afr., *46*: 301-310.

Erichson, W. F.
1848. Naturgeschichte der Insecten Deutschlands, Erste Abt., Colep., *3.* Nicolai, Berlin, 968 pp.

Essig, E. O.
1958. *Insects and Mites of Western North America.* Macmillan Company, N. Y., 1,050 pp.

Fabre, J. H.
1912. *Social Life in the Insect World,* Ch. 16: A trufflehunter: the *Bolboceras gallicus.* Trans. by Bernard Miall. The Century Company, N. Y., pp. 217-237.

Fall, H. C.
1928. A review of the genus *Polyphylla* (Scarabaeidae: Coleoptera). Proc. Ent. Soc. Wash., *30*(2):30-35.

Felt, E. P.
1906. Insects affecting park and woodland trees. N. Y. State Mus. Mem., *8:492.*

Fidler, J. H.
1936. Some notes on the morphology of the immature stages of some British chafer beetles. Ann. Appl. Biol., *23*: 114-132.

Fleming, W. E.
1962. The Japanese beetle in the United States. U. S. Dept. Agr., Agriculture Handbook 236, 30 pp.

Fleming, W. E., and F. W. Metzger.
1936. Control of the Japanese beetle and its grub in home yards. U. S. Dept. Agr. Circ. 401, 14 pp.

Fletcher, T. B.
1919. Second hundred notes on Indian Insects. (Pusa.)

Fluiter, H. J. de.
1941. Waarnemingen omtrent engerlingen (oerets) en hun bestrijding in *Hevea*-aanplantingen. (Observations on white grubs and their control in rubber plantations.) Arch. Rubber cult., *25*(2):167-270.

Fluke, C. L., and P. O. Ritcher.
1934. Essential factors in control of white grubs. Wisc. Agr. Expt. Sta. Bull. 428, pp. 97-100.

Franklin, H. J.
1950. Cranberry insects in Massachusetts. Mass. Agr. Expt. Sta. Bull. 445, part IV, pp. 36-53.

Friend, Rober B.
1927. The Asiatic beetle in Connecticut. Conn. Agr. Expt. Sta. Bull. 304, pp. 585-664.

Frost, C. A.
1946. *Dichelonyx canadensis* Horn. Psyche, *53*(1-2):20-21.

Garcia-Tejero, F.
1947. Escarabeidos horticolas (los gusanos blancos de las huertas). Estacion de Fitopatologia Agrícola de Madrid, Trabajos (Serie Fitopatologia) núm. 190, 36 pp.

Gardner, J. C. M.
1929. Immature stages of Indian Coleoptera (6). Indian Forest Rec., *14*:129-131.
1935. Immature stages of Indian Coleoptera (16) (Scarabaeoidea). Indian Forest Rec. (New Ser.) Ent., *1*(1): 1-33.
1944. On some coleopterous larvae from India. Indian Jour. of Ent., *6*:111-116.
1946. A note on the larva of *Trox procerus* Har. (Scarabaeidae, Col.). Indian Jour. of Ent., *8*(1):31-32.

Gerstaeker, C. E. A.
1883. Üeber die Stellung der Gattung *Pleocoma* Les. in System der Lamellicornier. Stettin Ent. Zeit., *44*:436-450. (Translation by J. B. Smith found in Ent. Amer., *3:* 202-211.)

Gilyarov, M. S.
1952. Larvae of beetles with pectinate antennae (Lucanidae) of the European part of the U.S.S.R. Zool. Zh., Moscow, *31*:253-256. (In Russian.)

Golovianko, Z. S.
1936. Les larves plus communes des Coléoptères lamellicornes de la partie européenne de l'U.R.S.S.-Tabl. Analyt. Fn. U.R.S.S., 20, Inst. Zool. Acad. Sci. U.R.S.S., 66 pp., 70 figs. (In Russian.)

Grandi, G.
1925. Contributo alla connescenea biologica e morfologica di alcuni Lamellicorni fillophagi. Bolletina del Laboratorio de Zoologia Generale e Agraria dells R. Scuola Superiore d'Agricoltura in Portici, *18*:159-224.

Gravely, F. H.
1916. XIV Some lignicolous beetle-larvae from India and Borneo. Indian Mus. Rec., *12*:137-175.
1919. XVII Descriptions of Indian beetle larvae III. Indian Mus. Rec., *16*:263-270.

Gray, I. E.
1946. Observations on the life history of the horned passalus. Amer. Midland Nat., *35*(3):728-746.

Gyrisco, G. G., W. H. Whitcomb, R. H. Burrage, C. Logothetis, and H. H. Schwardt.
1954. Biology of the European Chafer *Amphimallon majalis,* Razoumowsky (Scarabaeidae). Cornell Agr. Expt. Sta. Memoir 3281, 35 pp.

Habeck, D. H.
1962. Description of immature stages of the Chinese rose beetle, *Adoretus sinicus* Burmeister (Coleoptera :Scarabaeidae). Proc. Hawaiian Ent. Soc., *18*(2) :251-258.

Halffter, G.
1959. Etología y Paleontología de Scarabaeinae (Coleoptera, Scarabaeidae). Ciencia (Méx.), *19*(8-10) :165-178.
1961. Monografía de las especies norteamericanos del género *Canthon* Hoffsg. (Coleopt. Scarab.). Ciencia (Méx.), *20*(9-12) :225-320.

Halffter, G., F. S. Pereira, and A. Martinez.
1960. *Megathopa astyanex* (Olivier) y formas afines (Coleopt. Scarab.) Ciencia (Méx.), *20*(7-8) :202-204.

Hall, M. C.
1929. Arthropods as intermediate hosts of helminths. Smithsn. Inst. Misc. Collect., *81*(15) :1-77.

Hatch, M. H.
1946. Notes on European Coleoptera in Washington including a new species of *Megasternum.* Pan-Pac. Ent., *22*(2) : 77-80.

Hawley, I. M., and H. C. Hallock.
1936. Life history and control of the Asiatic garden beetle. U. S. Dept. Agr. Circ. 246, 20 pp. *Autoserica castanea* Arrow.

Hamilton, John.
1887. Natural history notes on Coleoptera—No. 3. Can. Ent., *19*(3) :62-67.

Hayes, W. P.
1917. Studies on the life-history of *Ligyrus gibbosus* DeG. (Coleoptera).Jour. Econ. Ent., *10*(2) :253-261.
1921. *Strigoderma arboricola* Fab.—Its life cycle (Scarab. Coleop.). Can. Ent., *53*(6) :121-124.
1925. A comparative study of the life-cycle of certain phytophagous scarabaeid beetles. Kans. Agr. Expt. Sta. Tech. Bull. 16, 146 pp.
1927. The immature stages and larval anatomy of *Anomala kansana* H. and McC. (Scarabaeidae, Coleop.) Ann. Ent. Soc. Amer., *20* :193-206.
1928. The epipharynx of lamellicorn larvae (Coleop.) with a key to common genera. Ann. Ent. Soc. Amer., *21*(2) : 282-306.
1929. Morphology, taxonomy and biology of larval Scarabaeoidea. Ill. Biol. Monog., *12*(2) :1-119.

Hayes, W. P., and J. W. McCulloch.
1924. The biology of *Anomala kansana.* Jour. Econ. Ent., *17*(5) :589-594.

Hayes, Wm. P., and Peh-I Chang.
1947. The larva of *Pleocoma* and its systematic position (Coleoptera, Pleocomidae). Ent. News, *58*(5) :117-127.

Hayward, K. J.
1936. Contribución al conocimiento de la langosta *Schistocerca paranensis* Burm. y sus enemígos naturales. Mem. Com. Centr. Invest. Langosta, Buenos Aires, pp. 205-216.

Heit, C. E., and H. K. Henry.
1940. Notes on the species of white grubs present in the Saratoga Forest Tree Nursery. Jour. Forestry, *38*(12) : 944-948.

Henney, H. J.
1942. Entomology. Colo. Agr. Expt. Sta. 55th Annual Rept., p. 25.

Hinton, H.
1930. Observations on two California beetles. Pan-Pac. Ent., *7*(2) :94-95.

Hintze, Anna L.
1925. The behavior of the larvae of *Cotinis nitida* (Burm.) Ann. Ent. Soc. Amer., *18*(1) :31-34.

Hiznay, P. A., and J. B. Krause.
1955. The structure and musculature of the larval head and mouthparts of the horned passalus beetle, *Popilius disjunctus* Illiger. Jour. Morph., *97*(1) :55-76.

Hoffman, C. H.
1935a. The biology and taxonomy of the genus *Trichiotinus* (Scarabaeidae-Coleoptera). Ent. Amer., *15*(4) :133-214.
1935b. Biological notes on *Ataenius cognatus* (Lec.), a new pest of golf greens in Minnesota. (Scarabaeidae-Coleoptera). Jour. Econ. Ent., *28*(4) :666-667.
1936. Additions to our knowledge of the biology of *Pelidnota punctata* Linn. (Scarabaeidae-Coleoptera). Jour. Kans. Ent. Soc., *9* :103-105.
1936. Additional data on the biology and ecology of *Strigoderma arboricola* Fab. (Scarabaeidae-Coleoptera). Bull. Brooklyn Ent. Soc., *31* :108-110.
1937. Biological notes on *Pseudolucanus placidus* Say, *Platycerus quercus* Weber and Ceruchus *piceus* Weber (Lucanidae-Coleoptera). Ent. News, *48*(10) :281-284.
1939. The biology and taxonomy of the nearctic species of *Osmoderma* (Coleoptera, Scarabaeidae). Ann. Ent. Soc. Amer., *32*(3) :510-525.

Horn, G. H.
1878. Revision of the species of *Listrochelus* of the United States. Trans. Amer. Ent. Soc., *7* :137-148.
1883. *Pleocoma* Lec. Its systematic position and indication of new species. Ent. Amer., *3* :233-235.
1884. Notes on the species of *Anomala* inhabiting the United States. Trans. Amer. Ent. Soc., *11* :157-164.
1888. Review of the species of Pleocoma with a discussion of its systematic position in the Scarabaeidae. Trans. Amer. Ent. Soc., *15* :1-18.
1888. Descriptions of the larvae of *Glyptus, Platypsylla* and *Polyphylla.* Trans. Amer. Ent. Soc., *15* :18-26.

Howden, H. F.
1952. A new name for *Geotrupes* (*Peltotrupes*) *chalybaeus* Le Conte, with a description of the larva and its biology. Coleopt. Bull., *6*(3) :41-48.
1954. The burrowing beetles of the genus *Mycotrupes* (Coleoptera: Scarabaeidae: Geotrupinae). Pt. III. Habits and life history of *Mycotrupes,* with a description of the larva of *Mycotrupes gaigei.* Univ. of Mich., Mus. of Zool., Misc. Pub. 84, pp. 52-59.
1955a. Cases of interspecific "parasitism" in Scarabaeidae (Coleoptera). Jour. Tenn. Acad. Sci., *30*(1) :64-66.
1955b. Biology and taxonomy of North American beetles of the subfamily Geotrupinae with revisions of the genera *Bolbocerosoma, Eucanthus, Geotrupes* and *Peltotrupes* (Scarabaeidae). Proc. U. S. Nat. Mus., *104*(3342) : 151-319.
1957. Investigations on sterility and deformities of *Onthophagus* (Coleoptera, Scarabaeidae) induced by gamma radiation. Ann. Ent. Soc. Amer., *50*(1) :1-9.
1958. Species of *Acoma* Casey having a three-segmented antennal club (Coleoptera: Scarabaeidae). Can. Ent., *90*(7) :377-401.

Howden, H. F., and O. L. Cartwright.
1963. Scarab beetles of the genus *Onthophagus* Latreille north of Mexico (Coleoptera: Scarabaeidae). Proc. U. S. Nat. Mus., *114*(3467) :1-135.

Howden, H. F., and A. Martinez.
1963. The new tribe Athyreini and its included genera Coleoptera: Scarabaeidae, Geotrupinae). Can. Ent., 95(4): 345-352.

Howden, H. F., and P. O. Ritcher.
1952. Biology of *Deltochilum gibbosum* (Fab.) with a description of the larva. Coleopt. Bull., 6(4):53-57.

Hubbard, H. G.
1894. The insect guests of the Florida land tortoise. Insect Life, 6(4):302-315.

Janssens, A.
1947. Contribution à l'étude des Coléoptères lamellicornes de la faune Belge. I. Table de determination generique des larves. Bull. Mus. Royal d'Hist. Belg., 23(6): 1-16.

Jerath, M. L.
1960. Notes on larvae of nine genera of Aphodiinae in the United States (Coleoptera: Scarabaeidae). Proc. U. S. Nat. Mus., 111(3425):43-94.

Jerath, M. L., and P. O. Ritcher.
1959. Biology of Aphodiinae with special reference to Oregon (Coleoptera: Scarabaeidae). Pan-Pac. Ent., 35(4): 170-175.

Jewett, H. H.
1943. Control of green June beetle larvae *Cotinis nitida* (L.), in tobacco beds. Ky. Agr. Exp. Sta. Bull. 445, 12 pp.

Johnson, J. P.
1941. *Cyclocephala* (*Ochrisidia*) *borealis* in Connecticut. Jour. Agr. Res., 62(2):79-86.
1942. White grubs during 1941. Conn. Agr. Expt. Sta. Bull. 461, pp. 523-530.

Johnson, P. C.
1954. A feeding record of the ten-lined June beetle. Jour. Econ. Ent., 47(4):717-718.

Korschefsky, R. von.
1940. Bestimmungstabelle der häufigsten deutschen Scarabaeidenlarven. Arb. physiol. angew. Ent. Berlin-Dahlem, 7(1):41-52. (Key for the identification of the most common German scarab larvae.)

Landis, B. J.
1959. The economic importance of *Pleurophorus caesus* Creutz. Jour. Econ. Ent., 52(6):1215.

Leng, C. W.
1920. *Catalogue of the Coleoptera of America, North of Mexico*. John D. Sherman, Jr., Mt. Vernon, N. Y., 470 pp.

Leng, C. W., and A. J. Mutchler.
1927. *Supplement to Catalogue of the Coleoptera of America, North of Mexico, 1919 to 1924* (*inclusive*). John D. Sherman, Jr., Mt. Vernon, N. Y., 78 pp.
1933. *Second and Third Supplements to the Catalogue of the Coleoptera of America, North of Mexico, 1925 to 1935* (*inclusive*). John D. Sherman, Jr., Mt. Vernon, N. Y., 112 pp.

Lindquist, A. W.
1933. Amount of dung buried and soil excavated by certain Coprini (Scarabaeidae) (1). Jour. Kans. Ent. Soc., 6(4):109-125.
1935. Notes on the habits of certain Coprophagous beetles and methods of rearing them. U. S. Dept. Agr. Circ. 351, 9 pp.

Linsley, E. G.
1938. Notes on the habits, distribution, and status of some species of *Pleocoma*. Pan-Pac. Ent., 14:49-58, 97-104.
1941. Additional observations and descriptions of some species of *Pleocoma*. Pan-Pac. Ent., 17(4):145-152.

1943. Notes on the habits of some beetles from the vicinity of Yosemite National Park. Bull. South. Calif. Acad. Sci., 41(3):164-166. (Habits of *Canthon simplex* var. *militaris*.)
1945. Further notes on some species of *Pleocoma* (Coleoptera, Scarabaeidae). Pan-Pac. Ent., 21(3):110-114.

Linsley, E. G., and Michener, C. D.
1943. Observations on some Coleoptera, from the vicinity of Mt. Lassen, California. Pan-Pac. Ent., 19(2):75-79. (Found larvae and pupae of *Odontaeus obesus* Lec.)

Linsley, E. G., and E. S. Ross.
1940. Records of some Coleoptera from the San Jacinto Mountains, California. Pan-Pac. Ent., 16(2):75-76.

Lockwood, L.
1868. The goldsmith beetle and its habits. Amer. Nat., 2(4): 186-192.

Loding, H. P.
1935. *Geotrupes ulkei* Blanchard. Bull. Brooklyn Ent. Soc., 30:108.

Lugger, O.
1899. Beetles (Coleoptera) injurious to our fruit-producing plants. Minn. Agr. Exp. Sta. Bull. 66, 248 pp.

Luginbill, Sr., P., and H. R. Painter.
1953. May beetles of the United States and Canada. U. S. Dept. Agr. Tech. Bull. 1060, 102 pp.

Madle, H.
1934. Zür Kenntnis der Morphologie, Okologie and Physiologie von *Aphodius rufipes* L. und einiger verwandter Arten. Zool. Jahrb. (Anat.) (Abt. f. Anatomie u. Ontogenie d. Tiere), 58(Heft 3):304-396.
1935. Die Larven der Gattung *Aphodius* I. Arb. Phys. angew. Entom. aus Berlin-Dahlem, 2:289-304 (1935) and 3: 1-20 (1936).

Manee, A. H.
1908. Some observations at Southern Pines, North Carolina. Ent. News, 19:286-288. (Biology of *Strategus antaeus* (Fab.).)
1908. Some observations at Southern Pines, North Carolina. Ent. News, 19:459-462. (Biology of *Eucanthus lazarus* (Fab.) and *Bolboceras ferrugineus* (Beauv.).)
1915. Observations in Southern Pines, North Carolina, (Hym., Col.) Ent. News, 26:266. (Biology of *Dynastes tityus* Linn.)

Mathews, E. G.
1961. A revision of the genus *Copris* Müller of the western hemisphere (Coleoptera, Scarabaeidae). Ent. Amer., 41:1-137.
1963. Observations on the ball-rolling behavior of *Canthon pilularius* (L.) (Coleoptera, Scarabaeidae). Psyche, 70(2):75-93.

McColloch, J. W.
1917. A method for the study of underground insects. Jour. Econ. Ent., 10(1):183-187.

McIndoo, N. E.
1931. Tropisms and sense organs of Coleoptera. Smithsn. Inst. Misc. Collect. 82(No. 18), 70 pp., Publ. 3113. (Sense organs of the larva of *Cotinis nitida* (L.), pp. 50-51.)

Medvedev, S. I.
1952. Larvae of Scarabaeid beetles of the fauna of the USSR. Opred. Faune SSSR, Moscow, 47:1-343. (In Russian.)
1960. Descriptions of the larva of eight species of lamellicorn beetles from the Ukraine and Central Asia. Zool. Zhurn., 39:381-393.

Meinert, F.
1895. Sideoganerne hos Scarabae-Larverne. D. Kg. Danske Vidensk. Selsk. Skr., 6. Raekke, naturvidenskabelig og mathematisk Afd., 8(1):3-72.

Metcalf, C. L., W. P. Flint, and R. L. Metcalf.
  1962. *Destructive and Useful Insects, Their Habits and Control.* McGraw-Hill Book Co. Inc., N. Y., 1,087 pp.
Miller, L. I.
  1943. A white grub injuring peanuts in eastern Virginia. Jour. Econ. Ent., *36*(1) :113-114.
Milne, L. J.
  1933. Notes on *Pseudolucanus placidus* (Say) (Lucanidae, Coleoptera). Can. Ent., *65*(5) :106-114.
Miyatake, M.
  1951. The immature stages of *Cylindrocaulus patalis* (Lewis, 1883) (Coleoptera: Passalidae). Trans. Shikoku Ent. Soc., *2*(2) :27-30.
Mulsant, E.
  1842. Histoire naturelle des coléoptères de France, Part II, Lamellicornes. Lyon (et Paris), Maison (later Paris, Magnen and Blanchard Co.), 624 pp.
Murayama, J.
  1927. Systematic description of the larva of *Hoplia aureola* Pall (Col. Lamell.). Jour. Chosen Nat. Hist. Soc., *5:* 1-8.
  1931. A contribution to the morphological and taxonomic study of larvae of certain May-beetles which occur in the nurseries of the peninsula of Korea. Forest Expt. Sta. (Japan) Bull. 11, 108 pp.
  1936. Rapport sur les moyens repressifs employes contre les hannetons. III. Recherches sur la vie et moers chez le *Phyllopertha pallidipennis* Reitter. Chosen Govt.-Gen., Forest Expt. Sta. Bull. 23, 164 pp.

Neiswander, C. R.
  1938. The annual white grub, *Ochrisidia villosa* Burm. in Ohio lawns. Jour. Econ. Ent., *31*(3) :340-344.
Nichol, A. A.
  1935. A study of the fig beetle, *Cotinis texana* (Casey). Ariz. Agr. Expt. Sta. Bull. 55, 42 pp.

Oberholzer, J. J.
  1959. A morphological study of some South African lamellicorn larvae. I. Descriptions of the third instar larvae. South African Jour. of Agr. Sci., *2*(1) :41-74.
  1959. A morphological study of some South African lamellicorn larvae. II. Comparative morphology. South African Jour. of Agr. Sci., *2*(1) :75-88.
Ohaus, F.
  1900. Bericht über eine entomologische Reise nach Zentralbrasilien. Stettin Ent. Zeit., *61*:164-273.
  1909. Bericht über eine entomologische Studienreise in Südamerika. Stettin Ent. Zeit., *70*:1-139.
  1934. Coleoptera Lamellicornia, Family Scarabaeidae, subfamily Rutelinae: Pt. I. Genera Insectorum, Fasc. 199A, 172 pp.
Olson, A. L., T. H. Hubbell, and H. F. Howden.
  1954. The burrowing beetles of the genus *Mycotrupes*. Univ. Mich. Mus. Zool. Misc. Publ. 84, 59 pp.
Osten Sacken, R.
  1874-76. Description of the larva of *Pleocoma* Lec. Trans. Amer. Ent. Soc., *5*:84-87.
Osterberger, B. A.
  1931. The importance of *Euetheola* (*Ligyrus*) *rugiceps* Lec., an enemy of sugar cane. Jour. Econ. Ent., *24*(4) :870-872.

Panin, S.
  1957. Fauna Republicii Populare Romine. Insecta: Coleoptera, Familia Scarabaeidae, *10*(4) :1-315, 35 plates. (In Roumanian.)

Paulian, R.
  1939. Les caractéres larvaires des Geotrupidae (Col.) et leur importance pour la position systématique du groupe. Bull. Soc. Zool. de France, *64*:351-360.
  1941. La position systematique du genre *Pleocoma* Le Conte (Col. Scarabaeidae). Rev. Franc. d'Ent., *8*(3) :151-155.
  1942. La Larve de *Rhyssemodes orientales* Muls. et God. (Col. Scarabaeidee). Bull. Soc. Ent. de France, *47*(7) : 129-131.
  1956. Atlas des larves d'insectes de France. Illus. Editions N. Boubee et Cie. Paris V/e, 222 pp.
  1959. Coléoptères Scarabeides. Faune de France, *63*:1-298.
Paulian, R., and Villiers, A.
  1939. Larves de Coléoptères 4. *Heptaulacus Peyerimhoffi* Paul et Vill. Rev. Franc. d'Ent., *6*(2) :49-50.
Pearse, A. S., M. T. Patterson, J. S. Rankin, and G. W. Wharton.
  1936. The ecology of *Passalus cornutus* Fabricius, a beetle which lives in rotting logs. Ecol. Monogr., *6*:455-490.
Pereira, F. S.
  1954. A new myrmecophilous scarabaeid beetle from the Philippine Islands with a review of *Haroldius*. Psyche, *61*(1) :1-8.
Perris, E.
  1877. Larves des Coleoptères: Lamellicorns et Pectinicorns. Ann. Soc. Linn. de Lyon, *22*:91-122.
Pessôa, S. B., and F. Lane.
  1941. Coléopteros necrófagos de interêsse médico-legal. (Sao Paulo) Arq. de Zool., *2*(17) :389-504.
Peterson, Alvah.
  1951. *Larvae of Insects, Part II.* Edwards Bros., Inc., Ann Arbor, Michigan, 416 pp.
Phillips, W. J., and H. Fox.
  1917. The rough-headed corn-stalk beetle in the southern states and its control. U. S. Dept. Agr. Farmer's Bull. 875, 10 pp.
  1924. The rough-headed corn-stalk beetle. U. S. Dept. Agr. Bull. 1267, 33 pp.
Potts, R. W. L.
  1945. A new *Coenonycha* from California (Coleoptera, Scarabaeidae). Pan-Pac. Ent., *21*(4) :141-143.
Pratt, R. Y.
  1943. Insect enemies of the scarabaeid *Polyphylla crinita* Lec. (Coleoptera: Scarabaeidae). Pan-Pac. Ent., *19* (2) :69-70.

Riegel, G. T.
  1942. *Cyclocephala abrupta* in Illinois. (Coleop: Scarab.). Trans. Ill. Acad. Sci., *35*(2) :215.
Ritcher, P. O.
  1938. A field key to Kentucky white grubs. Jour. Kans. Ent. Soc., *11*(1) : 24-27.
  1940. Kentucky white grubs. Ky. Agr. Expt. Sta. Bull. 401, pp. 71-157.
  1943. The Anomalini of eastern North America with descriptions of the larvae and key to species (Coleoptera: Scarabaeidae). Ky. Agr. Expt. Sta. Bull. 442, 27 pp.
  1944. Dynastinae of North America with descriptions of the larvae and keys to genera and species (Coleoptera: Scarabaeidae). Ky. Agr. Expt. Sta. Bull. 467, 56 pp.
  1945a. Rutelinae of eastern North America with descriptions of the larvae of *Strigodermella pygmaea* (Fab.) and three species of the tribe Rutelini (Coleoptera: Scarabaeidae). Ky. Agr. Expt. Sta. Bull. 471, 19 pp.
  1945b. Notes on *Phyllophaga barda* (Horn) with a description of the larva (Coleoptera: Scarabaeidae). Proc. Ent. Soc. Wash., *47*(4) :97-99.

1945c. North American Cetoniinae with descriptions of their larvae and keys to genera and species (Coleoptera: Scarabaeidae). Ky. Agr. Expt. Sta. Bull. 476, 39 pp.

1945d. Coprinae of eastern North American with descriptions of larvae and keys to genera and species (Coleoptera: Scarabaeidae). Ky. Agr. Expt. Sta. Bull. 477, 23 pp.

1947a. Description of the larva of *Pleocoma hirticollis vandykei* Linsley (Coleoptera: Scarabaeidae). Pan-Pac. Ent., *23*(1):11-20.

Ritcher, P. O.
1947b. Larvae of Geotrupinae with keys to tribes and genera (Coleoptera: Scarabaeidae). Ky. Agr. Expt. Sta. Bull. 506, 27 pp.

1948. Description of the larvae of some ruteline beetles with keys to tribes and species. Ann. Ent. Soc. Amer., *41*(2):206-212.

1949a. Larvae of Melolonthinae with keys to tribes, genera and species (Coleoptera: Scarabaeidae). Ky. Agr. Exp. Sta. Bull. 537, 36 pp.

1949b. May beetles and their control in the inner bluegrass region of Kentucky. Ky. Agr. Exp. Sta. Bull. 542, 12 pp.

1958. Biology of Scarabaeidae. Ann. Rev. of Ent., *3*:311-334.

1962. Notes on *Phyllophaga sociata* (Horn) with a description of the larva (Coleoptera: Scarabaeidae). Pan-Pac. Ent., *38*(3):163-166.

Ritcher, Paul O., and Frank M. Beer.
1956. Notes on the biology of *Pleocoma dubitalis dubitalis* Davis (Coleoptera: Scarabaeidae). Pan-Pac. Ent., *32*(4):181-184.

Robinson, M.
1941. Notes on some rare Scarabaeidae with the description of one new species (Coleoptera). Ent. News, *52*:227-232.

Roffey, Jeremy.
1958. Observations on the biology of *Trox procerus* Har. (Coleoptera, Trogidae), a predator of eggs of the desert locust, *Schistocerca gregaria* (Forsk.). Bull. Ent. Res., *49*(3):449-465.

Sanderson, M. W.
1940. Arkansas Cyclocephalini with notes on Burmeister types. Ann. Ent. Soc. Amer., *33*(2):377-384.

1958. Faunal affinities of Arizona *Phyllophaga* with notes and descriptions of new species (Coleoptera, Scarabaeidae). Jour. Kansas Ent. Soc., *31*(2):158-173.

Saunders, W.
1874. On some of our common insects. No. 18: The spotted Pelidnota—*Pelidnota punctata* Linn. Can. Ent., *6*:141-142.

1879. The goldsmith beetle (*Cotalpa lanigera*). Can. Ent., *11*:21-22.

Say, Thomas.
1823. Descriptions of Coleopterous insects collected in the late expedition to the Rocky Mountains, performed by order of Mr. Calhoun, Secretary of War, under the command of Major Long. Jour. Acad. Nat. Sci. Phila., *3*:139-216.

Saylor, L. M.
1933. Attraction of beetles to tar. Pan-Pac. Ent., *9*(4):182.

1937. Revision of California *Cyclocephala*. Pomona Coll. Jour. Ent. and Zool., *29*(3):67-70.

1938. A new *Phyllophaga* from Nevada (Scarab.). Proc. Ent. Soc. Wash., *40*.(5):129-131.

1939. Revision of the beetles of the melolonthine subgenus *Phytalus* of the United States. Proc. U. S. Nat. Mus., *86*(3048):157-167.

1940. Synoptic revision of the beetle genera *Cotalpa* and *Paracotalpa* of the United States with description of a new subgenus. Proc. Wash. Ent. Soc., *42*(9):190-200.

1942. Notes on beetles related to *Phyllophaga* Harris, with descriptions of new genera and subgenera. Proc. U. S. Nat. Mus., *92*(3145):157-165.

1945. Revision of the scarab beetles of the genus *Dichelonyx*. Bull. Brook. Ent. Soc., *40*(5):137-158.

1946. Synoptic revision of the United States scarab beetles of the subfamily Dynastinae, No. 3: Tribe Oryctini (Part). Jour. Wash. Acad. Sci., *36*(2):41-45.

Schaerffenberg, Bruno.
1941. Bestimmungsschlüssel der wichtigsten deutschen Scarabaeiden-larven. Zeitschr. Pflanzenkr., *51*(1):24-42. (U. Rostock.)

Schenkling, S.
1921. Cetoniidae: Coleoptorum Catalogus Pars 72, 431 pp. Berlin.

1922. Trichiinae: Valginae. Coleoptorum Catalogus Pars 75, 58 pp. Berlin.

Schiødte, J. C.
1844. De metamorphosi eleutheratorum observationes: bidrag til insekternes udviklings-histoire. Pars 8, Scarabaei. Saetryk of Naturh. Tidsskr. (Ser. 3), *9*:227-376.

Schmidt, A.
1911. Coleoptera lamellicornia, family Aphodiidae. Genera Insectorum, Fasc. 110, 155 pp.

Schwarz, E. A.
1878. The coleoptera of Florida. Proc. Amer. Phil. Soc., *17*:353-466.

Shenefelt, R. D., and H. G. Simkover.
1950. White grubs in Wisconsin forest tree nurseries. Jour. Forestry, *48*(9):429-434.

Sim, R. J.
1930. Scarabaeidae, Coleoptera; observations on species unrecorded or little known in New Jersey. Jour. N. Y. Jour. Ent. Soc., *38*(2):139-147. (Biologies of *Odontaeus, Bolbocerosoma,* and *Eucanthus*).

1934. Characters useful in distinguishing larvae of *Popillia japonica* and other introduced Scarabaeidae from native species. U. S. Dept. Agr. Circ. 334, 20 pp.

Smith, R. F., and Potts, R. W. L.
1945. Biological notes on *Pleocoma hirticollis vandykei* Linsley (Coleoptera: Scarabaeidae). Pan-Pac. Ent., *21*(3):115-118.

Spaney, A.
1910. Beitrage zür unserer einheimischen Rosskäfer. Deutsche. Ent. Zeitschr., *1910*:625-632.

Sweetman, H. L., and Hatch, M. H.
1927. Biological notes on *Osmoderma* with a new species of Ptiliidae from its pupal case (Coleoptera). Bull. Brooklyn Ent. Soc., *22*(5):264-265.

Smyth, E. G.
1916. Report of the south coast laboratory. Puerto Rico Dept. Agr. Rpt., pp. 45-50. (Life cycles of *Strategus titanus, S. quadrifoveatus, Ligyrus tumulosus* and *Dyscinetus barbatus*.)

1920. The white grubs injuring sugar cane in Puerto Rico. II. The rhinoceros beetle. Puerto Rico Dept. Agr. Jour,. *4*(2):1-29. (Life histories and habits of *Strategus titanus* and *S. quadrifoveatus*.)

Stein, W. I.
1963. *Pleocoma* larvae, root feeders in western forests. Northwest Science, *37*(4):126-143.

Thomas, I., and Heal, G. M.
1944. Chafer damage to grassland in north Wales in 1942-43 by *Phyllopertha horticola* L. and *Hoplia philanthus* Fuess. I. Notes on population, life history and morphology. Ann. App. Biol., *31*(2):124-131.

Tilden, J. W., and G. S. Mansfield.
1944. Notes on three species of the genus *Coenonycha* Horn (Coleoptera: Scarabaeidae). Pan-Pac. Ent., *20*(3): 115-117.

Ting, P. C.
1934. Back-crawling Scarabaeid grubs intercepted in quarantine at San Francisco. Calif. Dept. Agr. Monthly Bull., *23*(7-9):185-191. (Description of the Asiatic species *Potosia affinis* And.)

Titus, E. S. G.
1905. Some miscellaneous results of the work of the Bureau of Entomology. The Sugar cane beetle (*Ligyrus rugiceps* Lec.). U. S. Dept. Agr., Bur. Ent. Bull. 54, pp. 7-18.

Van Dyke, E .C.
1928. Notes and descriptions of new species of Scarabaeidae from western North America. Pan-Pac. Ent., *4*(4): 151-162.

Vaurie, P.
1955. A revision of the genus *Trox* in North America (Coleoptera, Scarabaeidae). Bull. Amer. Mus. Nat. Hist., *106*:1-89.

Viado, G. B.
1939. External anatomy and diagnostic characters of some common Philippine white grubs. Philippine Agr., *28*(5):339-410.

Wallis, J. B.
1928. Revision of the genus *Odontaeus* Dej. (Scarabaeidae, Coleoptera). Can. Ent., *60*:119-128, 151-156, 168, 176.

Warren, J. C.
1917. Habits of some burrowing Scarabaeidae (Col.). Ent. News, *28*:412-413.

Wassman, E.
1918. Myrmecophile und termitophile Coleopteren aus Ostindien, hauptsachlich gesammelt von P. v. Assmuth S. J. Wiener Ent. Ziet., *37*:1-23.

Weiss, H. B.
1921. Notes on the larval and pupal stages of *Xyloryctes satyrus* (Col: Scarabaeidae). Ent. News, *32*(7):193-198. (Detailed description of the larva.)

Wheeler, Wm. M.
1908. Studies on Myrmecophiles, I. *Cremastochilus*. Jour. N. Y. Ent. Soc., *16*:68-79.
1908. The ants of Casco Bay, Maine, with observations of two races of *Formica sanguinea* Latr. Bull. Amer. Mus. Nat. Hist., *24*:619-645. *Cremastochilus castaneae*, pp. 625-627.

Wickham, H. F.
1894. The Coleoptera of Canada. IV. The pleurostict Scarabaeidae of Ontario and Quebec. Can. Ent., *26*(9):259-264.

Woodruff, R. E.
1961. A Cuban May beetle, *Phyllophaga* (*Cnemarachis*) *bruneri* in Miami, Florida (Coleoptera: Scarabaeidae). Florida Div. of Plant Industry Bull., *1*(1):3-31.

Wray, D. L.
1950. Insects of North Carolina, Second Supplement. North Carolina Dept. of Agr., 59 pp.

Index